THE NATURE
OF ENZYMOLOGY

THE NATURE OF ENZYMOLOGY

R. L. FOSTER

A HALSTED PRESS BOOK

JOHN WILEY & SONS
NEW YORK

Published in the U.S.A.
by Halsted Press, a Division of
John Wiley & Sons, Inc., New York
ISBN: 0-470-26860-3

Printed and bound in Great Britain

Contents

44.95

Acknowledgements

The author would like to thank the following who have given their permission to reproduce figures and tables.

Professor D. C. Phillips, Figures 1.2 and 1.4; Professor D. Blow, Figures 1.6 and 4.9; Dr J. E. Walker, Figure 1.7; Dr G. C. K. Roberts, Figure 1.8; Professor M. G. Rossman, Figures 4.11 and 4.12; Professor J. H. Wilkinson, Figures 6.3 and 6.4; Professor E. Katchalski-Katzir, Figure 7.3; Professor F. W. Schmidt, Figure 6.2; Professor K. K. Kannan, Figures 5.15 and 5.16.

Nature (London), Figures 1.2, 1.4, 1.7 and 4.9; Elsevier (Amsterdam), Table 5.11 and Appendix I; Academic Press (London and New York), Figures 1.6, 1.8, 4.11, 5.15, 5.16, 6.2 and 7.3; The Institute of Medical Laboratory Sciences, Figures 6.3 and 6.4; International Union of Biochemistry—International Union of Pure and Applied Chemistry (IUB-IUPAC), Table 5.11 and Appendix I.

To   Hetty, Les, Margaret, Han and Anneke (for their enzymes!)

'Come forth into the light of things
Let Nature be your Teacher'
                    —Wordsworth.

Preface

The study and application of enzymes are profoundly influencing the development of biochemistry and related disciplines. X-ray crystallographic investigations have illuminated many structural features underlying their functions, their selectivities of action are beginning to be applied to the treatment of thrombo-embolic and neoplastic diseases and their high efficiencies are increasingly utilised to improve the quality and economic supply of naturally occurring molecules.

Although there are many advanced treatises dealing with these different aspects the majority have been written for specialists. This book has been designed to serve the following three purposes (a) to introduce undergraduates to the broad subject of enzymology by explaining the principal concepts underlying its modern ideas and applications, (b) to give an account for interested persons in other fields and (c) to present to both a picture of the extraordinary diversification which the subject has now attained.

The first half of the book is devoted to the general characteristics and properties of enzymes, placing emphasis on their structures and active site chemistry, and in the second half their important physiological roles and current applications in clinical medicine and industry are outlined. In all the chapters the object has been to bring to the attention of the reader as large a part of the field of enzymology as possible within a reasonable volume. Thus specific enzymes have been used as examples since there are far too many to describe each in detail (approximately 2,000 have been classified to date). For the same reason, more references, especially to key articles and reviews, are given than is perhaps usual in this type of book, but it is hoped that they will provide bridges to further more detailed and advanced study.

My thanks are due to many present colleagues, past teachers and questioning students who knowingly or unknowingly have helped the writing of this book. In particular I should like to thank Dr T. Wileman for his thorough reading of the manuscript, the publishers for their patience and above all, my wife, whose constancy and secretarial help ensured its fulfilment.

R. L. F.

1   The Character of Enzymes

1

Enzymes are the catalysts of living organisms. Virtually all the chemical reactions occurring in plants, microorganisms and animals proceed at a measurable rate as a direct consequence of enzymic catalysis. The phenotypes of these living systems are functions of the molecules synthesised enzymically and the morphologies, organisations and functions of their cells are effected through enzymic agency.

Many of the commonly encountered substances which make palatable man's existence are also the products of enzymic action. Cheese, for example, is produced from milk by the action of 'rennin' a protein digesting enzyme derived from the stomachs of weaning calves and everyday alcohol, ethanol, is one product of the consecutive action of twelve enzymes elaborated by yeast cells growing on glucose.

Enzyme catalysis is therefore one of the most important, ubiquitous but at the same time, enigmatic expressions of evolution.

Biological catalysis has been known for nearly 150 years. In 1837 Berzelius recognised that there were naturally occurring 'ferments' which promoted chemical reactions and which fulfilled the criteria of catalysis he had proposed a few years earlier. These ferments he classified as 'organised' or 'unorganised' depending on the presence or absence of intact cells. But although the origins of enzymology can be traced so far back in time, the first significant advances were made a century ago by W. Kühne through his investigations into the nature of trypsin catalysed reactions. Kühne is also responsible for introducing the name 'enzyme', transliterated from the classical Greek 'in yeast', to describe the naturally occurring catalysts present in unorganised ferments.

Except for Fischer's attempt to explain the specificity of an enzyme for its substrate (reacting substance transformed during catalysis) by the analogy of complementarity between a lock and its key, little further progress was made until the turn of the present century. Considerable impetus to the development of enzymology was then provided by the mechanistic formulations of V. Henri, L. Michaelis and others, who rationalised the kinetic manifestations of enzymic catalysis by postulating the intermediate formation of an

3

active enzyme-substrate complex which reacts to form products and release active enzyme. This kinetic advance was followed soon afterwards in 1926 by the demonstration by Sumner that 'urease', an enzyme catalysing the hydrolysis of urea ($NH_2 \cdot CO \cdot NH_2$) to ammonia plus carbon dioxide, was a protein. By showing that an enzyme had a distinctive chemical identity and belonged to an established class of molecules, Sumner tolled the knell of the 'vitalistic force' and similar theories and paved the way for the laboratory handling and analysis of enzymes.

During the succeeding half century enzymology has developed apace (Table 1.1). As the steps comprising the major metabolic pathways, for example, glycolysis and the tricarboxylic acid cycle, were elucidated, knowledge grew of the multifarious reactions catalysed by enzymes, and as the biochemical events underlying the physiological processes of digestion, muscular contraction, endocrine function, coagulation and biosynthesis were unravelled, their important roles in the maintenance, control and integration of complex metabolic processes came to be recognised. Simultaneously during this fruitful period, kinetic frameworks to rationalise the observations of enzyme action and inhibition, and procedures specifically designed to analyse the structures of functionally sensitive proteins were developed. By the time the structural explorations culminated in the 1960s with the resolution of the three-dimensional structures of the oxygen transporting 'honorary enzymes', myoglobin and haemoglobin, and more especially with the resolution of the structure of the bacteriolytic enzyme, lysozyme, by Phillips *et al.*, the basic conceptual and molecular bases of enzymology had been laid down. This groundwork is covered in the first five chapters of this book.

More recently, attention has increasingly been directed to the applications of enzymes. Their high efficiencies make them potentially valuable as catalysts in manufacturing industry. For example, glucamylase and amylase (polysaccharide degrading enzymes) are gradually replacing strong acids in the depolymerisation of starch to glucose. Both enzymes can act on a wide variety of glucose polymers so making available potentially more supplies of the sweetener. In a more general industrial context, the other main advantages offerred by enzymic catalysis are the mild conditions needed for efficient operation. These requisites thus facilitate process design and maintenance. Their other remarkable capability, that is, specificity of action, is proving of benefit in clinical medicine.

Table 1.1: Chronology of Enzyme Studies

1833	Payen and Persoz. Alcohol precipitation of thermolabile 'diastase' from malt.
1835	Berzelius. Concept of *catalysis*.
1837	Berzelius. Recognition of biological catalysis.
1850	Wilhelmy. Quantitative evaluation of the rates of sucrose inversion.
1878	Kühne. Investigations of trypsin catalysed reactions and introduction of the word 'enzyme'.
1894–5	Fischer. 'Lock and key' simile of enzyme specificity.
1896–7	Bertrand. 'Coenzyme' or 'coferment' (now called cofactors).
1898	Duclaux. Nomenclature—substrate plus suffix 'ase'.
1901–3	Henri. General procedures for the derivation of kinetic rate laws: principle of the enzyme-substrate complex.
1906	Harden and Young. 'Cozymase' (NAD).
1913	Michaelis and Menten. Extension of the kinetic theory of enzyme catalysis.
1925	Briggs and Haldane. Derivation of enzyme rate equations using the steady-state approximation.
1926	Sumner. Crystallisation of urease—a 'protein' not a 'property'.
1930–3	Northrop and Kunitz. Crystallisation of proteolytic enzymes.
1937–9	Cori and Cori. Muscle phosphorylase.
1940	Beadle and Tatum. 'One gene-one enzyme' hypothesis.
1943	Chance. Spectrophotometric techniques.
1953	Koshland. Induced fit hypothesis.
1956	Umbarger, Yates and Pardee. Control of enzyme activity through feedback inhibition.
1956	Sutherland. Cyclic AMP, adenyl cyclase.
1956–8	Anfinsen. Amino acid sequence determines folding pattern and activity of ribonuclease.
1961	Jacob, Monod and Changeux. Allosterism.
1965	Phillips, Johnson and North. Three-dimensional structure of lysozyme obtained at 1.5 Å resolution.

The accuracy with which a disease state can be diagnosed often depends on the activity measurements of tissue specific enzymes. Once diagnosed the treatment of the pathological disorder may rely on the selective toxicity of chemotherapeutic agents. The differences between normal and abnormal or foreign cells are in many instances slight and one way in which they may be differentiated is to utilise the specificity particularly extant in enzymes. These more applied areas of enzymology are discussed in Chapters 6 and 7.

1.1   ENZYME STRUCTURE AND PROPERTIES

To date nearly two thousand enzymes have been catalogued, together catalysing all the known types of chemical reaction. Oxidation, reduction, hydrolysis, elimination, polymerisation, isomerisation and transfer are all catalysed in biological systems by enzymes that have evolved to respond to local physiological conditions. However, even though so many enzymes exist and catalyse a large variety of reactions, they do possess several characteristics in common.

All enzymes behave as catalysts, that is, only small quantities, relative to the concentrations of their substrates, are needed to considerably increase the rate of chemical reactions, while they themselves undergo no net change. In addition the total amount of substrate transformed per mass of enzyme is often very large. Like all true catalysts, an enzyme does not change the final equilibrium position of a reaction, which is thermodynamically determined, and only the *rate* of attainment of equilibrium of a feasible reaction is increased. The reaction rate may, of course, be unobservably slow in its absence.

A characteristic of considerable importance is that all enzymes are proteins. Sumner's first announcement to this effect was confirmed soon afterwards for several proteolytic enzymes by Northrop and Kunitz (Table 1.1). Therefore in addition to their catalytic properties, enzymes exhibit the chemical and physical behaviour of proteins: their electrolytic behaviours, solubilities, electrophoretic properties and chemical reactivities in addition to their catalytic activities depend on the L-α-amino acid sequence and peptide bonds constituting the protein molecule (see Table 1.2 and Figure 1.1).

Table 1.2: L-Amino Acid Side Chains (R) Commonly Found in Enzymes.

	R
Aliphatic amino acids	
1. Glycine (gly)	$-H$
2. Alanine (ala)	$-CH_3$
3. Valine (val)	$-CH(CH_3)_2$
4. Leucine (leu)	$-CH_2 \cdot CH(CH_3)_2$
5. Isoleucine (ile)	$-CH(CH_3) \cdot CH_2 \cdot CH_3$
Acidic amino acids	
6. Aspartic acid (asp)	$-CH_2 \cdot CO_2H$
7. Glutamic acid (glu)	$-CH_2 \cdot CH_2 \cdot CO_2H$
Acid amides	
8. Asparagine (asn)	$-CH_2 \cdot CO \cdot NH_2$
9. Glutamine (gln)	$-CH_2 \cdot CH_2 \cdot CO \cdot NH_2$
Basic amino acids	
10. Lysine (lys)	$-(CH_2)_4NH_2$

11. Histidine (his)

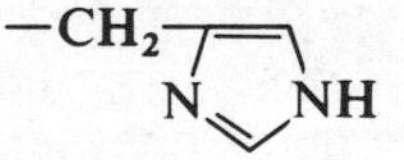

12. Arginine (arg) $-(CH_2)_3 \cdot NH \cdot C(:NH)NH_2$

Hydroxy amino acids
13. Serine (ser) $-CH_2 \cdot OH$
14. Threonine (thr) $-CH(OH) \cdot CH_3$

Sulphur-containing amino acids
15. Cysteine (cys) $-CH_2 \cdot SH$
16. Cystine $-CH_2 \cdot S \cdot S \cdot CH_2 -$
17. Methionine (met) $-(CH_2)_2 \cdot S \cdot CH_3$

Aromatic amino acids

18. Phenylalanine (phe)

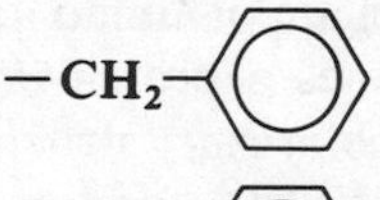

19. Tyrosine (tyr)

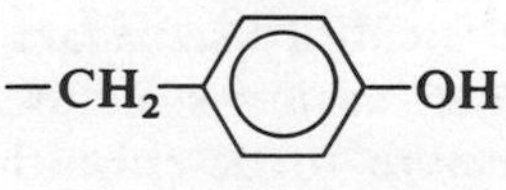

20. Tryptophan (try)

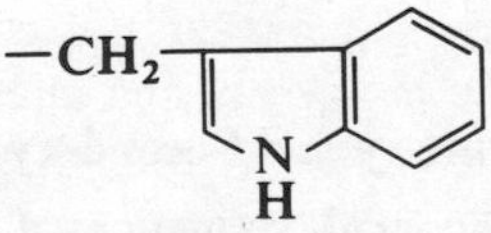

Imino acid

21. Proline (pro)

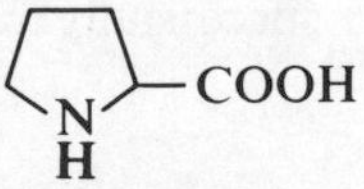

Figure 1.1: Diagrammatic Representation of the Polypeptide Backbone of an Enzyme

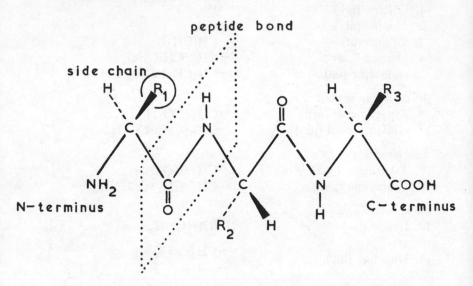

Being proteins, enzymes differ considerably from traditional chemical catalysts such as hydrogen ions, hydroxyl ions, heavy metals or metal oxides. Whereas these are most effective in organic solvents, at high temperatures or at extreme pH values, enzymes operate most efficiently under very mild conditions. Departure from homogeneous, aqueous solutions, physiological pH and temperature rapidly destroys their activities, but under normal conditions the rate increases achieved are rarely matched by their non-protein counterparts.

The linear chain of amino acid residues joined by peptide bonds, which constitutes a protein molecule has been called by Linderström Lang the *primary* structure. Localised folding of the primary structure is entitled *secondary* structure, and the overall folding of the molecule *tertiary* structure. To these Bernal later added *quaternary* structure to describe the agglomeration of several folded chains.

1.1.1 Primary and Secondary Structure

The amino acid sequence of hen's egg white lysozyme, the first enzyme to be successfully analysed three dimensionally,[1] illustrates

Figure 1.2: The Primary Structure of Lysozyme showing the Residues found in α-Helix Conformations (continuous lines) and those in the Apparent Binding Site (underlined)

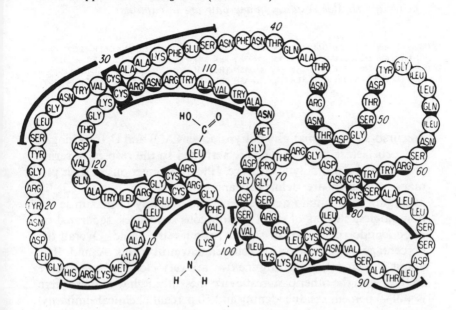

Source: Reproduced with permission, from reference 1.

several features found in primary structures (see Figure 1.2). The order of amino acid residues is the same in all molecules from the same source and appears to be random with, as yet, no obvious predictability. No branching occurs, although many enzymes are intramolecularly crosslinked through the disulphide bridges of cystine. By this means widely separated sections along the chain may be covalently joined: for example residues 5 and 129 form a disulphide bond which brings together both ends of the lysozyme molecule. Present evidence indicates that few of the amino acids are superfluous and most are 'functional', i.e. most co-operatively determine the higher orders of structural organisation and hence catalytic activity.

Comparisons of the primary structures of enzymes performing similar functions have demonstrated extensive structural homologies in their sequences, particularly in the patterns of their non-polar residues.[2] For example, pancreatic juice contains five inactive

Figure 1.3: Sequence Homologies in the first 21 Residues of the B Chains of Mammalian Serine Proteases (CA, *bovine chymotrypsinogen* A; CB, *bovine chymotrypsinogen* B *and* E, *porcine elastase; chemically similar residues in any pair are in capitals*)

	16	17	18	19	20	21	22	23	24	25	26	27	28	29	30	31	32	33	34	35	36
CA:	ILE	VAL	Asn	GLY	GLU	GLU	ALA	Val	Pro	GLY	SER	TRP	PRO	TRP	GLN	VAL	SER	LEU	GLN	Asp	LYS
CB:	ILE	VAL	Asn	GLY	GLU	ASP	ALA	Val	Pro	GLY	SER	TRP	PRO	TRP	GLN	VAL	SER	LEU	GLN	Asp	Ser
T:	ILE	VAL	GLY	GLY	Tyr	Thr	Cys	Gly	Ala	ASN	THR	Val	PRO	TYR	GLN	VAL	SER	Leu	ASN	–	–	–
E:	VAL	VAL	GLY	GLY	Thr	GLU	ALA	GLN	Arg	ASN	SER	TRP	PRO	Ser	GLN	ILE	SER	LEU	GLN	Tyr	ARG

Source: Reference 2.

precursors (zymogens), chymotrypsinogens A, B and C, trypsinogen and proelastase, all of which are activated to the respective serine proteases* by proteolytic cleavage. Illustrative sections of their primary structures are delineated and compared in Figure 1.3. By arranging their amino acid sequences in this way to maximise the homologies, clusters of short homologous sequences, separated by oligopeptide regions of dissimilar residues are found. Overall the percentage residue identity between chymotrypsins A and B is about 78 percent, indicating the two are very closely related, more so than with the other pancreatic enzymes although with these there is still 40 percent residue identity and 50 percent chemical similarity. Calculations of the mutation values of amino acid pairs in CA versus CB and the chymotrypsins versus trypsin, indicate that the former pair possess a more recent evolutionary ancestor in common than do either with trypsin, i.e. the mammalian serine proteases appear to have undergone divergent evolution.

The overall folding of the amino acid sequences to give the functional enzyme appears to be unorganised (Figure 1.4), but on closer inspection, it can be seen that regions are organised into structures of definable symmetry. For example, residues 24–34 and 41–54 in lysozyme are folded into elements of secondary structure, α-helix and β-sheet respectively. These are named after the corresponding structures of the α- and β-keratins and were first predicted by Pauling and Corey using X-ray diffraction data obtained from crystalline short polypeptides. The peptide bond was found to be shorter by approximately 0.01 nm than the 0.14 nm of a single carbon-nitrogen bond, and to have more double bond character. Rotation

* Protein degrading enzymes each requiring an unmodified seryl residue in its active site for activity.

Figure 1.4: Schematic Drawing of the Main Chain Conformation of Hen's Egg White Lysozyme

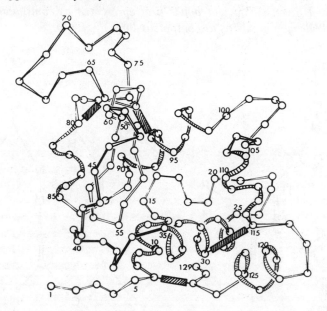

Source: Reproduced with permission, from Reference 1.

around this bond is restricted at normal temperatures, causing the peptide bond and the two adjacent α-carbon atoms to lie in one plane, with the carbonyl oxygen and amino hydrogen in the energy minimum of a *trans* configuration, as shown in Figure 1.1. Rotation can then only occur around those bonds to the α-carbon atoms, and the possible types of stable structure which maximise the number of hydrogen bonds, are restricted to the α-helix and β-pleated sheet. In closer atomic detail these structures are shown in Figure 1.5.

1.1.2 Tertiary Structure

The overall folding of the polypeptide chain, which incorporates the organised secondary structure as well as random stretches, is entitled the tertiary structure (Figure 1.4). At present the most reliable and powerful method for determining the three-dimensional structure of a protein, and in combination with the primary sequence, for portraying the relative stereochemical positions of the atoms is X-ray crystallography.

Figure 1.5: Diagrammatic Representation of (A) One Turn of a Right-handed α-Helix and (B) Two Strands of an Anti-parallel β-Pleated Sheet. (Covalent bonds are represented as continuous lines and hydrogen bonds as discontinuous lines.)

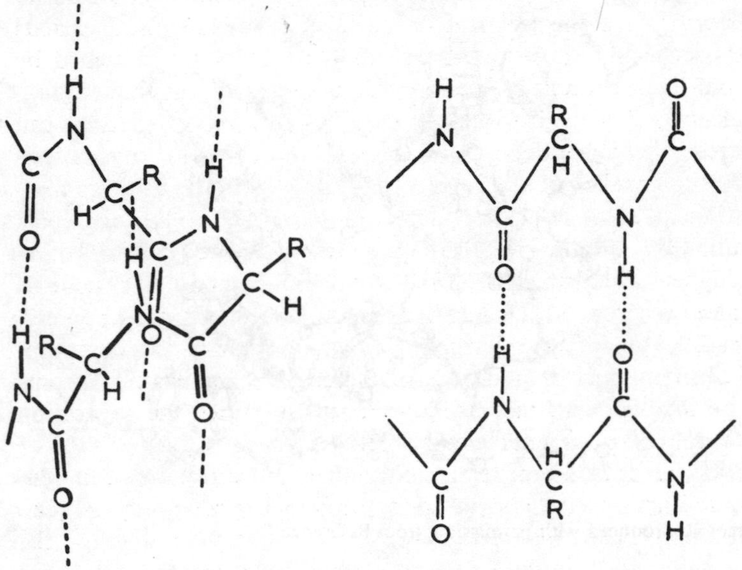

A   B

X-rays are used because their wavelengths are the same order of size as the molecular dimensions in the protein crystal. Crystals are comprised of orderly arrangements of molecules—the crystal lattice or space lattice—which is an almost infinite number of parallel planes. The lattice then acts as a three-dimensional diffraction grating. When monochromatic beams of X-irradiation bombard the crystal they are transmitted through and scattered by the atoms as a number of diffracted beams. If a photographic plate is placed behind the crystal these beams make their appearance on the plate as spots which represent the diffractions at the various planes. The complete array of spots is called the diffraction pattern. From a consideration of the intensity of the spots in the pattern the distribution of electron density in the crystal and thus an image of the protein can be calculated by applying the Fourier mathematical series. For this to be possible however, the phases of the scattered beams must be known. These are obtained by diffusing heavy metal atoms (e.g. uranium) into specific sites in the crystal but without distorting the lattice. Metal ions scatter X-rays more strongly than do the protein atoms, thus by measuring the differences in intensities a solution to the phase problem is approached. The Fourier transforms give the electron densities at many, regularly spaced points, enabling those points of equal intensity to be joined by contour lines. Models of the amino acid residues are then fitted into the electron density distribution. The accuracy with which these can be fitted in however depends on the resolution of the X-ray analysis i.e. by the number of scattered intensities used in the Fourier computations. At low resolution, 4–6 Å, in which a few hundred spots are utilised, only the overall shape of the molecule is obtained. At the higher resolutions of 2.8–3 Å the polypeptide backbone is revealed, whereas at 2.5 Å, amino acid side chains are apparent. From 2 Å down to 1.4 Å resolution (which requires the computation of many tens or hundreds of thousands of intensities) atoms can be located with increasing accuracy, provided the precession photographs are of a sufficiently high standard.

In its current state of refinement not only the primary convolutions and amino acid residue configurations in protein crystals can be determined but also detailed information about binding in the active site may be gathered from crystals, isomorphous with those of the native protein into which cofactors, inhibitors or substrate analogues have been infused. Most of the information discussed in Chapter 4 has been obtained in this way.

Like most physical techniques X-ray crystallography does have limitations. The experimental technique is slow whereas enzyme catalysis is rapid, so that little can be ascertained about the dynamics of catalysis. The models are constructed using data obtained from crystalline derivatives, and are subject to the criticism that the structures in homogeneous solution may be different. For the majority of proteins studied so far however, this concern has to a large extent been dispelled. In the crystalline state most have been found to exhibit catalytic activity and retain the capacity to stoichiometrically bind inhibitors or cofactors in specific orientations and with binding constants similar to those in solution. Several, e.g. ribonuclease, also possess the same conformation in different crystalline forms, indicating that forces in the crystal do not deform the enzyme to a significant degree. The environment enclosing the enzyme molecules in the crystalline state is usually between 30–60 percent solvent. The agreement between crystallographic data and the known physico-chemical properties of the enzymes in solution also suggest that for most of those analysed their enzymically active structures are probably not too different from those calculated. Table 1.3 is a compilation of the enzymes analysed crystallographically to mid-1978.

The enzymes listed in Table 1.3 catalysing dissimilar reactions have been found to possess unique tertiary structures, but the accumulated atomic detail does permit several generalisations to be made. All are compactly folded molecules which, although some regional flexibility is retained, have very little space inside to accommodate even small water molecules. This interior consists mainly of hydrophobic side chains (of tryptophan, phenylalanine, valine, leucine etc.), surrounded by the more polar amino acids (arginine, aspartic acid etc.). One corollary of this is that enzymes catalysing similar reaction types and possessing homology in their primary structures (Figure 1.3), especially in the sequences of their non-polar residues (the aggregation of which provides the driving force for protein folding) also possess three-dimensional homology. For example, members of the serine protease family all have tertiary conformations with a common hydrophobic core; that for chymotrypsin is shown in Figure 1.6.

A second type of localised three-dimensional homology, but one which is not evident from amino acid sequence data, has emerged from X-ray studies on enzymes binding similar cofactors. The tertiary structures of dogfish-muscle lactate DH (dehydrogenase), beef-heart soluble malate DH, horse-liver alcohol DH and lobster and *B. stearothermophilus* glyceraldehyde-3-phosphate DH, all of which require NADH as coenzyme, can be divided into two separate *domains*, i.e. polypeptide regions associated with particular functions (see Figure 1.7). These are a catalytic domain which is structurally unique to each individual enzyme and a coenzyme binding domain whose construction is remarkably similar to all. The latter is comprised of six parallel strands of β-pleated sheet and four α-helices (Figure 1.7), which are found arranged in the sequence β-α-β along the primary structure. (Part of this domain binding the adenosine monophosphate half of NADH has also been found in the adenosine triphosphate dependent phosphoglycerate kinase and adenylate kinase.) The continuous amino acid sequences constituting the domain are not located identically in the overall

Table 1.3: X-Ray Crystallography of Enzymes

Enzyme	Source	Molecular weight	Resolution (Å)
Adenylate kinase	Porcine muscle	22,000	3
Alcohol DH	Horse liver	80,000	2.9
Aspartate transcarbamylase	*E. coli*	310,000	5.5
Carbonic anhydrase B	Human erythrocyte	30,000	2.2
Carbonic anhydrase C	Human erythrocyte	30,000	2
Carboxypeptidase A	Bovine pancreas	34,600	2, 2.8
Carboxypeptidase B	Bovine pancreas	34,000	5.5
α-Chymotrypsin	Bovine pancreas	25,000	2
γ-Chymotrypsin	Bovine pancreas	25,000	2.7
δ-Chymotrypsin	Bovine pancreas	25,000	5
Chymotrypsinogen	Bovine pancreas	27,000	2.5
Elastase	Porcine	25,900	3.5
Glyceraldehyde-3-phosphate DH	Lobster	143,000	3
Hexokinase	Yeast	51,000	2.3
Lactate DH	Pig heart	140,000	6
	Dogfish	140,000	2, 2.8
Lysozyme	Bacteriophage T$_4$	18,800	2.5
	Hen's egg white	14,600	2
	Human	14,500	2.5
Malate DH	Porcine	72,000	2.5
Papain	Papaya latex	23,000	2.8
Phosphoglycerate kinase	Horse muscle	38,000	3
	Yeast	46,000	3.5
Phosphoglycerate mutase	Yeast	111,000	3.5
Phosphorylase b	Rabbit muscle	200,000	6, 5.2
Protease	*Rhizopus chinensis*	35,000	5.5
Pyruvate kinase	Cat muscle	240,000	6
Rhodanase	Bovine liver	37,000	3
Ribonuclease A	Bovine	13,600	2
Ribonuclease S	Bovine	13,600	2
Staphylococcal nuclease	*S. aureus*	16,800	2
Subtilisin BPN′	*B. amyloliquefaciens*	27,500	2.5
Subtilisin novo	*B. subtilis*	27,500	2.8
Superoxide dismutase	Bovine	16,000	3, 5.5
Thermolysin	*B. thermoproteo-lyticus*	34,600	2.3
Thioredoxin	Bacteriophage T$_4$	10,050	4.5
Triose phosphate isomerase	Chicken muscle	49,000	2.5
β-Trypsin	Bovine pancreas	24,000	1.8, 2.7

Figure 1.6: A View of the Complete Polypeptide Chain of α-Chymotrypsin. (The chain is represented by a ribbon folded at each α-carbon atom. Disulphide linkages are indicated by shaded bars.)

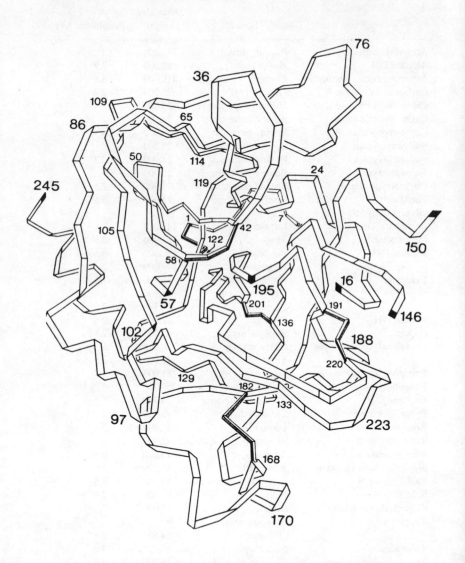

Source: Reproduced with permission from J. J. Birktoft and D. M. Blow *J. Mol. Biol.*, vol. 68 (1972), p. 187.

Figure 1.7: The Catalytic Domain and Coenzyme Binding Domain in Glyceraldehyde-3-phosphate Dehydrogenase from B. stearothermophilus. NAD⁺ *is represented by bold lines.*

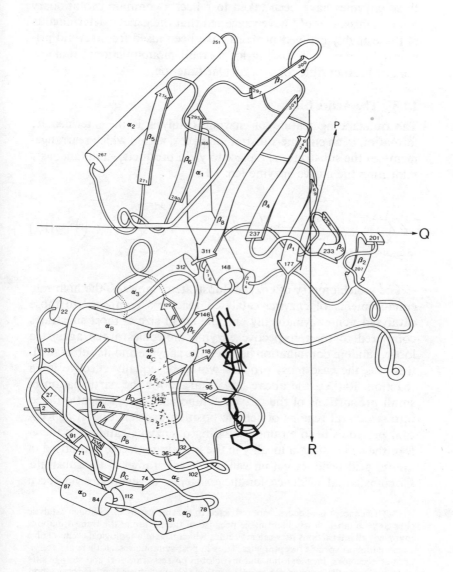

Source: Reproduced with permission, from Reference 3.

primary structures however; that in glyceraldehyde-3-phosphate DH is formed from the first 147 residues, from 22–164 in lactate DH and 193–318 in alcohol DH.

The common structural organisations of sheets and helices in these enzymes have been taken to reflect a common evolutionary origin. Ohlsson *et al.*[4] have suggested that the genetic determinants of these highly evolved proteins have been fused from several primordial genes, each coding for a small monofunctional unit (or domain) responsible for a particular function.

1.1.3   The Active Site

The outstanding feature of enzymic catalysis is the intermediate formation of an enzyme-substrate complex, within which rearrangements of the substrate take place to yield products, simultaneously reforming the native enzyme (eq. 1.1).

$$\left(E\right) + \langle S \rangle \rightleftharpoons \left(E\langle S \rangle\right) \rightarrow \left(E\right) + \langle P \rangle \qquad (1.1)$$

The stoichiometry of enzyme-substrate complexes, the high rate enhancements of enzyme catalysed reactions compared to those catalysed non-enzymatically and the small size of most substrates compared to their respective enzyme molecules argue against a loose, random combination between the enzyme and its substrate. If this was the case most proteins would be equally effective for a reaction. Rather, the above factors argue for the participation of small proportions of the enzyme amino acid complement ordered into structural regions of definite position and function. A conceptual prerequisite to an understanding of enzyme catalysis is therefore the *active site*,* a locality in the protein molecule comprised of amino acid residues within van der Waals range of the substrate. Circumstantial evidence for the existence of such sites has been

---

* This concept is of course, not restricted to enzyme catalyses, although catalytic sites are the most clearly identifiable because of the nature of the transformations involved; all interactions between molecules which trigger a biological event involve combination at specific receptor sites. Thus in post-synaptic vessels there are acetylcholine receptors, haemoglobin and myoglobin possess transport and storage sites respectively, antibody sites are complementary to their antigens and endocrine hormones interact with specific receptor proteins.

provided by the stoichiometries of binding observed spectrophotometrically and by the kinetic experiments described in the next
chapter. More direct evidence is provided by X-ray crystallography.

Information on the structure and relative stereochemical orientations of amino acid residues involved in binding and catalysis is
therefore fundamental to an understanding of the chemical and
physiological action of an enzyme. Figure 1.8, for example, shows a
schematic picture, based on nuclear magnetic resonance data, of the
active site of ribonuclease S containing the inhibitor

*Figure 1.8: Schematic Drawing of the Cytidine-3'-monophosphate-
ribonuclease Complex as seen from the Back of the Active-site Cleft*

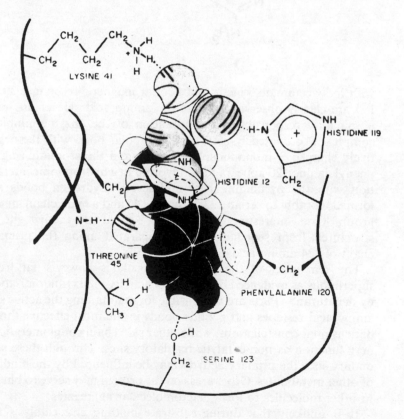

cytidine-3′-monophosphate (I).[5] (Ribonuclease catalyses the endo-nucleolytic cleavage of ribonucleic acid at the 3′-position of a pyrimidine nucleotide residue.) The view is from the back of the active site cleft. Histidines 12 and 119 (implicated in catalysis by modification

$$HO \quad OPO_3^{2-}$$

$$CH_2OH$$

$$O=\ \ N$$

$$N$$

$$NH_2$$

I

with iodoacetamide, kinetic and nuclear magnetic resonance studies) are clearly observable as are the amino acid side chains predominantly responsible for stabilisation of the enzyme-inhibitor complex. Protonated lysine 41 forms a polar bond with the negatively charged 3′-monophosphate group, and the aromatic ring of phenylalanine 120 appears to be involved in a hydrophobic interaction with the pyrimidine base and several hydrogen bonds are formed, notably to serine 123, threonine 45 and a main chain amido group. One interesting observation is that the active site is structured from sequences widely separated along the primary chain of 124 amino acids.

The remainder of the protein molecule is however far from superfluous as evidenced by the loss of activity on its randomisation by denaturants. Parts are responsible for maintaining the active site amino acid residues and peptide bonds in the most effective three-dimensional constellations, while other side chain conglomerations may function as non-catalytic regulatory sites. Through these secondary sites the primary activity may be influenced by the binding of other metabolites. Other areas on the protein may serve to bind it to other molecules to form macromolecular aggregates.

It is unlikely that during substrate binding and catalysis the enzyme and substrate remain in rigid conformations but that one or both undergo conformational alterations before electronic rearrangement. It was Koshland[6] who proposed that the substrate, by inducing an altered geometry in the active site, rearranged the amino acid side chains into catalytic orientations. According to this 'induced fit' hypothesis 'good' substrates would be those that induced the correct alignment of catalytic groupings, while 'poor' substrates would not. There is much physico-chemical and X-ray evidence to support Koshland's view, some of which will be described in Chapter 4. The hypothesis does not however say anything specific about the substrate but it is reasonable to hold that its configuration is also strained on enzyme binding. Distortion of the substrate and/or enzyme will of course necessitate the expenditure of energy and it may be this that is, to some extent, responsible for the specific increases in rate.

The active-site concept provides a rational basis for another characteristic of enzymic catalysis–substrate *specificity*. For example, mammalian lactate dehydrogenases catalyse the NAD^+ linked oxidation of L-lactate, eq. 1.2, but not the oxidation of its structural isomer, β-hydroxyproprionate or its stereoisomer D-lactate; yeast lactate dehydrogenase on the other hand is a flavo-protein (see Section 1.4) and converts solely D-lactate. Totally different enzymes have evolved which catalyse the conversion of substrates almost identical in their chemical reactivities.

$$ (1.2) $$

NAD^+ L-Lactate   NADH   Pyruvate

The stereochemical requirement of the lactate dehydrogenases for D- or L-lactate, is readily discernible; less obvious however is the capability of the enzymes to distinguish between the two apparently identical hydrogen atoms on the tetrahedral carbon atom, C-4, eq. 1.2. If the atoms are differentiated and the structure represented as below (II), it is seen that association of the NADH molecule with an enzyme could cause the two atoms to experience different microenvironments. Both D-lactate DH and L-lactate DH select H_R for transfer and are referred to as A-side specific, in common with several others (see Table 1.4). Some other dehydrogenases are B-face specific, reversibly transferring H_S to their substrates (Table 1.4).

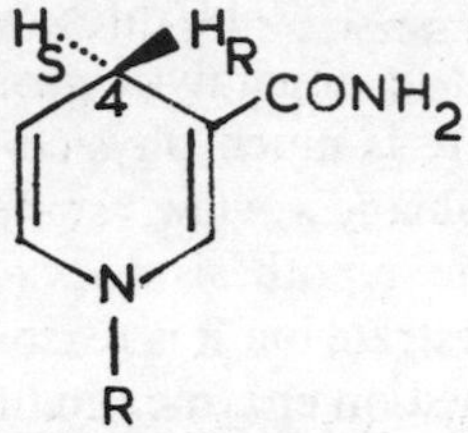

II

Table 1.4: Dehydrogenase (DH) Stereospecificities

A-side specific:
   Alcohol DH, cytochrome b_5 DH, dihydrofolate DH, glycerate DH, glyoxylate DH, isocitrate DH, D-lactate DH, L-lactate DH, L-malate DH, orotate DH.

B-side specific:
   Cytochrome c DH, dihydrolipoate DH, glucose-6-phosphate DH, L-glutamate DH, glutathione DH, α-glycerolphosphate DH, L-β-hydroxybutyryl-CoA DH, hydroxysteroid DH, 6-phosphogluconate DH, triose phosphate DH.

By means of specificity, enzymes are able to exert considerable metabolic control. In the cell or organism, one substrate is often catalytically transformed into several different products by different enzymes; since metabolic pathways involve the consecutive action of enzyme teams wherein the product of one becomes the substrate of the next, the specificity of each serves to maintain and control the directional supply of intermediates. Specificity of reaction towards both substrate and product also increases catalytic efficiency by eliminating side reactions, and thus ensures the economic utilisation of valuable cellular energy.

1.1.4   Quaternary Structure
Most enzymes of intracellular origin also possess quaternary structure, the agglomeration of several units of tertiary structure. In such

enzymes each of the contributing tertiary structures is termed a *subunit* or *monomer* and the complete complex is then an *oligomer*, and a dimer, trimer, tetramer, etc. depending on the number of subunits it contains. Oligomeric proteins are classified as *homologous* if they contain identical subunits and *heterologous* if they contain different subunits. A short list of such enzymes together with their subunit complements is compiled in Table 1.5. This table dem-

Table 1.5: Subunit Composition of some Enzymes possessing Quaternary Structure

	Molecular weight	Number of subunits
Identical (homologous) subunits		
Superoxide dismutase (*E. coli*)	39,500	2
Chorismate mutase (*Streptococcus*)	63,000	4
Chorismate synthetase (*N. crassa*)	110,000	2
Nucleoside diphosphokinase (yeast)	102,000	6
Ornithine transcarbamylase (bovine)	108,000	3
Phosphoglycerate mutase (yeast)	110,000	4
Pyruvate kinase (yeast)	161,000	8
Carboxylesterase (human liver)	186,000	3
Lactate oxygenase (*Mycobacteriun phlei*)	350,000	6
Glutamine synthetase (*B. stearothermophilus*)	592,000	12
Non-identical (heterologous) subunits		
Superoxide dismutase (*Photo. lecognathi*)	33,700	AB
Procarboxypeptidase A (bovine)	88,000	ABC
Histidine decarboxylase (Micrococcus sp.)	190,000	A_5B_5
Aspartate transcarbamylase (*E. coli*)	310,000	A_6B_6
RNA polymerase (*E. coli*)	400,000	A_2BB'

Source: Reference 7.

onstrates that the number of subunits comprising an enzyme cannot be predicted from a knowledge of its molecular weight. It has also been found that the number of active sites per enzyme cannot be equated with the subunit complement, for example, only six of the aspartate transcarbamylase subunits are catalytically active. The other, non-catalytic, subunits perform a regulatory function critical to the physiological role of the enzyme (see Chapter 3).

Quaternary structure is not stabilised by covalent bonds, the subunits associate through combinations of the weaker forces, electrostatic, hydrophobic and hydrogen bonds. Consequently this level of organisation is readily dissociated by denaturing agents such as urea, guanidine hydrochloride or sodium dodecyl sulphate into the contributing subunits. These reagents do not cleave covalent bonds, so that after dissociation, gel filtration or electrophoresis can be used to estimate the number, types and sizes of the subunits.

1.2   ENZYME ASSEMBLAGES

From the foregoing it could be assumed that an enzyme molecule, whether or not it possesses quaternary structure, catalyses a single type of reaction. For example, *E. coli* aspartate transcarbamylase, although it is a heterologous protein comprised of twelve subunits, catalyses only the carbamylation of aspartate. However protein aggregates have been isolated, and purified to homogeneity, that catalyse more than one step in a metabolic pathway. Of course the difficulty here is to be absolutely sure that the aggregate under investigation is a physiologically distinct assemblage and not an artifact of isolation. However, copurification of the several activities and their spontaneous recombination under physiological conditions after dissociation are strong pieces of evidence that the complexes are autonomous assemblies. Enzyme assemblages containing multiple enzyme activities can be classified into two types—multienzyme complexes and multifunctional enzymes.

1.2.1   Multienzyme Complexes

These consist of non-covalently associated subunits each one of which catalyses one distinct and different reaction in a sequence. Possibly the most extensively documented multienzyme complex is pyruvate dehydrogenase,[8] which catalyses the oxidation of pyruvate to acetyl-CoA.

$$CH_3 \cdot CO \cdot CO_2H + NAD^+ + CoA \cdot SH \rightleftharpoons CH_3 \cdot CO \cdot S \cdot CoA$$
$$+ NADH + CO_2 + H^+ \quad (1.3)$$

The complex is constructed from multiple copies of three distinct polypeptides: pyruvate decarboxylase (E_1, MW 10^5 daltons), lipoate acetyltransferase (E_2, MW 8×10^4 daltons) and lipoamide dehydrogenase (E_3, MW 5.6×10^4 daltons). Estimates vary of the number of each in the complex. A model for that from *E. coli* consistent with electron microscopy envisages a 'core' containing eight trimers of E_2 arranged at the corners of a cube (III), with a total of 12 or 24 molecules of E_1 arranged along its edges and 12 or 24 molecules of E_3 on its faces (IV).

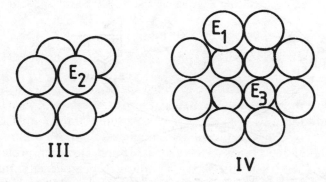

III

IV

In the presently accepted sequence of reactions (Figure 1.9) pyruvate decarboxylase catalyses the elimination of carbon dioxide from pyruvate and transfers the acetyl group to thiamine pyrophosphate (TPP, see V). Mediated by lipoate acetyltransferase, the acetyl group is then carried by a lipoic acid side chain from β-hydroxyethyl-TPP to coenzyme A. Transfer of the acetyl group to coenzyme A leaves lipoic acid in the reduced form, so that before it can re-enter the cycle in order that only co-catalytic quantities are

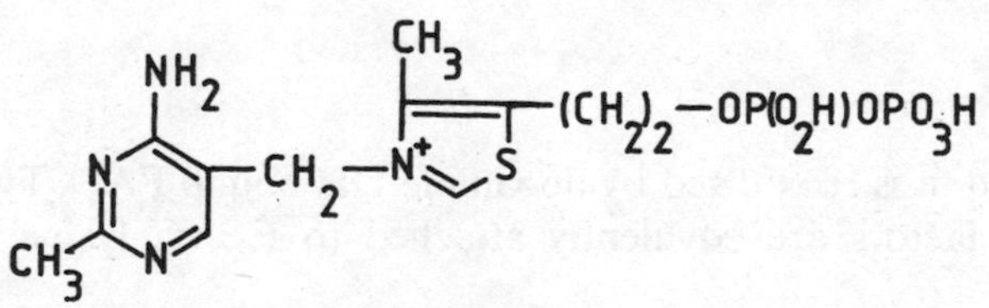

4-Amino-2-methyl pyrimidine	4-Methyl-pyrazole	Pyrophosphate

V

Figure 1.9: Scheme for Oxidation of Pyruvate to Acetyl-CoA by Pyruvate Dehydrogenase

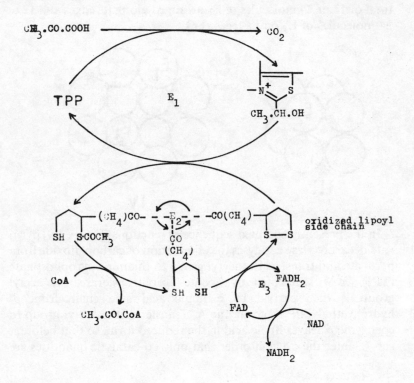

involved, it is reoxidised by lipoamide DH bound FAD. Two lipoic acid cofactors are covalently attached to E_2 via lysine ε-amino groups.

It is possible that the long, centrally placed lipoyl-lysyl side chains act as 'rotating arms' in the complex, accepting the hydroxy-ethyl group from pyruvate DH bound TPP, delivering it to the active site of lipoate acetyltransferase and then conveying the disul-phide terminus to lipoate DH for oxidation by FAD. The final reaction is recycling of the FADH by oxidation with NAD^+.

Table 1.6: Examples of Polycephalic Proteins

Enzyme	Enzyme Commission No.*	Source	Subunit Structure (MW)	Reaction catalysed**
Chorismate mutase: prephenate DH	5.4.99.5: 1.3.1.12	*E. coli*	A_2 A = 40,000	Chorismate $\downarrow$ 5.4.99.5 Prephenate $\downarrow$ NAD 1.3.1.12 4-Hydroxyprephenate
Anthranilate synthase: anthranilate phosphoribosyltransferase	4.1.3.27: 2.4.2.18	*E. coli*	$A_2 B_2$ A = 62,000 B = 62,000	Glutamate + Chorismate $\downarrow$ 4.1.3.27 Pyruvate + Glutamate + Anthranilate $\downarrow$ PRPP** 2.4.2.18 P-ribosylanthranilate
Tryptophan synthase	4.2.1.20	*N. crassa*	A_2 A = 75,000	Indoleglycerol-P G3P** $\nearrow$ + Serine $\searrow$ G3P Indole $\xrightarrow{+ \text{Serine}}$ Tryptophan
Chorismate mutase: prephenate dehydratase	5.4.99.5: 4.2.1.51	*E. coli*	A_2 A = 40,000	Chorismate $\downarrow$ 5.4.99.5 Prephenate $\downarrow$ 4.2.1.51 Phenylpyruvate + CO_2

* See Section 1.6.
** PRPP = Phosphoribosylpyrophosphate, G3P = Glyceraldehyde-3-phosphate.

1.2.2 Multifunctional Enzymes

These enzymes may or may not possess subunits but each polypeptide chain has multiple catalytic functions, so that they are variously described as multiheaded, chimeric or polycephalic proteins, whose binding sites are generated by 'folding of contiguous stretches of chains to yield autonomous domains'.[9] Table 1.6 gives a short list of well documented multifunctional proteins; two others, the fatty acid synthetase (FAS) complex and aspartokinase : homoserine dehydrogenase are described in more detail in Chapter 5.

1.2.3   Physiological Significance of Enzyme Assemblages

What survival advantage has an organism possessing multienzyme complexes rather than separate enzymes possessing the same contributing activities? Detailed answers to this question arise from considerations of their protein structures and their kinetic properties.[7]

First, the information necessary for protein folding, including that necessary for quaternary aggregation is probably contained in the primary sequence. This structure is encoded by the bases in the nuclear DNA, thus less genetic information and hence less storage space will be needed. Secondly, in a large molecule wrong incorporation of one or more amino acids could render the whole protein inactive. In a smaller polypeptide, though the frequency of mutation will be the same, there is a greater probability that the effect of any mutation will be minimised: the oligomers exist in monomer-oligomer dynamic equilibria, thus the presence of defective subunits could cause the oligomer to dissociate and allow the correctly encoded subunits to reorganise and associate. Thirdly, subunit association is concentration dependent, giving an extra facility for control of enzyme function. Smaller particles will be more easily transported from the ribosome to the site of action, but at the site, as the concentration builds up, association and hence activity will increase. Fourthly, oligomeric proteins are possible targets for molecular evolution, since any mutation causing a slight change in tertiary structure may be amplified through subunit interactions.

Control of several enzyme activities in a metabolic sequence by the same combination of activators, inhibitors or inducers would be more effective if they were part of the same complex or even on the same polypeptide chain. This is well illustrated by the regulation of the aspartokinase : homoserine DH enzyme complex discussed in Chapter 5. A complex such as fatty acid synthetase (FAS) which catalyses as many as eight consecutive reactions would also be more efficient than a loose mix of separate enzymes of similar reactivities, since the encounter frequencies of the enzymes with their substrates would be considerably improved as a result of increased local concentrations (the effect of limited diffusion is discussed in Chapter 7). It is interesting that FAS of bacteria are essentially loose agglomerations of separate enzymes whereas in higher organisms e.g. liver and yeast, the FAS systems are oligofunctional enzymes. Maintaining the catalytic activities within a restricted region will also reduce the opportunities for intermediates to undergo competing reactions.

1.3 PROTEIN FOLDING

Proteins are synthesised by the ribosomal machinery with amino acid sequences translated from the base sequences in the nucleic acids, but after leaving the ribosome they must be organised into specific three-dimensional structures before becoming functional molecules. Although considerable detail is known concerning translation, much less is known about the post-translational sequence of folding events. Do enzymes assume the conformation which is thermodynamically the most stable under the conditions? Or is the unique three-dimensional structure imparted *after* synthesis of the molecule?

Two experiments point to the former of these alternatives, to the important role of the primary sequence in determining the overall structure and hence catalytic activity of an enzyme. In 1969 two groups of workers, Gutte and Merrifield[10] and Hirschman et al.,[11] simultaneously reported the total synthesis of an enzyme with the polypeptide structure of ribonuclease from individual amino acids. After oxidation of the cysteine groups with atmospheric oxygen, the synthetic molecule exhibited ribonuclease activity. A few years previously, Anfinsen[12] and his colleagues had completely unfolded bovine pancreatic ribonuclease by reduction of the four disulphide links with β-mercaptoethanol in 8M urea. This treatment effectively destroyed its structural organisation and activity. But removal of the denaturants by dialysis followed by air oxidation of the cysteines regenerated 95 to 100 percent activity. Moreover the four disulphide bridges were shown to be correctly paired. If reoxidation had occurred at random only one percent of the original activity would have been expected.

Anfinsen's experiment supports the hypothesis that thermodynamic stabilisation is the main driving force for protein organisation and that no additional genetic information is necessary. No conclusions can be made however about the nature of the folding process but some form of epigenetic (post-translational) promotion of folding must operate, as a simple calculation shows.[13] An amino acid residue can rotate around two bonds, thus for 100 amino acids the number of conformations is 3^{200} (approx. 10^{100}); the time scale for rotation around a single bond being about 10^{-13} seconds, the time needed for all conformations to be tried is 2×10^{89} seconds or about 10^{80} centuries! The time taken for Staphylococcal nuclease (149 amino acids) to fold is only a few seconds.

Evidence has accumulated indicating that protein folding (and unfolding) is a highly co-operative process, i.e. that folding of a few residues greatly facilitates the rate and degree of subsequent folding. This is seen experimentally as a sudden and large transition in physico-chemical properties such as viscosity or optical rotation over a narrow range of change in denaturant concentration. The initial folding could therefore be considered analogous to the nucleation of crystallisation of a saturated solution. Anfinsen and Scheraga,[14] based on experiments such as those described above for ribonuclease renaturation, have argued that the native three-dimensional system is the one with the lowest Gibbs free energy, in other words, the one in which the free energy of all interactions is minimised. On this basis they visualise a system in which the unfolded polypeptide (U), a partially folded form containing segments of native structure called 'nucleation' centres (P) and the fully folded native protein (N) are in equilibrium:

$$U \rightleftharpoons P \rightleftharpoons N \tag{1.4}$$

The driving force to the native conformation is then the negative free energy of stabilisation afforded by long range forces between the nucleation centres and the amino acid residues. The pathway to folding is proposed to be, firstly, formation of backbone nucleation centres from the randomised polypeptide, organised through nearest neighbour, short range interactions between the side chain of the residue and the atoms of its backbone. Secondly, these nucleation centres are stabilised and several brought into close proximity by medium range interactions between neighbouring residues. Finally the whole native structure is stabilised by long range interactions between the organised sites and the other residues.

The nucleation sites may be situated anywhere along the polypeptide chain and thus could be formed at any step of its ribosomal synthesis. Little is known concerning protein folding patterns *in vivo* but partially folded intermediates have been found and characterised during the *in vitro* renaturation of reduced bovine pancreatic trypsin inhibitor (PTI).[15] Native PTI has fifty-eight residues and three disulphide bridges, between half-cystines 5–55, 14–38 and

30–51. The reoxidation of PTI promoted by dithiothreitol was periodically stopped by acidification or alkylation and the intermediates separated. Only a few were observed and from the kinetics of their formation and breakdown, it was proposed that they were formed in an obligatory order. It is interesting to find in view of the discussion above that the half-cystines of the first disulphide bond formed, residues 30 to 51, are both in secondary structural regions, antiparallel β-pleated sheet (16–36) and α-helix (47–56) respectively. This supports the proposal that an early step in protein folding may be the formation of secondary structural elements.

The next question to be asked is what is the driving force under which the protein folds up to attain its specific and unique orientation? The experimental evidence suggests that protein organisation is the consequence of a small difference in entropy as the hydrophobic side chains (phe, try, val, etc.) withdraw from the aqueous environment and into the interior of the protein molecule.[7] This 'hydrophobic force' is considered the main driving force for folding but it alone is not entirely responsible for maintaining the three-dimensional structure. The strength of this force increases with temperature whereas most enzymes are denatured at high temperature and the effectiveness of chemical denaturants do not always parallel their hydrophobicity. In addition the hydrophobic force is nondirectional whereas enzymes certainly have distinctive structures. Thus directional forces superimposed on the non-directional must operate.

The other types of non-covalent bond responsible for the maintenance of protein structure are the interpeptide hydrogen bond (VI), side chain hydrogen bond (VII) and polar or electrostatic bond (VIII).

$$\text{C=O} \cdots \text{H-N} \qquad \text{—} \langle \bigcirc \rangle \text{—OH} \cdots \text{OOC—} \qquad \text{—COO}^- \cdots \text{H}_3\overset{+}{\text{N}} \text{—}$$

VI   VII   VIII

Since most polar side chains are on the surface of the enzyme they are bound to solvent and thus their contribution to the overall stability is minimal. However isolated instances of polar bonding have been observed, for example in active chymotrypsin, an electrostatic bond exists between the α-amino group of isoleucine 16 and the 'buried' β-carboxylate group of aspartate 194. In an aqueous environment the hydrogen bonding contribution to stability will also be small but will increase with the apolar nature of the protein interior.

Many proteins, particularly those of extracellular function, are additionally crosslinked by covalent disulphide bridges, eq. 1.5.

$$
\underset{\substack{| \\ \text{NH} \\ |}}{\overset{\substack{| \\ \text{CO} \\ |}}{\text{C}}}-\text{CH}_2-\text{SH} \;+\; \text{HS}-\text{CH}_2-\underset{\substack{| \\ \text{NH} \\ |}}{\overset{\substack{| \\ \text{CO} \\ |}}{\text{CH}}} \longrightarrow \underset{\substack{| \\ \text{NH} \\ |}}{\overset{\substack{| \\ \text{CO} \\ |}}{\text{HC}}}-\text{CH}_2-\text{S}-\text{S}-\text{CH}_2-\underset{\substack{| \\ \text{NH} \\ |}}{\overset{\substack{| \\ \text{CO} \\ |}}{\text{CH}}} \qquad (1.5)
$$

Originally it was thought that the disulphides only added to stability after the protein had already attained its thermodynamically favoured configuration. But some experimental evidence, such as that of Creighton,[15] is consistent with a kinetic influence on the folding processes, in which the cysteine residues play important roles in determining the folding pathway and hence the final three-dimensional configuration. The evidence adduced includes the observation of a ten-fold faster rate of disappearance of free SH groups than rate of recovery of ribonuclease activity from its scrambled state, and the identification of non-native disulphide pairs during its refolding. Therefore a pathway involving rapid formation of disulphide bonds followed by their intramolecular rearrangement is possible, eq. 1.6.

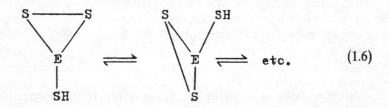

$$ (1.6) $$

The rate of this disulphide interchange, even in the presence of added thiols, eq. 1.7, is slow compared to the biological rate, thus it is likely that *in vivo* this process is also enzyme catalysed.

$$R \cdot SH \ + \ \begin{array}{c} S \!\!-\!\!-\!\!-\!\! S \\ \diagdown \quad \diagup \\ E \\ | \\ SH \end{array} \ \rightleftharpoons \ \begin{array}{c} HS \quad S\!\!-\!\!S\!\!-\!\!R \\ \diagdown \quad \diagup \\ E \\ | \\ SH \end{array} \qquad (1.7)$$

Preparations containing 'protein disulphide isomerase' activity have been isolated from endoplasmic reticula, i.e. from locations close to the machinery of protein synthesis, which accelerate intra-molecular disulphide interchange and recovery of activity of 'scrambled' ribonuclease.[16] The enzyme is widely distributed in mammalian tissues and is relatively nonspecific, catalysing the disulphide exchange of a wide variety of oxidised proteins. However its involvement in protein synthesis has not been firmly established and no evidence for its definite requirement in a physiological process has yet been obtained.

1.4 PROSTHETIC GROUPS

Although all enzymes have been found to be proteins, for activity some require the additional presence of a non-protein *prosthetic* group, thus:

$$\text{Holoenzyme} \quad = \text{Apoenzyme} + \text{Prosthetic group} \quad (1.8)$$

(Total complex)   (Protein)   (Non-protein)

As pointed out by Dixon and Webb,[17] prosthetic groups may be categorised functionally into two classes, *coenzymes* and *cofactors*. The first may be considered biosynthetically related to the vitamins. For example, the coenzyme nicotinamide adenine dinucleotide (NAD) important for cellular energy metabolism incorporates the vitamin niacin into its chemical make-up. In addition coenzymes may be regarded as co-substrates, undergoing a chemical transformation during the enzyme reaction (NAD is reduced to NADH) reversal of which requires a separate enzyme, possibly in a different cellular location. They may therefore travel intracellularly between apoenzymes and by transferring chemical groupings integrate several metabolic processes. By comparison, cofactors, for example pyridoxal phosphate or haem groups, remain with one enzyme molecule and in conjunction complete a cycle of a chemical change within one enzyme turnover. Other enzymes, for example carboxypeptidase, require metal ions as cofactors, the divalent cations Mg^{2+}, Zn^{2+}, Mn^{2+} being most common; these are often termed enzyme *activators*. Table 1.7 gives a list of the more common coenzymes, their metabolic or coenzymic roles and the related vitamins. It has been established experimentally that many of the coenzymes are derived from vitamins and viewing Table 1.7 it is

Table 1.7: Coenzymes Participating in Enzyme Catalysed Processes

Coenzyme	Vitamin	Metabolic roles
Adenosine Triphosphate (ATP)	—	Transphosphorylation
Biotin	Biotin	Transcarboxylation
Cobamide	B_{12}	Transmethylation
Coenzyme A	Pantothenic acid	Transacylation
Flavin mononucleotide (FMN) Flavin adenine dinucleotide (FAD)	Riboflavin (B_2)	Hydrogen carrier
Nicotinamide adenine dinucleotide (NAD)	Niacin	Hydrogen carrier
Non-haem iron	—	Hydrogen peroxide hydrolysis
Lipoic acid	—	Transacylation
Pyridoxal phosphate	B_6	Amino transfer, decarboxylation, racemisation.
Tetrahydrofolate (THF)	Folic acid	One carbon transfer
Thiamine pyrophosphate (TPP)	Thiamine	Aldehyde transfer, oxidative decarboxylation

apparent that the majority of these belong to the class described as water soluble. By definition, a vitamin is an essential growth factor which cannot be synthesised by the organism because the requisite enzymes are absent. A reduction in its dietary supply will therefore cause a decreased concentration of the derived coenzyme and will lead to a reduction in its dependent enzymic activity, even though the tissue concentration of the apoenzyme may be normal. One medical consequence of this could be a deficiency disease. However, not all coenzymes have a vitamin origin, ATP and lipoic acid for example being synthesised *de novo* by most mammalian cells.

1.5 ENZYME PURIFICATION BY AFFINITY CHROMATOGRAPHY

The classical methods of isolation, separation and purification, developed for proteins in general can also be applied to enzymes. Ammonium sulphate precipitation, ion-exchange chromatography and the less preparative, more analytical techniques such as electrophoresis and isoelectric focusing, all utilise differences in the physico-chemical properties of the individual protein molecules. Although the majority of purifications have been performed using these techniques their relative unselectivities often result in impaired resolution of enzymes possessing similar physico-chemical properties, necessitating a balance to be made between yield and purification at each stage.

The development of 'affinity chromatography', which avoids some of the above disadvantages, has therefore had a considerable impact on enzymology. Introduced by Cuatrecasas, Wilchek and Anfinsen[18] and developed initially for enzyme purification, it is readily adaptable to the isolation of either component of a receptor-ligand system. Where affinity chromatography departs from the methods above, is in employment of the biological characteristic of specificity. Interactions between a macromolecule and its ligand are usually tight, but non-covalent, and a complex between the two can often be dissociated by a non-denaturing change in the environmental conditions. Based on this principle, if the ligand is covalently attached to an insoluble support and then loaded into a chromatography column, when a mixture containing the required receptor macromolecule is passed through the material, under the right conditions it will be retained by the matrix while non-specific molecules will pass through unretarded. Because the interaction is non-covalent, suitable adjustment of the eluting buffer composition will dissociate the adsorbed enzyme from the matrix and enable its collection (Figure 1.10).

The main factor determining the success of the affinity chromatography method is the degree to which the experimental conditions used approach those of the analogous enzyme-ligand association in homogeneous solution. Contributing to this are the nature of the supporting matrix, which must be inert yet capable of activation to a ligand binding form, and the steric relationships between enzyme, ligand and matrix. Included in the latter are the matrix-ligand distance, point of attachment on the ligand and its conformation. When

Figure 1.10: The Principle of Affinity Chromatography. E, *ligand-bound enzyme,* P_1 *and* P_2, *noncomplementary proteins*

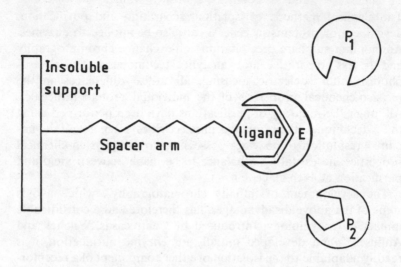

the association constant is low (below $5 \times 10^{+3} M^{-1}$) its attached concentration becomes important. Yield and purity are optimised by suitable adsorption and desorption conditions.

Gel particles of agarose have been used successfully in many cases as the support matrix for affinity chromatography although cellulose derivatives, crosslinked dextrans, polyacrylamide gels and glass beads have also proved effective. Agarose is a polysaccharide of D-galactose β(1-4) linked to 3,6-anhydrogalactose. Beads formed from this linear carbohydrate possess porous but fairly rigid, chemically and mechanically stable structures. In addition they exhibit good flow characteristics and allow free access of both macromolecules and chemical modifiers. This latter property is important for homogeneous chemical derivatisation of the matrix.

If it possesses suitably reactive groups the ligand may either be attached directly to the agarose matrix or to a reactive derivative. The latter, 'matrix activation', is usually preferred since the polyhydroxyl groups possess limited reactivity, the concentration of valuable ligand is usually the limiting factor and the attachment of ligand to a reactive intermediate may more easily be controlled. Cyanogen bromide is rapidly becoming the standard reagent for agarose activation. The procedure is straightforward and gives reproduceable substitution of the galactose hydroxyl groups. After reaction, the agarose beads contain a high proportion of cyclic carbamic acid imide groupings (eq. 1.9) although side reactions giving unreactive carbamates have been found.

$$(1.9)$$

To avoid possible matrix-restricted access of enzyme to a directly attached ligand a 'spacer arm' is usually first coupled to the activated support and the ligand attached to this. Initially connecting arms of the type $NH_2 \cdot (CH_2)_n \cdot R$ ($n = 2$ to 12) were interposed but because of their hydrophilicity and fewer degrees of rotational freedom, oligoglycines of the type

$$NH_2 \cdot (CH_2 \cdot CO \cdot NH)_n \cdot CH_2 \cdot CO_2H$$

are now preferred. The resulting matrix is then probably substituted with some or all of the species below,

$$X = -(CH_2 CO \cdot NH)_n \cdot CH_2 COOH$$

Attachment of the ligand to the spacer arm depends on the chemical natures of both, which are usually chosen to complement each other. For example, if the ligand has a non-essential (from the viewpoint of enzyme association) amino group, its attachment to the spacer carboxylic acid groups promoted by a water soluble carbodiimide may be feasible.

The choice of ligand is often fairly wide. Substrate analogues, inhibitors, allosteric modifiers, coenzymes or cofactors may all exhibit specific interactions with the enzyme of interest, and the problem is to successfully attach the ligand to the activated matrix via a position on the former which does not substantially decrease the enzyme-ligand association constant. For example, attachment of adenosine monophosphate to 6-aminohexyl sepharose via its purine amino group proved successful for the purification of alcohol dehydrogenase and glycerokinase but not for glyceraldehyde-3-phosphate dehydrogenase or hexokinase; glyceraldehyde-3-phosphate dehydrogenase but not the others could however be attached to adenosine diphosphate bound to sepharose via its diphosphate group. These observations reflect the different binding requirements of the different enzymes for the nucleotide.[18]

An illustration of the comparative efficiency of this technique is its application to the isolation and purification of alkaline phosphatase from a crude sucrose homogenate of calf intestine.[19] The data, shown in Table 1.8, are also used to illustrate the terms usually used as indices of success in purification procedures. *p*-Arsanilic acid, (a competitive inhibitor of alkaline phosphatase i.e. a molecule which combines specifically at the active site but which is not a substrate), was attached to Sepharose by the three different side arms A, B, C, below:

$$\text{—NH(CH}_2\text{ CO NH)}_2\ \underset{\displaystyle \text{CH·CH}_2}{\overset{\displaystyle \text{COOH}}{|}}\ \text{—}\!\!\underset{\text{OH}}{\bigcirc}\!\!\text{—N=N—}\bigcirc\text{—AsO}_3\text{H}_2 \qquad \text{A}$$

$$\text{—NH·(CH}_2)_6\cdot\text{NH·CH}_2\text{·CO NH—}\bigcirc\text{—AsO}_3\text{H}_2 \qquad \text{B}$$

$$\text{—NH (CH}_2)_2\ \text{—}\!\!\underset{\text{OH}}{\bigcirc}\!\!\text{—N=N—}\bigcirc\text{—N=N—}\bigcirc\text{—AsO}_3\text{H}_2 \qquad \text{C}$$

The enzyme adsorbed to all three affinity matrices, and unattached protein was washed out by tris buffer pH 8.4. Desorption and elution were carried out in a phosphate buffer of the same pH; phosphate also competitively inhibits the enzyme but with a larger association constant than *p*-arsanilic acid. Matrix B was found to be twelve-fold less effective than C for the isolation and purification of alkaline phosphatase, probably because its shorter length elicited steric restriction of enzyme approach by the support and the flexibility of the hydrophobic chain may additionally have allowed a population of conformers in which the chain was folded back towards

Table 1.8: Purification of Calf-intestinal Alkaline Phosphatase

Procedure	Specific activity[a]	Yield[b]	Fold purification[c]
Butanol treatment	1.8	78	1
Acidification	6	71	3
Ammonium sulphate precipitation	22	68	12
DEAE cellulose ion exchange[d]	130	23	72
Affinity chromato-graphy[d] (Matrix C)[e]	900	56	500

[a] Units of activity per mg protein (see Section 2.1).
[b] Percentage of total original activity.
[c] Calculated from specific activity; the first step is arbitrarily assigned unity.
[d] Ion exchange and affinity chromatography were performed separately on aliquots of the same salt precipitate.
[e] See text.
Source: Reference 19.

Table 1.9: Affinity Chromatography Purification of Enzymes

Enzyme	Affinity matrix
Alcohol DH	NAD-glass
Carbonic anhydrase	Sulphanilamide-sepharose
DNA polymerase	DNA-acrylamide
Dopamine hydroxylase	Concanavalin A-sepharose
Hydroxysteroid DH	Blue dextran-sepharose
Isoleucyl-tRNA synthetase	Aminoacyl-tRNA-agarose
Luciferase	FMN-sepharose
Phosphodiesterase	ATP-agarose
Pyruvate kinase	Cibacron blue-sephadex
Xanthine oxidase	Allopurinol-agarose

Source: Reference 20.

the polysaccharide. Supports A and C are both less flexible than B, and longer; although C is the more hydrophobic, the probable cause of its three-fold lower efficiency, it enabled collection of a product less contaminated with phosphate binding enzymes and so was considered more selective.

The selectivity and efficiency afforded by this technique compared to ion-exchange chromatography are apparent from Table 1.8. A representative list of enzymes purified using affinity chromatography is compiled in Table 1.9.[20]

1.6   ENZYME CLASSIFICATION AND NOMENCLATURE

A reaction catalysed in a particular manner is the specific property of an enzyme and this provides a rational basis for its characterisation, nomenclature and classification. Accordingly the enzymes fall into six main classes.[21] These are Class 1, oxidoreductases; Class 2, transferases; Class 3, hydrolases; Class 4, lyases; Class 5, isomerases; Class 6, ligases. Every enzyme known to catalyse a specific reaction has been assigned to one of these classes, and as new enzymes are discovered they are also included in the scheme. Within these classes, each enzyme is identified in three ways, by (a) a 'systematic' name which is as exact as possible stressing the type of reaction, (b) a 'recommended' or 'trivial' name which is a shortened form intended for general use, and (c) by a code number of four elements, the first of which denotes the main class, the second the sub-class, the third the sub-sub-class and the fourth element is the serial number of the enzyme in the sub-sub-class. The subdivisions identify the type of substrate molecule, cofactor or bond cleaved.

The key to the numbering and classification of the enzymes, recommended by the Enzyme Commission (EC) of the International Union of Biochemistry,[21] is shown in Appendix I, and the main classes are briefly described below:

CLASS 1. *Oxidoreductases* are the systematic names given to proteins catalysing oxidation-reduction reactions. For example, glycerol : NAD^+ 2-oxidoreductase (EC code number 1.1.1.6) catalyses the reaction:

$$CH_2 \cdot OH$$
$$|$$
$$CH \cdot OH + NAD \rightleftharpoons C{=}O + NADH \quad (1.10)$$
$$|$$
$$CH_2 \cdot OH \qquad CH_2 \cdot OH$$

Glycerol   Dihydroxyacetone

The recommended names for these enzymes are 'dehydrogenase' if NAD^+ or FAD are the hydrogen acceptors, 'oxidase' if O_2 is the electron acceptor, 'monooxygenase' if O is incorporated and 'peroxidase' if electrons are donated to H_2O_2.

CLASS 2. *Transferases* catalyse the transfer of a group from a donor molecule to an acceptor. The systematic names are donor: acceptor grouptransferases, for example, L-aspartate : 2-oxoglutarate aminotransferase (EC 2.6.1.1) catalyses transfer of the $-NH_2$ group between L-aspartate and L-glutamate:

$$CO_2H \cdot CH_2 \cdot CH(NH_2) \cdot CO_2H + CO_2H \cdot CH_2 \cdot CH_2 \cdot CO \cdot CO_2H$$

L-Aspartate   2-Oxoglutarate

$$(1.11)$$

$$CO_2H \cdot CH_2 \cdot CO \cdot CO_2H + CO_2H \cdot CH_2 \cdot CH_2 \cdot CH(NH_2) \cdot CO_2H$$

Oxaloacetate   L-Glutamate

The recommended names for these enzymes are acceptor grouptransferases or donor grouptransferases, e.g. L-aspartate aminotransferase (occasionally transaminase is encountered). Enzymes transferring the orthophosphate group from ATP are often termed 'kinases'.

CLASS 3. *Hydrolases* catalyse the hydrolysis of covalent C—O, C—N, C—C, C—P, C—S, C—halide, P—O and P—N bonds. Strictly they should be included in Class 2 as transferases since they use a water or other suitable molecule as group acceptor, but from a historical viewpoint and because hydrolysis is their main physiological function they are classed separately. Their systematic names are substrate hydrolases except for certain enzymes such as lysozyme, rennin, plasmin and others whose historical names are retained. An example from this group is EC 3.1.1.7, acetylcholine hydrolase or acetylcholinesterase which catalyses the hydrolysis of acetylcholine to choline plus acetate:

$$CH_3 \cdot CO \cdot O \cdot CH_2 \cdot CH_2 \cdot \overset{+}{N}(CH_3)_3 \longrightarrow$$

$$CH_3 \cdot COO^- + (CH_3)_3 \overset{+}{N} \cdot CH_2 \cdot CH_2 \cdot OH \quad (1.12)$$

CLASS 4. *Lyases* catalyse the cleavage of covalent bonds by removal of a group to leave a double bond. Their systematic names are substrate group-lyases or, if the reverse reaction is predominant, synthases. The recommended names are usually those adopted by usage, thus aldolase, dehydratase and decarboxylase are common. For example fumarate hydratase (EC 4.2.1.2) catalyses the reaction:

$$HO_2C \cdot CH(OH) \cdot CH_2 \cdot CO_2H \rightleftharpoons$$

L-Malic acid   (1.13)

$$HO_2C \cdot CH \overset{trans}{=\!=\!=} CH \cdot CO_2H + H_2O$$

Fumaric acid

CLASS 5. *Isomerases* catalyse structural rearrangements within a molecule. Their nomenclature is based on the type of isomerism, thus these enzymes are identified as racemases, epimerases, isomerases, mutases, tautomerases or anomerases. For example:

$$CHO \cdot CH(OH) \cdot CH_2 \cdot OPO_3H_2 \underset{\text{isomerase EC 5.3.1.1}}{\overset{\text{Triosephosphate}}{\rightleftharpoons}}$$

D-Glyceraldehyde-3-phosphate

$$CH_2(OH) \cdot CO \cdot CH_2 \cdot OPO_3H_2 \quad (1.14)$$

Dihydroxyacetone phosphate

CLASS 6. *Ligases* (synthetases) are enzymes catalysing the conjunction of two molecules coupled to the hydrolysis of a nucleoside triphosphate. The systematic names are A : B ligases, for example, pyruvate carboxylase (EC 6.4.1.1) or systematically, pyruvate : carbon dioxide ligase (ADP-forming) catalyses the carboxylation of pyruvate to oxaloacetate:

$$ATP + CH_3 \cdot CO \cdot CO_2^- + CO_2 + H_2O \longrightarrow$$
$$ADP + HPO_3^{2-} + HO_2C \cdot CH_2 \cdot CO \cdot CO_2^- \quad (1.15)$$

1.7 REFERENCES

1. C. C. F. Blake, D. F. Koenig, C. A. Mair, A. C. T. North, D. C. Phillips and V. R. Sharma, *Nature*, vol. 206 (1965), p. 757.
2. B. S. Hartley and D. M. Shotton, in P. D. Boyer (ed.), *The Enzymes* (Academic Press, New York, 1971), vol. 3, p. 345.
3. G. Bresecker, J. I. Harris, J. C. Thierry, J. E. Walker and A. J. Wonacott, *Nature*, vol. 266 (1977), p. 328.
4. I. Ohlsson, B. Nordstrom and C. I. Branden, *J. Mol. Biol.*, vol. 89 (1974), p. 339.
5. D. H. Meadows, G. C. K. Roberts and O. Jardetzky, *J. Mol. Biol.*, vol. 45 (1969), p. 491.
6. D. E. Koshland, *Adv. Enzymol.*, vol. 22 (1960), p. 45.
7. I. M. Klotz, D. W. Darnall and N. R. Langerman, in H. Neurath and R. L. Hill (eds.), *Proteins*, 3rd edn (Academic Press, New York, 1975), vol. 1, pp. 294–402.
8. D. L. Bates and R. N. Perham, *Nature*, vol. 268 (1977), p. 313.
9. K. Kirschner and H. Bisswanger, *Ann. Rev. Biochem.*, vol. 45 (1976), p. 143.
10. B. Gutte and R. B. Merrifield, *J. Am. Chem. Soc.*, vol. 91 (1969), p. 501.
11. R. Hirschman et al., *J. Amer. Chem. Soc.*, vol. 91 (1969), p. 502.
12. C. B. Anfinsen, C. J. Epstein, R. F. Goldberger, *Cold Spring Harbour Symp. Quant. Biol.*, vol. 28 (1963), p. 439.
13. A. L. Lehninger, *Biochemistry*, 2nd edn (Worth, New York, 1975), p. 144.
14. C. B. Anfinsen and H. A. Scheraga, *Adv. Prot. Chem.*, vol. 29 (1975), p. 205.
15. T. E. Creighton, *J. Mol. Biol.*, vol. 95 (1975), p. 167.
16. C. B. Anfinsen, *Science*, vol. 181 (1973), p. 223.
17. M. Dixon and E. C. Webb, *Enzymes*, 2nd edn (Longmans, New York, 1964).
18. C. R. Lowe and P. D. G. Dean, *Affinity Chromatography* (Wiley, London, 1974).
19. O. Brenna, M. Perella, M. Pace and P. G. Pietta, *Biochem. J.*, vol. 151 (1975), p. 291.
20. M. Wilchek and C. S. Hexter, *Methods of Biochem. Analysis*, vol. 23 (1976), p. 345.
21. *Enzyme Nomenclature: Recommendations of the International Union of Pure and Applied Chemistry and the International Union of Biochemistry* (Elsevier, Amsterdam, 1972).

2 Enzyme Activity

46

The relationships that have proved to exist between the structures of enzymes and their specific functions are beginning to be deciphered as more enzymes are prepared in high states of purity and in forms suitable for investigation by X-ray crystallographic methods. As yet however this has been realised for only a few enzymes; even for these, interpretation of their catalytic actions in terms of their molecular organisations is proving far from an easy task. Because of this complexity, the actual reaction an enzyme catalyses is of prime importance to its identity. This reaction is the main distinguishing feature of the enzyme and accordingly is adopted as a basis for its nomenclature and classification (see Chapter 1). Thus in addition to the molecular properties described in the last chapter, for a full understanding of an enzyme, the reaction it catalyses and the mechanism by which it operates must also be analysed.

By mechanism is meant the detailed molecular, atomic and ionic processes that take place in the conversion of substrate to product. Included in this are the bonds which are formed, the bonds cleaved and the stereochemical relationships between them.* For this purpose, techniques have been developed which are peculiar to enzymology and which are essential tools in the unravelling of the catalytic behaviour of an enzyme.

2.1 METHODS OF INVESTIGATING THE MECHANISMS OF ENZYME CATALYSED REACTIONS

2.1.1 Substrate and Product Identification

Substrates are the substances enzymically transformed into reaction products and are one of the major determinants of the reaction. Some enzyme catalysed isotope exchange reactions have been described in which the substrates and products were not identified,

---

* In enzymology the term 'mechanism' is also used to describe the order of binding and release of substrates and products, see Section 2.2.3.

but elucidation of their molecular structures is an essential first step to an understanding of the catalytic mechanism of an enzyme, for which they provide circumstantial evidence.

The methods used for the separation and determination of the molecular structures of enzyme substrates and products are essentially those of analytical and physical chemistry.[1]

2.1.2   Isotope Labelling

Knowledge of the structures of the substrates and end-products can give some information on the bond(s) cleaved during reaction but not in all cases. Where this information cannot be obtained, isotope labelling (or tracer) experiments have proved effective. For example, hydrogen transfer between lactate or alcohol and NAD, catalysed by the appropriate dehydrogenase, was monitored by the use of deuterium and tritium isotopes (Section 1.1.3). By this means the nicotinamide carbon atom C-4 was identified and the side specificity of each dehydrogenase deduced.

Enzyme catalysed ether, ester and glycoside hydrolyses, like oxidation-reduction, are also commonly observed reactions. The substrates may be cleaved at either of the two bonds 'a' or 'b' below.

$$R_1 \overset{a}{-} O \overset{b}{-} R_2 \xrightarrow{+H_2O} R_1OH + R_2OH \tag{2.1}$$

where R_1 and R_2 are monosaccharide, aryl-, alkyl-, acyl-, phosphate or sulphate groups.

In water of normal isotopic distribution, the products would be identical whichever bond was broken, but if either the substrate or the water is labelled and the location of the label in the products then ascertained, the position of cleavage can be determined, as shown below.

$$R_1 \overset{a}{-} {}^{18}O \overset{b}{-} R_2 + H_2O \begin{cases} \overset{a}{\nearrow} R_1OH + R_2{}^{18}OH \\ \overset{b}{\searrow} R_1{}^{18}OH + R_2OH \end{cases} \tag{2.2}$$

$$R_1 \overset{a}{-} O \overset{b}{-} R_2 + H_2{}^{18}O \begin{cases} \overset{a}{\nearrow} R_1{}^{18}OH + R_2OH \\ \overset{b}{\searrow} R_1OH + R_2{}^{18}OH \end{cases} \tag{2.3}$$

Koshland[2] has made two statements concerning the patterns generally to be expected from the enzyme catalysed cleavage of these bonds: (a) the weaker bond is the one usually split, for example esters of general formula, $R_1CO \cdot OR_2$ are usually cleaved at the weaker, ester, bond and (b) if the enzyme displays more selectivity for a certain group in the molecule then it is the bond between that and the oxygen that is normally ruptured. For example acid phosphatase catalyses the hydrolysis of most organophosphates, breaking the phosphate-oxygen bond in preference to that joining the organic moeity to oxygen.

2.1.3   Kinetic Methods*

The foregoing are all essential to working out the catalytic mechanism of a particular enzyme, but they provide little information regarding its activity–the rate of reaction of substrate attributable to catalysis. For this, kinetic investigations must be applied; these are the most general means of determining reaction mechanisms.[3] Experiments are designed to yield information on the steps of a reaction and to furnish facts describing the precise influence of substrate concentration, enzyme concentration and physical and chemical modifiers on enzymic activity. From this knowledge, a mechanistic model is built up to account, as closely as possible, for the observed catalytic behaviour. The consistency of the proposed hypothesis is then tested by further experimentation under varying limiting conditions.

Kinetic procedures also provide information concerning the involvement of intermediate complexes, whether a particular form of the substrate is involved and whether structural or conformational changes occur in the enzyme. They cannot however give much information on the chemical nature of these intermediates or changes, nor provide final proof of a mechanism, for which experiments similar to those described in the preceding sections are necessary.

The majority of enzyme kinetic experiments have been performed *in vitro* in isothermal, homogeneous, closed systems. In closed systems there is no exchange of material with the surroundings and in homogeneous solution, with the concomitant absence of concentration gradients, more precise measurements and theoretical

---

* In this and subsequent sections, upper case characters will represent both species and their concentrations.

analyses may be made than are feasible in heterogeneous or open systems. Isothermal systems are of advantage because the temperature can then be independently varied.

2.1.3.1   *Enzyme Velocity*

Traditionally, kinetic investigations consist of mixing the enzyme and substrate and then monitoring the substrate or product concentration as a function of time. A typical curve portraying the time dependence of the product concentration corresponding to the reaction, eq. 2.4, is illustrated in Figure 2.1.

$$\text{Substrate (S)} \xrightarrow{\text{Enzyme (E)}} \text{Product (P)} \tag{2.4}$$

Figure 2.1 is entitled the *progress curve* for the reaction eq. 2.4. From this curve the kinetic law obeyed by the system may be derived. A common analytical procedure is the 'differential method' which involves drawing tangents to the curve at frequent time or concentration intervals, Figure 2.1. The slope of the tangent at any time, t, then represents the rate of change, v, of the substrate or product concentration with time, at that point, i.e.

$$v = -\frac{dS}{dt} \tag{2.5}$$

or

$$v = +\frac{dP}{dt} \tag{2.6}$$

The positive and negative signs indicate respectively an increase and a decrease in concentration. For the single substrate-single product reaction above, eqs. 2.5 and 2.6 are identical.

As can be seen from Figure 2.1 the tangent of maximum slope, hence the highest rate, is that passing through t_0. This maximum rate, the *initial velocity* of the enzyme catalysed reaction, has proved most useful in the analysis of enzyme kinetic behaviour, for three main reasons. First, on mixing, the reactants are at their maximal, set concentrations but as the reaction proceeds substrate molecules are continually being converted, causing the rate to fall. The initial velocity thus reflects more accurately the known commencing conditions. Secondly, at t_0, the enzyme protein molecules have had insufficient time to denature or inactivate significantly, so that their maximum effect will be exerted. Thirdly, close to zero time very

Figure 2.1: Progress Curve for the Enzyme Catalysed Reaction S → P Illustrating the Differential Method for Determining Reaction Rates

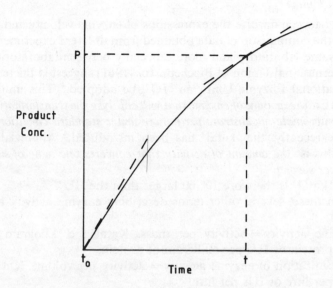

little product will have been formed, and this is important since some products inhibit the enzymes forming them and the product of a forward reaction is often enzymically catalysed back to its respective substrate. As the reaction progresses the extent of the reverse step will also increase.

Initial velocities are usually measured at several substrate concentrations with enzyme concentration, temperature, etc. kept constant, in order to give quantitative expression to the general rate law

$$v = -\frac{dS}{dt} = k \cdot S^n \qquad (2.7)$$

The value of n indicates the extent to which the observed rate depends on the substrate concentration, i.e. the order of reaction, and the proportionality constant, k, is the rate constant of the reaction, with units $(\text{mol dm}^{-3})^{n-1} \text{ s}^{-1}$. Graphical analysis of the variation of velocity with substrate concentration is a very useful procedure for estimating the parameters which characterise the enzyme reaction. These in turn give valuable clues to the reaction mechanism. The theoretical and mechanistic significance of such plots and parameters is discussed below.

2.1.3.2   *Units*

In order to standardise the expressions of enzyme velocity and so enable the comparison of data obtained from different experiments in the same laboratory and more especially between laboratories, the International Union of Biochemistry (1961) suggested the term 'International Enzyme Unit', an 'IU', be adopted.[4] This unit is defined as: *the amount of enzyme that will catalyse the transformation of one micromole of substrate per minute under standard conditions.*

More recently, the 'katal' has been introduced.[4] Abbreviated 'kat', this is: *the amount of activity that converts one mole of substrate per second.*

The 'katal' is therefore $10^6/60$ larger than the 'IU'.

From these, several other terms describing enzyme activity are derived:

Specific activity = activity per mass. Katals per kilogram of protein or IUs per milligram of protein.

Concentration of enzyme activity = activity per volume. Katals per litre or IUs per litre.

Catalytic constant = moles product produced per minute per mole of pure enzyme.

Turnover number = catalytic constant divided by the number of active sites.

2.1.4   Principles Involved in the Practical Measurement of Enzyme Activity

Kinetic methods can be classified according to the time scale over which measurements are made. In terms of decreasing half lives $(t_{1/2})$, the time for fifty percent completion of reaction, these are:

Steady-state methods $t_{1/2} > 10$ s
Fast reaction or flow techniques $t_{1/2} \sim 10^{-3}$ s
Relaxation procedures $t_{1/2} < 10^{-3}$ s

Steady-state kinetics, the theory of which is treated in some detail later, is overall the easiest of the three to perform. The time scale, ranging from a few seconds to hours and even days, means that the manipulatory protocol, adding, mixing reagents, and measurement procedures do not require the complexity of instrumentation required by flow and relaxation techniques. In its simplest form, steady-state kinetics may be performed with commonly available laboratory equipment, test-tubes, pipettes, colorimeters or spectrophotometers. This type of measurement has therefore constituted the bulk performed throughout the developmental history of enzymology, and has been crucial to the support of the basic theories of enzyme catalysis. It continues today in clinical, teaching and analytical laboratories. Steady state measurements are often termed *assays of enzyme activity*, or in short *enzyme assays*.

Greater insight into the mechanism of enzyme catalysis is however obtained by investigating those processes which occur before the steady state has been attained.[5] For these circumstances fast reaction techniques have been developed which follow the progress of the reaction over the first few milliseconds. Rapid mixing of the reagents then becomes the first priority, and special mixing chambers have been developed into which the reactants are injected. The types of flow technique are then described as continuous flow and stopped flow depending on the treatment received by the reacting solution after this prior mixing. In continuous flow the solution is passed along a tube and the extent of reaction monitored at strategic positions along this tube. The progress curve can then be drawn and analysed to determine the kinetic behaviour. The stopped flow principle requires the solution to flow from the mixing chamber to a special reaction cell in which the substrate and product concentrations are measured, usually spectrophotometrically, as a function of time.

Proton transfer and conformational transitions in enzyme molecules have rate-determining steps with half-lives less than a millisecond. Thus in order to analyse such reactions, techniques which avoid slow reagent mixing are necessary. Eigen and his colleagues pioneered such methods.[5] The reaction is first allowed to reach its equilibrium position and is then very rapidly disturbed by a change in temperature or pressure, which is made to take place within a time period shorter than that for the system to adjust to a new equilibrium position. The time course of this response, the relaxation, is monitored and analysed.

2.1.4.1 Steady-state Methods

The progress curve can be obtained experimentally by discontinuous or continuous practical procedures. In the discontinuous or sampling method, accurately measured aliquots are withdrawn from the reaction pool at frequently recorded time intervals, the reaction in the sample is terminated by a rapid change in pH, precipitation of the enzyme protein or rapid freezing, and the substrate or product concentration therein is estimated. A series of time-concentration points are then obtained from which the progress curve can be constructed. Concentrations are determined by the usual analytical procedures of spectroscopy, colorimetry, fluorimetry, titration, radioactivity, etc.

Preferred however is the continuous procedure. Immediately after mixing, changes in a physical property of the substrate or product are monitored continuously without interfering with the system. Changes in extinction coefficient, fluorescence, hydrogen ion concentration or optical rotation have proved the most reliable, especially when the output from the instrument has been linked to an external recording system to enable the progress curve to be recorded automatically. This method gives a closer approximation to the true progress of the reaction since in effect an infinite number of points is outputted, and any irregularity, such as a lag period, which could occur between successive sampling points, is readily seen. No physical interference with the reaction mixture occurs, obviating contamination, and no time lag is interposed between sample withdrawal and reaction cessation. For these reasons initial rates determined by this method are more precise, provided the proportionality between the physical property and concentration is adequately known.

For many enzyme catalysed reactions the first part of the progress curve is linear so calculation of the initial velocity is simple; for others however the data handling is less straightforward. A practice commonly employed is to statistically fit to the experimental data a series polynomial such as:

$$y = a_0 + a_1 t + a_2 t^2 + \cdots + a_n t^n \qquad (2.8)$$

where y is concentration, t is time and a_i $(i = 0, 1, \ldots, n)$ are adjustable coefficients.

Several different statistical methods have been put forward for limiting the polynomial order, but for many practical purposes the quadratic

$$y = a_0 + a_1 t + a_2 t^2 \qquad (2.9)$$

is sufficient and can be fitted by the usual least squares procedures.

At any time, t, the rate is given by the differential

$$\frac{dy}{dt} = a_1 + 2a_2 t \qquad (2.10)$$

which when $t = 0$ (initial velocity conditions) reduces to a_1; the initial velocity is given therefore by the second coefficient of the polynomial.

Such methods can be adapted to give the standard error in the initial velocity. These errors can be reduced by designing the experiment to measure appearance of product rather than disappearance of substrate. Measurement of a small change in a large amount will give more quantitative uncertainty than measuring a small increase from nothing.

2.1.4.2 Precautions and Controls

Three sets of precautions are necessary when setting up a series of enzyme kinetic experiments. First, except under certain circumstances, initial velocities are measured. Secondly, regulation of the physical variants of temperature, hydrogen ion concentration and ionic strength is crucial to the interpretation of the enzyme assay. A small change in temperature can affect the reaction in three different ways: by altering the rate constant in accordance with Arrhenius' law (Section 3.1), by inactivating or denaturing the enzyme, or by causing a temperature dependent conformational change in the enzyme and hence changing the mechanism. The pH of the reaction medium is maintained by a suitable buffer system which does not otherwise interfere with the reacting molecules and a concentration of buffering components is chosen to compensate for any changes in the hydrogen ion concentration that may occur during the reaction. The reasons for this are twofold. The ionisation of side chains in the enzyme is strongly pH dependent and those at the active site contribute to the enzyme activity (see Section 3.2.2). Other ionisable side chains may be responsible for maintaining the three-dimensional integrity of the protein molecule, so that pH constancy must be maintained to prevent denaturation during reaction. Ionic strength, μ ($= 0.5\Sigma cz^2$, where c is the concentration of an ionised species possessing a charge z) affects the reactant activity coefficients (γ_i)—the primary salt effect, and the dissociation of weak acids and bases—the secondary salt effect.[3] Thus from a consideration of the macromolecular structure and the local structure of the active site it is clear that the ionic strength of the buffering system should also be controlled; not only for each assay at the same pH but also for those performed over a range of pH values. Thirdly, in addition to the external controlling factors above, a set of internal controls is also usually performed. To discover if contributions to the time dependent change in the concentrations of substrate or product are made by substances in the assay medium other than the enzyme itself, the system is set up without the enzyme preparation. Some substrates, for example, *p*-nitrophenyl acetate, sometimes used in the assay of hydrolytic enzymes, also react measurably with water molecules and buffer components. A separate control containing a heat inactivated aliquot of the enzyme preparation is also usually performed. Quantitative allowance can then be made for any non-specific catalytic effect of the protein molecule or of a heat stable component. Finally, to detect contamination of the enzyme preparation by endogenous substrate, cofactor or coenzyme, assay systems are composed with each of these missing.

2.1.4.3 Continuous Methods

These may be divided into methods which record changes in a spectral property of the substrate or product and those that monitor changes in the activities of these substances.

Ultraviolet and Visible Spectrophotometry. Spectroscopy is the method of choice for the determination of enzyme activities, and several commercial instruments have been specially developed for this purpose. These permit a considerable number of assays to be performed rapidly and with minimal manipulation. Because spectrophotometric methods have thus proven readily adaptable for rate determinations much effort has been directed to the design of chromophoric substrates. These can be discussed under three broad headings, naturally occurring substrates and cofactors, synthetic substrates and coupled reactions.

Naturally occurring substrates which undergo changes in their absorption spectra on enzymic conversion are the most useful indicators of the reaction. Commonly encountered examples of this type are the nicotinamide dinucleotides, NAD and NADP, the coenzymes for many dehydrogenases. The main difference between the reduced and oxidised forms of these coenzymes is at 340 nm. (Figure 2.2), where the latter shows almost no peak, thus the majority of NAD(P) linked dehydrogenases are most readily assayed at

Figure 2.2: Ultraviolet Spectra of Nicotinamide Dinucleotide Coenzymes

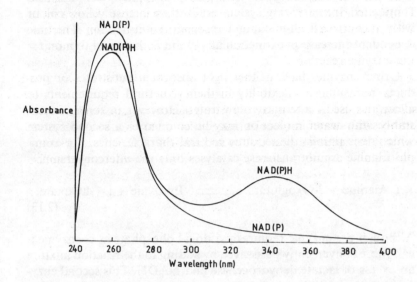

this wavelength. The reaction of the coenzyme with the substrate is stoichiometric (eq. 2.11), and thus the progress curve for disappearance or appearance of NADH is also that for the substrate.

$$S + NADH + H^+ \rightleftharpoons SH_2 + NAD^+ \qquad (2.11)$$

Not all enzymes however act on natural substrates or give products that contain suitably absorbing chromophores. Yet many do exhibit sufficient latitude in their specificities to enable the alternative use of synthetic chromophoric substrates. For example alkaline phosphatase catalyses the specific hydrolysis of phosphate esters, eq. 2.12.

$$R \cdot O \cdot PO_3^{2-} \longrightarrow R \cdot OH + HPO_4^{2-} \qquad (2.12)$$

Those that occur naturally are non-chromophoric but the enzyme does not discriminate between organic groups, R; so for its assay, chemically synthesised chromogenic substrates can be used without eliciting a reduction in catalytic efficiency. One used routinely for the clinical assay of alkaline phosphatase in blood serum is *p*-nitrophenyl phosphate. This substrate is not highly coloured but the main reason for its choice is that under alkaline assay conditions it is cleaved to phosphate plus *p*-nitrophenolate ion. This conjugated aromatic ring system exhibits an intense yellow colour with a wavelength of maximum absorption 400–410 nm. The time dependent increase in extinction at 410 nm is thus used to monitor the enzyme reaction.

Other enzymes have neither light absorbent substrates or products nor sufficient flexibility in their structural requirements to allow the use of alternative substrates. However in some circumstances the enzymic reaction may be coupled to a second system which does possess the required spectral characteristics. For example, alanine aminotransferase catalyses only the interconversion,

$$\text{L-Alanine} + \text{2-Oxoglutarate} \rightleftharpoons \text{Pyruvate} + \text{L-Glutamate} \tag{2.13}$$

Although the enzyme requires pyridoxal phosphate as coenzyme it can more conveniently be assayed by adding to the reaction mixture an excess of lactate dehydrogenase and NADH. This second enzymic system immediately converts any pyruvate formed in the first step to lactate, simultaneously oxidising NADH to NAD. The time dependent optical density change at 340 nm then represents the progress curve of the alanine aminotransferase reaction since the excess of NADH plus lactate dehydrogenase ensures that the production of pyruvate is the limiting rate.

The coupled reaction may be represented more generally by eq. 2.14,

$$A \xrightarrow{E_1} B \xrightarrow{E_2} C \tag{2.14}$$

where E_1 and E_2 represent the two enzymes, and X and Y two forms of a co-substrate of the second enzyme catalysed reaction. This second reaction is termed the 'indicator' reaction and either the concentration of final product (C) or co-substrate (X/Y) can be monitored.

The concentration ranges over which rates may be determined by ultraviolet spectroscopy are set partly by the extinction coefficients of the chromophores and the size of the instrument cell holders. The Beer-Lambert law states that the observed absorption (A) depends on the extinction coefficient (ε), concentration (C) and length of absorbing solution (l) as $A = \varepsilon \times C \times l$. A total absorbance change of at least 0.1 optical density units is normally required for accurate rate determinations and a path-length of 10 cm is usually the maximum accommodated by most spectrophotometers. The lower limit of concentration change that can comfortably be measured is thus $0.01/\varepsilon$; for NADH with an extinction coefficient of approximately 6×10^3 mole^{-1} cm^{-1}, this is approximately 2 micromolar. A lower limit is therefore set on the use of this type of spectrophotometry. However, measurements at lower concentrations are often necessary for the determination of kinetic parameters and where this cannot be met using absorption spectroscopy, fluorescence spectroscopy offers distinct advantages.

Fluorescence Spectroscopy.[6]   Some molecules on irradiation with ultraviolet or visible light at certain wavelengths emit radiation at longer wavelengths. This lower energy emission is entitled fluorescence. Fluorescence depends on the concentration of the species in solution and thus if the substrate and product differ in their emission wavelengths fluorescence spectrophotometry may be used for rate measurements.

The main advantage offered by fluorimetry over absorption spectroscopy is its greater sensitivity. This allows much lower concentrations of substrate to be used and importantly also enables quantitative measurements at low enzyme concentrations. Its main disadvantage however is the lack of a direct relationship between fluorescence reading and concentration, and thus calibration curves which take into account instrument geometry, etc., are necessary, thereby introducing additional errors. Fluorescence is also highly temperature dependent and measurements are sensitive to light scattering particles in the sample. Per unit time, however, the fluorescence response is larger than the equivalent in absorbance terms. Therefore this type of spectroscopy is finding increasing application for the estimation of very small enzyme concentrations and for the rapid analysis of large quanties of samples.

Spectropolarimetry.   Enzymes catalysing the interconversions of molecules which rotate the plane of polarised light can in principle be assayed spectropolarimetrically. Thus the reactions of glycosidases such as invertase, which catalytically cleaves sucrose, specific rotation, $(\theta)_D^{25} = +67°$, into D-glucose $(\theta)_D^{25} = +53°$ plus D-fructose $(\theta)_D^{25} = -92°$, or the enzymatic racemisation of amino acids may be followed. Polarimetry is however less sensitive than the other spectrophotometric techniques, the observed optical rotation (θ) depends on concentration, C grammes per litre as:

$$(\theta)_D^{25} = \frac{(\theta) \cdot 100}{C \cdot L} \qquad (2.15)$$

where L is the length in decimeters of the rotating solution. Thus to follow the invertase reaction, the concentration change needed to alter the rotation of plane polarised light at 589 nm by 0.5° would be about 0.013M. This concentration is at least two orders of magnitude larger than that required to make a spectroscopic measurement of similar precision (provided of course a convenient assay could be designed). A further factor which tends to decrease the precision of optical rotatory measurements is the high specific rotations of protein molecules which may change during the enzyme reaction.

Electrode Methods.[7]   Instrumental techniques that permit selective nondestructive measurements to be made are of great value in assays of enzyme activity especially if direct and continuous. Electrode methods are becoming of increasing value in this respect. From an enzymological viewpoint those most commonly used are the hydrogen ion responsive glass electrode, the oxygen electrode and enzyme electrodes.

The pH electrode is usually incorporated together with a calomel reference electrode into a pH-stat, an apparatus for maintaining constant pH in a reaction involving the release or uptake of hydrogen ions. For example many esterases catalyse the hydrolytic release of acids from esters:

$$R \cdot CO_2 R' \xrightarrow{H_2O} R \cdot CO_2 H + R'OH$$
$$(R \cdot CO_2 H \rightleftharpoons R \cdot CO_2^- + H^+) \qquad (2.16)$$

In an unbuffered medium, the release of hydrogen ions would tend to lower the pH with resultant detrimental effects on esterase activity. In the pH-stat the change in electromotive force experienced by the electrode triggers the addition of sufficient base (e.g. potassium hydroxide solution) from a micrometer syringe to maintain the pH at a constant value. The amount added per unit time, stoichiometrically related to the utilisation of substrate, is recorded and is the time course of the reaction.

Molecular oxygen is the co-substrate for many enzymes such as glucose oxidase, the amino-acid oxidases, cytochrome oxidase and xanthine oxidase, and the quantitation of oxygen uptake and release is central to the investigation of the mitochondrial and photosynthetic enzyme systems. The standard procedure for measurement of oxygen utilisation, as that of the other gases, carbon dioxide and nitrogen, was Warburg manometry, but the development of a potentiometric method for oxygen estimation (the Clark electrode) has enabled more sensitive, rapid and convenient measurements of oxygen utilisation to be made.

The electrochemical reactions of the oxygen electrode are:

$$O_2 + 2H_2O + 4e = 4OH^-$$

$$(O_2 + 2H_2O + 2e = H_2O_2 + 2OH^- \qquad (2.17)$$

$$H_2O_2 + 2e = 2OH^-) \quad \text{Formation of } H_2O_2 \text{ not}$$

$$\text{established conclusively.}$$

Thus at the negative polarisable platinum electrode (cathode), oxygen is reduced; at the silver anode, in potassium chloride as supporting electrolyte

$$4Ag + 4Cl^- = 4AgCl + 4e \qquad (2.18)$$

and the resulting current flow through the solution from the silver anode to the platinum cathode is thus proportional to the oxygen activity. Oxygen is of course consumed in the process but the concentration is negligible compared to the total amount in solution. Charlton *et al.*[8] have described an oxygen electrode apparatus which can be built in the laboratory. Enzymes other than oxidases which convert oxygen directly can also be assayed using this electrode by the coupled enzyme principle outlined earlier. For example glucose mutarotase catalyses the reaction

$$\alpha\text{-D-Glucose} \rightleftharpoons \beta\text{-D Glucose} \qquad (2.19)$$

and glucose oxidase

$$\beta\text{-}D\text{-Glucose} + O_2 \rightarrow D\text{-Gluconolactone} + H_2O_2 \quad (2.20)$$

Thus the mutarotation of glucose or the concentration of each anomer in solution can be assayed.
Enzyme electrodes are discussed in Chapter 7.

2.1.4.4 *Radioassays*[9]

Radiochemical techniques utilise substrates radioactively labelled with, most commonly, carbon-14, hydrogen-3, iodine-125 or iodine-131. The enzyme catalysed reaction of substrates containing these isotopes in withdrawn samples is terminated and the substrate and product molecules separated in order to count their individual radioactivities.

The advantages of radioassays compared to other assays are sensitivity, the wide range of substrate concentrations that can be used, and more particularly the measurement of concentration directly rather than via a concentration dependent parameter. This last characteristic is useful for *in vivo* enzyme assays. Against these advantages, the separation procedures which must be performed for every sample are often tedious and time consuming, the storage and handling of radioactively labelled substrates and products requires special care and the radioactive decomposition and inactivation of substrates and enzymes may cause problems.

Radiochemical assays have however proved useful for those enzymes whose substrates and products can easily be separated. For example decarboxylases catalyse the uptake and release of gaseous carbon dioxide which can easily be collected, and aminoacyl-tRNA synthetase assays are facilitated by acid precipitation of the ^{14}C-amino acyl-tRNA macromolecule after incorporation of the radioactively labelled amino acid.

Exchange reactions, for example the submitochondrial phosphate exchange reaction, $ADP + P_i \rightleftharpoons ATP$, can as yet only be monitored by such labelling methods. Based on such observations, the enzymes which catalyse these exchange reactions have been considered to belong to a separate (seventh) class.[9]

2.2 STEADY-STATE ENZYME KINETICS

The central axiom of enzyme kinetics is that an enzyme acts by first combining with its substrate to form an enzyme-substrate complex.

Within this complex, rearrangement of the substrate to product takes place; subsequently the enzyme-product complex breaks down to reform the native enzyme and release the products into solution, Figure 2.3.

Figure 2.3: The General Mechanism of a One Substrate-One Product Enzyme Catalysed Reaction

$$E + S \rightleftharpoons ES \rightleftharpoons EP \rightleftharpoons E + P$$

This concept of an enzyme-substrate intermediate, besides leading to that of the enzyme active site (the region on the enzyme where the substrate is attached), provides a rationalisation of the effects of substrates, inhibitors and activators on the rates of enzyme catalysed reactions. In the first instance therefore the observed velocities depend on the initial concentrations of enzyme and substrate present.

2.2.1 Influence of Enzyme Concentration on Initial Velocity

The initial velocity of an enzyme catalysed reaction is directly dependent on the amount of enzyme, i.e. the reaction order with respect to total initial enzyme concentration (E_0) is unity. Apparent exceptions to this generalisation have been observed but have been found to be due to impurities in either the substrate or enzyme preparations or to inadequate substrate levels at high enzyme concentrations.

2.2.2 Effect of Substrate Concentration on Initial Velocity

The initial velocity is also dependent on the concentration of substrate in the medium but the dependency observed is not as straightforward as that found on changing enzyme concentration. The influence of substrate concentration on the rate is shown in Figure 2.4. As the concentration is raised, initially, the rate is also found to increase in direct proportion i.e. the reaction order with respect to substrate is unity. However the rate cannot be increased indefinitely by raising the substrate levels. The kinetic order gradually decreases and the rate is found to attain a maximum, V_{max}.

A rationalisation of this behaviour is the achievement of Henri[10a] and Michaelis and Menten,[10b] who proposed the intermediate formation of an enzyme-substrate complex, since called in their honour the Michaelis-Menten-Henri complex. The reasoning can

Figure 2.4: Effect of Substrate Concentration on Initial Velocity

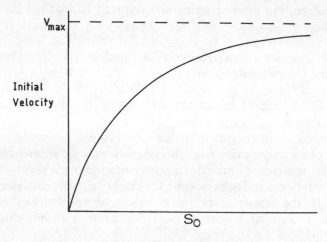

be simplified as follows: while free enzyme is available, an increase in substrate concentration will result in a higher rate; but when all enzyme molecules are saturated with substrate their total effect will be exerted, giving the maximum rate, V_{max}. The rate can then only be increased further by adding more enzyme.

2.2.2.1   Henri-Michaelis-Menten Hypothesis

A mathematical expression describing the observed hyperbolic dependence of initial velocity on substrate concentration was developed by the same authors[10] from the assumptions that (a) the enzyme-substrate complex was in equilibrium with the free species in solution and (b) only by means of this complex could products be formed at an observable rate. Their mechanism can be represented:

$$E + S \rightleftharpoons ES \quad \text{equilibrium formation of the complex} \quad (2.21)$$

$$ES \longrightarrow E + P \quad \text{breakdown of complex to products}$$

$$\text{with regeneration of enzyme} \quad (2.22)$$

The equilibrium, eq. 2.21, can be described by an apparent equilibrium constant, K_s, defined as,

$$K_s = \frac{E \cdot S}{ES} \quad (2.23)$$

where E and S are the free concentrations of enzyme and substrate respectively and ES is the concentration of enzyme-substrate complex. The rate of product formation, v, is $k_{cat} \cdot ES$, where k_{cat} is the first order rate constant for breakdown of the enzyme-substrate complex. In order to obtain an expression for this rate in terms of measurable quantities and adjustable parameters the unknown quantity, ES, is eliminated. When substrate concentrations are high compared to the catalytic concentrations of enzyme, the free concentration of enzyme will be substantially decreased but that of the substrate not significantly so.

$$K_s = (E_0 - ES) \cdot S_0 / ES \qquad (2.24)$$

$$\therefore \ ES = \frac{E_0 \cdot S_0}{K_s + S_0} \qquad (2.25)$$

where E_0 and S_0 are total initial concentrations. Substitution of eq. 2.25 into $v = k_{cat} \cdot ES$ yields

$$v = \frac{k_{cat} \cdot E_0 \cdot S_0}{K_s + S_0} \qquad (2.26)$$

The maximum velocity is attained when all enzyme molecules are saturated with substrate, i.e.

$$V_{max} = k_{cat} \cdot E_0 \qquad (2.27)$$

$$\therefore \ v = \frac{V_{max} \cdot S_0}{K_s + S_0} \qquad (2.28)$$

This is the Henri or Michaelis-Menten equation and describes the hyperbola of Figure 2.4.

At low substrate concentrations, $K_s \gg S_0$ and the initial rate becomes equal to $V_{max} S_0 / K_s$, i.e. the rate is first order dependent on substrate concentration; conversely at high substrate levels $S_0 \gg K_s$ and eq. 2.28 predicts that the initial rate will approach the maximum attainable, V_{max}.

2.2.2.2 *Briggs-Haldane Hypothesis*

Although eq. 2.28 adequately described the observed substrate dependence of initial velocity, Briggs and Haldane[11] derived a more general expression from application of the steady-state approach (a technique proven useful for the derivation of equations describing more complex kinetic processes). In the steady-state or stationary-state approximation, the net rate of change of the concentrations of the intermediates are taken as zero, i.e. their formation and breakdown are considered dynamically balanced. The concentration of any intermediate is also assumed to be small, which for a very reactive complex such as ES is probably true. The validity of the steady-state approximation both for organic reactions and enzyme catalysed reactions has been discussed at length,[3] and for the Michaelis-Menten mechanism above, it is probably justified provided the concentrations of substrates used are at least a thousand fold greater than those of the enzyme.

For the Michaelis-Menten mechanism:

$$E + S \underset{k_{-1}}{\overset{k_1}{\rightleftharpoons}} ES \overset{k_2}{\longrightarrow} E + P \qquad (2.29)$$

Rate of complex formation $= k_1 \cdot E \cdot S$
Rate of complex dissociation $= k_{-1} \cdot ES$
Rate of complex breakdown $= k_2 \cdot ES$

and
$$\frac{dES}{dt} = k_1 \cdot E \cdot S - k_{-1} \cdot ES - k_2 \cdot ES$$

Under the steady-state assumption, $dES/dt = 0$, thus $k_1 \cdot E \cdot S = k_{-1} \cdot ES + k_2 \cdot ES$, i.e. Rate of complex formation = Rate of complex disruption.

As before $E = E_0 - ES$, and $S = S_0$

$$\therefore k_1 \cdot S_0 \cdot (E_0 - ES) = ES(k_{-1} + k_2)$$

and
$$(E_0 - ES) \cdot S_0 = ES \frac{(k_{-1} + k_2)}{k_1} \qquad (2.30)$$

If the combination of rate constants is written K_m, rearrangement of eq. 2.30 yields

$$ES = \frac{E_0 \cdot S_0}{S_0 + K_m} \qquad (2.31)$$

and since $v = k_2 \cdot ES$

$$v = \frac{k_2 \cdot E_0 S_0}{S_0 + K_m} = \frac{V_{max} \cdot S_0}{S_0 + K_m} \qquad (2.32)$$

This Briggs-Haldane equation is identical in form to the Michaelis-Menten equation but is more rigorously derived and more general. The Michaelis equation is a special case of eq. 2.32 since:

$$K_m = (k_{-1} + k_2)/k_1 = k_{-1}/k_1 + k_2/k_1 = K_s + k_2/k_1$$

thus if $k_1 \gg k_2$ the numerical values of the constants K_m and K_s are identical. Their meanings however are different. Originally K_m was described operationally as the substrate concentration which yielded a rate equal to half the maximum,* whereas K_s is the dissociation constant of the enzyme-substrate complex and is an index of their mutual affinity (the larger the K_s the lower the affinity and vice versa). Although in many enzyme-substrate systems K_m equals K_s, this equality cannot be assumed without further experimental verification, which is obtained from pre-steady state studies, the effect of modifiers or from binding experiments.

The numerical values of the parameters K_m and V_{max} depend on the reaction conditions and the presence of enzyme modifiers, thus quantitative analysis of changes in their apparent values caused by pH, temperature, inhibitors, etc. provide valuable insights into the enzyme mechanism (see Chapter 3). They also provide a means of comparing the properties of enzymes prepared from different sources.

2.2.2.3 Determination of K_m and V_{max}

As discussed above the Michaelis-Menten parameter, K_m, and maximum velocity, V_{max}, are valuable and important indices of a particular enzyme/substrate system, and thus their accurate numerical evaluation is essential. Although in principle they can be obtained directly from the hyperbolic dependence of initial velocity on substrate concentration, it is not always possible to dissolve sufficient quantities of the substrate to enable determination of the maximum velocity. Even if the substrate is highly water soluble the resulting

* If in eq. 2.28 or 2.32, v is set equal to $V_{max}/2$ then K_s or K_m become equal to S_0.

environmental change, formation of non-productive substrate-enzyme complexes or salt effects may invalidate the results obtained. In addition, the evaluation of these constants depends on a few points only, in the middle and the high end of the substrate range, and it is obviously sounder practice to use as much of the experimental data as possible. For these reasons Lineweaver and Burk[12] proposed plotting the data according to eq. 2.33, the reciprocal of eq. 2.32.

$$\frac{1}{v} = \frac{K_m}{V_{max}} \cdot \frac{1}{S_0} + \frac{1}{V_{max}} \qquad (2.33)$$

This is the equation of a straight line with slope K_m/V_{max}, abscissal intercept $-K_m^{-1}$ and ordinate intercept V_{max}^{-1} (Figure 2.5), and enables the parameters to be estimated more readily from data collected over a wide range and at substrate concentrations less than saturating—see Example 2.1 and Figures 2.6 and 2.7.

Example 2.1
In order to evaluate K_m and V_{max} for the fumarase catalysed conversion of malate into fumarate at pH 8.0 and 30° the following data were collected.

S (mM)	0.02	0.033	0.05	0.1	0.2	0.3	0.4	0.5
v (10^{-5} mol min^{-1})	1.67	2.5	3.3	5	6.7	7.7	8.0	8.3

Reciprocation gives

S^{-1} (mM^{-1})	50	30	20	10	5	3.3	2.5	2
v^{-1} (10^5 mol^{-1} min)	0.6	0.4	0.3	0.2	0.15	0.13	0.125	0.12

These data points are plotted in Figures 2.6 and 2.7 from which the Michaelis parameters can be estimated as follows.

$$1/K_m = 10^{-4}M^{-1} \qquad\qquad K_m = 10^{-4}M$$

$$1/V_{max} = 0.1 \times 10^5 mol^{-1}\ min \qquad V_{max} = 10^{-4} mol\ min^{-1}$$

Figure 2.5: Lineweaver-Burk Graphical Plot for Evaluation of K_m and V_{max}

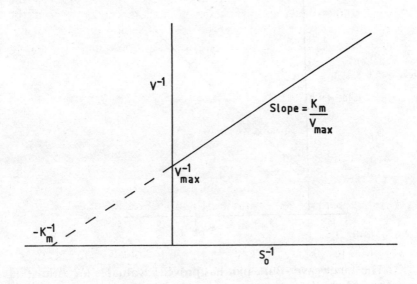

Figure 2.6: Plot of Data from Example 2.1

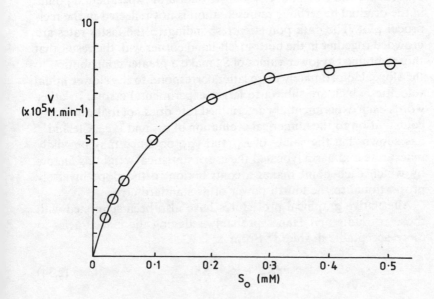

Figure 2.7: Plot of Data from Example 2.1

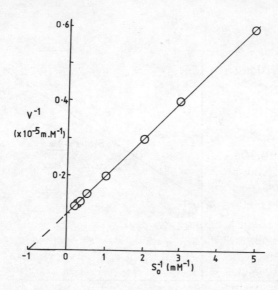

The Lineweaver-Burk plot has proved invaluable for estimation of the Michaelis parameters and an improvement on the unaltered hyperbolic method, but it can still lead to inaccuracies. As is seen in Figures 2.6 and 2.7 a fairly even distribution of experimental points in the original hyperbolic representation is not reflected in the reciprocal plot. The data points corresponding to the faster rates are crowded together in the bottom left-hand corner with the result that those obtained at lower values of S_0 make a greater contribution to the slope. Unfortunately these latter correspond to the slower initial velocities which are subject to larger experimental errors. In other words each experimentally determined rate does not make an equal contribution to the numerical evaluation of K_m and V_{max}. Cleland[13] has shown that the values of K_m and V_{max} obtained may be wildly inaccurate and has advocated the use of statistical weighting factors by which each point makes a contribution to the slope inversely proportional to the fourth power of its standard error.

Alternative graphical procedures have also been suggested with those of Eadie and Hanes probably reducing the inaccuracies to more acceptable degrees.[14] From eq. 2.32

$$v \cdot K_m + vS = S \cdot V_{max} \qquad (2.34)$$

division by $S \cdot K_m$ followed by rearrangement gives:

$$\frac{v}{S} = -\frac{v}{K_m} + \frac{V_{max}}{K_m} \tag{2.35}$$

Alternatively division of eq. 2.34 by $v \cdot V_{max}$ and rearranging gives:

$$\frac{S}{v} = \frac{S}{V_{max}} + \frac{K_m}{V_{max}} \tag{2.36}$$

Both eq. 2.35 and 2.36 are straight line equations. (It is instructive to plot the data of Example 2.1 in these two ways.)

2.2.2.4 Comments on the Michaelis-Menten Mechanism

Many enzymes exhibit Michaelis-Menten kinetic behaviour, i.e. exhibit the hyperbolic substrate-rate profiles consistent with the minimal mechanism (eqs. 2.21, 2.22), but this is not sufficient proof that an enzyme acts via this mechanism. The reversible reaction (eq. 2.37), involving more than one intermediate, also gives a rate expression (eq. 2.38), with a form identical to eq. 2.32, simplifying to the latter when either $S = 0$ or $P = 0$.

$$E + S \rightleftharpoons ES \mid \rightleftharpoons EP \mid \rightleftharpoons E + P \tag{2.37}$$

$$v = \frac{V_{max}S/K_m^S - V_{max}P/K_m^P}{1 + S/K_m^S + P/K_m^P} \tag{2.38}$$

where K_m^S, V_{max}^S and K_m^P, V_{max}^P are the Michaelis parameters for the forward and reverse directions respectively. The hyperbolic expression does not therefore provide any information on the number of intermediate complexes on the catalytic pathway.

A special case of eq. 2.37 is that in which only the initial formation of the Michaelis complex is reversible, viz.:

$$E + S \rightleftharpoons ES \xrightarrow{k_2} ES' \xrightarrow{k_3} E + P_2 \tag{2.39}$$
$$+$$
$$P_1$$

In this case the single substrate is cleaved by the enzyme to give two products, released consecutively. This type of mechanism is followed particularly by proteolytic enzymes such as trypsin and elastase, in which the intermediate ES' is an acyl-enzyme and by phosphatases in which it is a phosphorylated enzyme.

The steady state equations describing the time dependences of ES and ES' are

$$\frac{d \cdot ES}{dt} = 0 = k_1 \cdot E \cdot S - (k_{-1} + k_2) \cdot ES$$

$$\frac{d \cdot ES'}{dt} = 0 = k_2 \cdot ES - k_3 \cdot ES'$$

and since $E_0 = E + ES + ES'$

$$ES = \frac{E_0 \cdot S_0}{\dfrac{k_{-1} + k_2}{k_1} + \dfrac{k_2 + k_3}{k_3} \cdot S_0}$$

The observed rate, v, is:

$$k_2 \cdot ES = k_3 \cdot ES' \tag{2.40}$$

$$\therefore v = \frac{V_{max} \cdot S_0}{K_m + S_0} \tag{2.41}$$

where $V_{max} = \dfrac{k_2 k_3}{k_2 + k_3} \cdot E_0$ and $K_m = \dfrac{k_{-1} + k_2}{k_1} \cdot \dfrac{k_3}{k_2 + k_3}$.

When $k_3 > k_2$, the second step is rate limiting and the intermediate complex rapidly breaks down; in this circumstance $V_{max} = k_2 E_0$ and $K_m = (k_{-1} + k_2)/k_1$, which is identical to that for the mechanism involving a single intermediate. When however $k_2 > k_3$, then $V_{max} = k_3 E_0$ and the maximum rate of product formation is dependent on the rate constant for the last step, the breakdown of the intermediate ES'. Under this condition, the experimentally derived Michaelis constant will be lower than that for the case $k_3 > k_2$ by the ratio $k_3/(k_2 + k_3)$. Proteolytic enzymes following this latter mechanism are discussed further in Section 2.3.

K_m and V_{max} may not always be composed of rate constants associated with the active formation of reaction products. They may include, for example, terms relating to the formation of nonproductive enzyme-substrate complexes. Lysozyme obeys the hyperbolic rate law in its catalysed hydrolysis of oligosaccharide substrates. Since, however, its binding site contains six sub-sites, the population of enzyme-substrate complexes formed with substrates with degrees of oligomerisation less than six will include a proportion in which the glycosidic bond does not traverse the cleavage site, eq. 2.42. The presence of these non-productive complexes will cause the measured Michaelis parameter, K_m to be smaller than that obtaining in their absence. This can easily be seen from a steady state analysis of a Michaelis-Menten mechanism expanded to include a non-productive enzyme-substrate complex, ES', eq. 2.43:

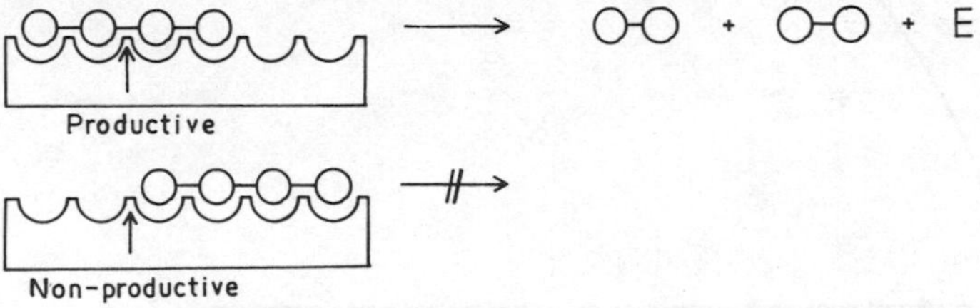

Productive

Non-productive

$$(2.42)$$

$$ES' \rightleftharpoons E + S \rightleftharpoons ES \longrightarrow E + P \qquad (2.43)$$

If
$$K = \frac{k_2 + k_{-1}}{k_1} \quad \text{and} \quad K_u = \frac{E \cdot S}{ES'}$$

then
$$v = \frac{S_0 \cdot V_{max}}{K_m + S_0} \qquad (2.44)$$

where
$$V_{max} = \frac{k_2 E_0}{1 + K/K_u} \quad \text{and} \quad K_m = \left(\frac{1}{K} + \frac{1}{K_u}\right)^{-1}$$

2.2.2.5 Deviations from Hyperbolic Michaelis-Menten Behaviour

Substrate Inhibition. This is different from non-productive complex formation, arising predominantly at high substrate concentrations. The general form of the substrate-rate profile is illustrated in Figure 2.8, curve a, which shows the rate decreasing as substrate levels increase instead of reaching a maximum asymptotic value. The corresponding Lineweaver-Burk plot is shown in Figure 2.9, curve a. Several mechanisms are consistent with these observations, for example, high substrate concentrations could cause the removal

Figure 2.8: a-Substrate Inhibition. b-Substrate Activation

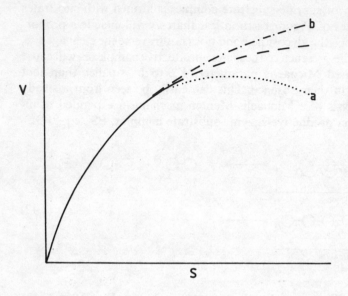

Figure 2.9: a-Substrate Inhibition. b-Substrate Activation

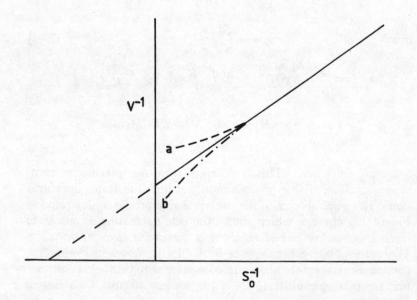

of an essential cofactor such as a metal ion, or result in a proportion of the enzyme molecules engaged in multiple binding patterns of the type:

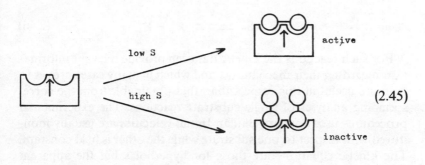

low S

high S

active

inactive

(2.45)

Substrate Activation.   Illustrated by Figures 2.8b and 2.9b, this type of behaviour usually also occurs at high substrate concentrations. One mechanism that can account for this behaviour is the removal of an impurity from the enzyme by the substrate. Alternatively the substrate may induce a favourable conformational change in the enzyme.

Sigmoidal Substrate-Rate Profiles.   At both high and low substrate concentrations, considerable departure from a hyperbolic to a significant 'S' shaped curve is a familiar observation. Sigmoidal substrate-rate profiles can result from two main mechanisms. (1) An impurity in the reaction medium may reversibly bind the substrate and thereby decrease the concentration available to the enzyme. As the substrate concentration is raised, proportionally more will become free hence increasing the rate, until finally all the impurity will be associated and the eventual maximum rate will be attained (reference 12, Chapter 3). (2) Sigmoidal catalytic behaviour may be the consequence of co-operation between subunits in the protein molecule. This response has an important relation to the regulation of enzyme activity and will be discussed in the next chapter.

2.2.3   Rate Equations for Two Substrate-Two Product Enzymes
The reactions considered above have all involved single substrates and are thus applicable to only a small proportion of the extant

enzyme catalysed reactions. The majority of enzymes can be considered to act simultaneously or consecutively on two substrates,* one of which may be a coenzyme:

$$S_1 + S_2 \overset{E}{\rightleftharpoons} P_1 + P_2 \qquad (2.46)$$

For such reactions the kinetic data can provide relevant information regarding their mechanisms and which in many cases proves to be more useful in this respect than that obtainable from the corresponding analyses of single substrate reactions. The experimental procedures used are very similar. Initial velocities are usually monitored with respect to one substrate while the other is held constant. The kinetic equations are those for hyperbolas but the apparent Michaelis parameters are functions of the substrate concentrations. Their interpretation often depends therefore on the reaction mechanism.

The number of different pathways by which three reacting components can form products is large and demands treatment in detail[15, 16] but they can be summarised into two main classes, sequential and non-sequential.

2.2.3.1   Sequential Mechanisms

In this class, both substrates add to the enzyme before the products are formed;

$$E + S_1 + S_2 \rightleftharpoons ES_1S_2 \longrightarrow E + P_1 + P_2 \qquad (2.47)$$

Depending on the order of addition of the substrates this class can be further subdivided into mechanisms in which substrate association and product dissociation are random and those in which they are ordered.

Random Sequential.   The ternary complex ES_1S_2 is formed via either of the binary complexes ES_1 or ES_2 as intermediates, and the enzyme therefore possesses distinct sites for both substrates. The random mechanism can be represented by the complexus:

---

* Hydrolytic enzymes are normally considered uni-substrate since molecules of the second substrate, water, are in excess.

$$
\begin{array}{c}
\text{ES}_1 \qquad\qquad \text{EP}_1 \\
{}^{+S_1}\nearrow \quad \searrow {}^{+S_2} \qquad {}^{-P_2}\nearrow \quad \searrow {}^{-P_1} \\
E \qquad\qquad \text{ES}_1\text{S}_2 \rightleftharpoons \text{EP}_1\text{P}_2 \qquad\qquad E \\
{}^{+S_2}\searrow \quad \nearrow {}^{+S_1} \qquad {}^{-P_1}\searrow \quad \nearrow {}^{-P_2} \\
\text{ES}_2 \qquad\qquad \text{EP}_2
\end{array}
\qquad (2.48)
$$

Examples of enzymes following a random sequential mechanism as in eq. 2.48 are creatine phosphokinase (R = creatine), hexokinase (R = glucose) and pyruvate kinase (R = pyruvate) where $S_1 = ROH$, $S_2 = ATP$, $P_1 = R \cdot PO_4^{2-}$ and $P_2 = ADP$.

Ordered Sequential. In an ordered sequential reaction the two substrates add in a compulsory order to the enzyme:

$$
E + S_1 \rightleftharpoons ES_1 \overset{S_2}{\rightleftharpoons} ES_1S_2 \rightleftharpoons
$$

$$
EP_1P_2 \underset{P_1}{\rightleftharpoons} EP_2 \rightleftharpoons E + P_2 \qquad (2.49)
$$

This mechanism is adopted by many enzymes exhibiting a coenzyme requirement. The coenzyme can be envisaged to first bind to a specific, complementary site on the enzyme to provide a reactive holoenzyme for the second substrate. Coenzyme binding sites have been demonstrated by separate binding experiments for enzymes such as lactate dehydrogenase and alcohol dehydrogenase which conform to this reaction sequence. In the case of horse liver alcohol dehydrogenase, for which $S_1 = NAD^+$, $S_2 = $ ethanol, $P_1 = $ acetaldehyde and $P_2 = NADH$, ES_1 is rapidly converted into EP_2 i.e. oxidation of ethanol is very fast, this is entitled the Theorell-Chance mechanism after the contributions made by these workers.

2.2.3.2 Non-sequential Mechanisms

This ordered reaction, entitled Ping-Pong by Cleland[13] because the enzyme oscillates between the different substrates, is comprised of two partial reactions. Initially the enzyme is modified by the first substrate which is simultaneously converted to product-1, and this is followed by reaction of the derived enzyme with the second substrate, eq. 2.50.

$$E + S_1 \rightleftharpoons ES_1 \rightleftharpoons E'P_1 \rightleftharpoons E' \overset{+S_2}{\rightleftharpoons}$$
$$+$$
$$P_1$$
$$E'S_2 \rightleftharpoons E'P_2 \rightleftharpoons E + P_2 \quad (2.50)$$

Hydrolytic enzymes where the second substrate is an acceptor water molecule can formally be considered to follow this sequence, in which $S_1 = R \cdot CO_2R'$, $S_2 = H_2O$, $E' = E-O \cdot CO \cdot R$, $P_1 = R'OH$ and $P_2 = R \cdot CO_2H$. But enzymes more conspicuously acting on two substrates and whose mechanism can be described by eq. 2.50 are phosphoglucomutase ($S_1 = \alpha$-glucose-1-phosphate, S_2 = glucose-1,6-diphosphate, P_1 = glucose-1,6-diphosphate, $P_2 =$ glucose-6-phosphate, $E' = E-O-PO_3^{2-}$) and transaminases such as L-aspartate : 2-oxoglutarate aminotransferase (S_1 = L-aspartate, S_2 = 2-oxoglutarate, P_1 = oxaloacetate, P_2 = L-glutamate, $E' = E-NH_2$).

2.2.4 Transient State Kinetics

There must be a period of time during which the enzyme-substrate complex concentration, even though it is small, builds up to reach a constant value. This period is entitled the 'transient phase'. In this phase the rate of formation of product also increases, attaining a constant value simultaneously with the steady state complex (Figure 2.10).

The theoretical treatment of this transient rate of formation of product and enzyme-substrate complex for the single intermediate Michaelis-Menten mechanism is given here. The experimental procedures for monitoring pre-steady state kinetics were briefly described in Section 2.1.4 and more fully in Reference 5.

As before, the Briggs-Haldane representation of the basic Michaelis-Menten mechanism is:

$$E + S \underset{k_{-1}}{\overset{k_1}{\rightleftharpoons}} ES \overset{k_2}{\longrightarrow} E + P$$

for which the rate of product formation is:

$$\frac{dP}{dt} = k_2 \cdot ES.$$

Figure 2.10: Curve for the Formation of the Enzyme-Substrate Complex

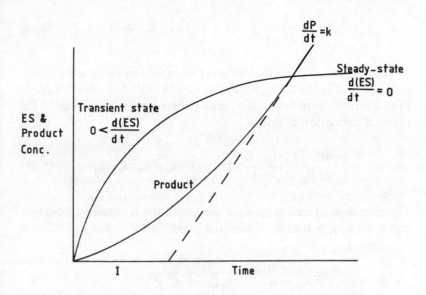

and the rate of formation of the enzyme-substrate complex:

$$\frac{d\,ES}{dt} = k_1 \cdot E \cdot S - k_{-1} \cdot ES - k_2 \cdot ES$$

$$= k_1(E_0 - ES)(S_0 - ES - P) - k_{-1} \cdot ES - k_2 \cdot ES \quad (2.51)$$

If the initial concentration of substrate, S_0, is much larger than that of the enzyme, E_0, and if relatively little has been converted, eq. 2.51 simplifies to:

$$\frac{d\,ES}{dt} = k_1 \cdot S_0 \cdot E_0 - k_1 \cdot S_0 \cdot ES - k_{-1} \cdot ES - k_2 \cdot ES \quad (2.52)$$

This can be integrated to give the time dependent concentration of ES:

$$ES = \frac{k_1 \cdot E_0 \cdot S_0}{(k_2 + k_{-1}) + k_1 S_0} \cdot [1 - \exp(k_2 + k_{-1} + k_1 S_0)t] \quad (2.53)$$

Since $\dfrac{k_2 + k_{-1}}{k_1}$ is K_m, then:

$$ES = \frac{E_0 \cdot S_0}{K_m + S_0} \cdot [1 - \exp k_1 t(K_m + S_0)] \qquad (2.54)$$

This describes the curve for formation of the enzyme-substrate complex in Figure 2.10. (When the steady state has been attained, the equation simplifies to the familiar first term.) The equation for product formation is thus:

$$\frac{dP}{dt} = \frac{k_2 \cdot E_0 \cdot S_0}{K_m + S_0} \cdot [1 - \exp k_1 t(K_m + S_0)] \qquad (2.55)$$

The time dependence of product concentration is obtained from this equation by integration. Under the condition $P = 0$ at $t = 0$

$$P = \frac{k_2 E_0 S_0}{K_m + S_0} \cdot t - \frac{k_2 E_0 S_0}{k_1 (S_0 + K_m)^2}$$
$$+ \frac{k_2 E_0 S_0}{k_1 (S_0 + K_m)^2} \cdot \exp[-k_1 t(S_0 + K_m)] \qquad (2.56)$$

This equation describes the transient build up to a constantly increasing product concentration. The resulting displacement or induction(I) obtained by extrapolation back to the time axis (see Figure 2.10) is:

$$I = \frac{1}{k_1 (S_0 + K_m)} = \frac{1}{k_1 S_0 + k_{-1} + k_2} \qquad (2.57)$$

Equation 2.57 predicts the displacement to be inversely proportional to the initial substrate concentration. A plot of I^{-1} versus S_0 has slope k_1, the rate constant for enzyme-substrate association, and intercept $(k_{-1} + k_2)$. Since K_m and k_2 can be evaluated from steady state analyses, then the rate constant for dissociation of the enzyme-substrate complex may also be evaluated. Some values of k_1 and k_{-1} for several different enzymes are presented in Table 3.2, in Chapter 3.

2.3 ANALYTICAL ENZYMOLOGY

The characteristics described in Chapter 1 make enzymes potentially valuable as analytical reagents. Their high rates of catalysis permit the rapid determination of activities in biological samples, leading to high rates of sample throughput. This latter aspect is particularly useful in hospital laboratories and in the food industry. Their second characteristic, specificity, permits the relatively easy detection and quantitation of small concentrations of a metabolite or inhibitor in a crude mixture under mild conditions. The main applications of analytical enzymology are shown in Table 2.1.

Table 2.1: Analytical Applications of Enzymes

A. Determination of enzyme mechanisms
B. Determination of enzyme amounts (see Chapters 6 and 7)
C. Metabolite, inhibitor and activator quantitation
D. Structural investigation of cells, organelles
   and complex molecules (see Chapter 5)

2.3.1 Measurement of Enzyme Levels

The level of an enzyme in a biological sample may be determined by measurement of either its catalytic activity or its molecular concentration.

2.3.1.1 Catalytic Activity

The principles involved in the measurement of catalytic activity have been described in Sections 2.1.4 and 2.2. For an enzyme that obeys Michaelis-Menten kinetics, at saturating substrate concentrations the maximum rate of reaction is equal to $k_2 \cdot E$; thus if either k_2 or E is known, the other may be determined. The occasions when either of these conditions are fulfilled however are infrequent and catalytic activities are usually compared, not by reference to a primary standard measurement, equivalent to molarity for example, but by the use of the empirical and relative quantities, 'International Units' or 'Katals' (see Section 2.1.3.2).

For many applications this is sufficient; the success of an enzyme purification procedure can be gauged by monitoring this relative activity at each stage, and in routine clinical analysis, measurements of the changes of a specific enzyme activity in blood plasma are usually sufficient to enable a contributory diagnostic statement to be made.

In both these cases, activity as a reflection of concentration is adequate, but other areas require more precise measurements of enzyme concentration.

2.3.1.2   Molecular Concentration

At present no enzyme preparation meets the kind of criteria that analytical chemistry requires, since there are no reference standards presently available. Rate assays are subject to many uncertainties which limit their accuracy, several variables, such as pH, ionic strength, and inhibitors, have to be controlled and it is not always possible to obtain zero-order kinetics, thus alternatives have to be considered. Activity depends on protein concentration but equally chemical constitution or physico-chemical properties are not always reliable guides. The protein may not be absolutely pure and even if it is pure, may be less than fully active. Also, a decreased activity may not be reflected in a change in its properties.*

Such limitations led Bender *et al.*[18] to examine the reactions of hydrolases with organophosphates. The stoichiometric reactions of these inhibitors with the enzymes led him to consider the 'normality' of active sites in solution and thus the possibility of 'titrating' the enzymes with suitable compounds. Many hydrolytic enzymes, e.g. trypsin, chymotrypsin, elastase and acetylcholinesterase follow the reaction pathway of eq. 2.39 in which ES' is a covalent acyl-enzyme complex, k_2 the rate constant for acylation and k_3 that for deacylation. If k_3 can be made very slow compared to k_2, before deacylation and regeneration of free enzyme will have occurred, an amount of P_1 will have been released in amounts stoichiometric with the active enzyme concentration.

If $S_0 \gg E$ both the pre-steady state acylation (k_2) and steady state deacylation (k_3) can be obtained. Although the former is often too fast to be measured by conventional steady state techniques, this

---

* A selective, sensitive method for the estimation of enzyme concentration has been developed based on radioimmunossay.[17] The difficulties that arise here are the possibly different topographical locations and independence of antigenic and catalytic sites; this means that inactivating modifications in one may not necessarily be transmitted to the other. In addition the presence of several different forms of the enzyme in a preparation may interfere both with the selectivity of the method and the raising of antisera.

Figure 2.11: Reaction Pathway of Acylation and Deacylation

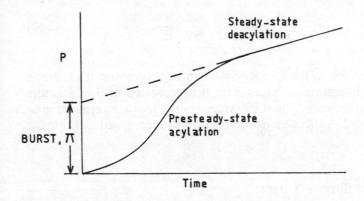

usually does not inconvenience the measurement of P_1 since the initial 'burst' (see Figure 2.11) yields the enzyme concentration. The time dependence of production of the first product, P_1, is given by an equation similar to eq. 2.56. Thus if $k_2 \gg k_3$ and t is large, this can be written,

$$P_1 = \pi + At \qquad (2.58)$$

where

$$\text{(a)} \quad \pi = \frac{E_0}{\left(1 + \dfrac{K_{m(app)}}{S_0}\right)^2} \quad \text{and} \quad \text{(b)} \quad A = k_{cat} \cdot \frac{E_0}{\left(1 + \dfrac{K_{m(app)}}{S_0}\right)} \qquad (2.59)$$

When $S_0 \gg K_{m(app)}$,

$$\pi = E_0 \qquad (2.60)$$

Thus if ΔA is the 'burst' absorbance change and $\Delta\varepsilon$ is the difference in extinction coefficient between products and starting materials, then the concentration of active enzyme in the preparation is

$$E_0 = \pi = \Delta A/\Delta\varepsilon \qquad (2.61)$$

As explained in Section 2.2.2.3, it may not always be possible to fulfil the condition $S_0 \gg K_{m(app)}$. In these circumstances $\pi^{-1/2}$ can be plotted against S_0^{-1} according to eq. 2.62, the reciprocal of eq. 2.59a.

$$\frac{1}{\pi^{1/2}} = \frac{1}{E_0^{1/2}} + \frac{K_m}{S_0} \cdot \frac{1}{E_0^{1/2}} \qquad (2.62)$$

By these means the operational normality of enzyme in a solution may be determined. Values have been reported in the literature indicating between 50 and 90 percent purity for the enzyme preparations tested. A short list of enzyme titrants is compiled in Table 2.2.

Table 2.2: Enzyme Titrants

Enzyme	Titrant
Acetylcholinesterase	N-Methyl-(7-dimethyl carbamoxy)-quinolinium iodide, 3,3'-bis (α-trimethyl-ammonium-methyl) azobenzene bromide.
Cholinesterase	o-Nitrophenyl dimethyl carbamate.
Chymotrypsin	p-Nitrophenylacetate, trans-cinnamoyl imidazole, p-nitrophenyl, p-trimethylaminocinnamate, 4-methylumbelliferyl, p-trimethylamino-cinnamate, 2-hydroxy-5-nitro-α-toluene sulphonic acid sultone.
Elastase	Diethyl-p-nitrophenylphosphate.
Papain	2,2-Dimethyl-5-phenyl oxazolinone, p-nitrophenyl N-acetyltryptophan.
Thrombin	N(2)-Carbobenzoxy-L-lysine, p-nitrophenylester.
Trypsin	p-Nitrophenyl, N(2)-benzyloxy carbonyl-L-lysine, 4-methylumbelliferyl p-guanidino-benzoate, N(2)-methyl-N(2) tosyl lysine, naphthyl ester, p-nitrophenyl p-guanidino benzoate.

2.3.2 Principles of the Measurement of Metabolite Concentrations

The concentration of a compound in a mixture may be determined enzymically if a specific enzyme is available that will rapidly convert it into products. Depending on the type of molecule, its physical properties and the laboratory context, several enzymological techniques have been designed for the quantitation of naturally occurring substances; these include end point methods, coupled assays, catalytic assays and enzyme immunoassays. Electrodes incorporating enzymes will be described in Chapter 7.

2.3.2.1 End Point or Total Change Assays

In an essentially irreversible enzyme catalysed reaction the substrate (S) is completely transformed into product (P). If the two differ in a measurable physical property insensitive to interference by other components in the mixture then the overall change in S or P can be quantitated.

For the irreversible reaction $S \xrightarrow{E} P$, the concentration of P, or metabolite S will be given by ΔA, the total change in the concentration dependent parameter used to monitor the reaction. An identical progress curve will also be obtained if the reaction has a coenzyme requirement, and provided this is added in excess so that the metabolite is in a limiting concentration the total change in its concentration can be used to determine directly the concentration of the metabolite. For example the concentration of alcohol in a mixture is easily measured by adding NAD^+ and alcohol dehydrogenase, when the overall increase in absorbance at 340 nm (due to NADH) gives the stoichiometric concentration of ethanol in the mixture.

2.3.2.2 Kinetic Assays

The concentration of a substance is also determinable from the rate of its transformation. If in the Michaelis-Menten equation, $S \ll K_m$, the rate, given by $S \cdot V_{max}/K_m$, will become first order dependent on the substrate concentration. To reduce the errors it is usual to construct a calibration curve from rates obtained at known concentrations under the particular reaction conditions, and to increase the accuracy, K_m and V_{max} are previously determined.

2.3.2.3   Catalytic Assays

Even if small amounts are present and satisfy the condition above $(S \ll K_m)$ the rate or extent of reaction may be too small to be measured with confidence. In certain cases it is possible to overcome this by regenerating the substrate by means of a second enzyme, eq. 2.63.

$$
\begin{array}{ccc}
 & E_1 & \\
S & \longrightarrow & P \\
 & & \\
B & \underset{E_2}{\longleftarrow} & A
\end{array}
\qquad (2.63)
$$

The additional enzyme will set up a constant steady-state concentration of substrate, so giving rise to a constant rate. Measurement of this rate is facilitated if the added enzyme has a cofactor requirement (A), the rate of its change will be proportional to P, which in turn depends on S; since S is constantly recycled, the rate and hence the concentration of S is amplified by the continuously increasing conversion of the indicator $A \rightarrow B$. By this technique 10^{-10} to 10^{-15} mole NADH have been estimated spectrofluorometrically using glutamate DH (E_1), lactate DH (E_2), lactate (A) and pyruvate (B).[19]

2.3.2.4   Enzyme Immunoassay[20] (EIA)

Introduced in 1971, the principle of this technique is similar to that of the more familiar radioimmunoassay for the quantitation of material in solution, being based on the principle of competitive binding. But in enzyme immunoassay an enzyme catalysed reaction is used as a label in place of the radioactive isotope. The method thus avoids possible health hazards from radioactivity, the short shelf lives of radioactively labelled compounds, expensive scintillation media and counters and the need for separation steps: but it still employs the selectivity and sensitivity of the antibody-antigen response.

Immunoassays whether isotope or enzyme labelled are described by the general reaction:

$$
B + B' + 2A \rightleftharpoons BA + B'A \qquad (2.64)
$$

where B is the antigenic substance to be measured, B' is labelled antigen and A is B-specific antibody. The equilibrium lies far to the right, favouring complexation. In the normal procedures B'A is isolated from the mixture and its concentration is determined by calibration. This concentration is reciprocally proportional to that of B and so its amount in solution may be quantitated. Therefore the accuracy of the technique depends on the successful separation of the complex from the unattached molecules. For enzyme immunoassay in particular two methods have been developed.

Enzyme-linked Immunosorbant Assay (ELISA). Antibody or antigen is first labelled with enzyme and then attached to an insoluble carrier. The advantage insolubilisation confers is ease of the separation of attached and unattached enzyme activity. This is important since the concentration of B is given by this activity ratio.

$$\overset{\text{B}}{\boxed{}} + \text{B} + 2\text{A} \rightleftharpoons \overset{\text{BA}}{\boxed{}} + \text{BA} \qquad (2.65)$$

Enzymes used as labels include the following: alkaline phosphatase (measurement of α-fetoprotein, haptoglobin, IgE, IgG), β-galactosidase (antiadenosine antibody, insulin), glucose oxidase (α-fetoprotein, IgG), malate dehydrogenase (thyroxine) and peroxidase (estrogens, α-fetoprotein, IgG, insulin, estradiol).

Homogeneous Enzyme Immunoassay (EMIT). This method in which the separation step is avoided, has proved to be more widely applicable, especially for the screening of patients' samples for therapeutic agents and drugs of abuse.

The enzyme is covalently coupled to the antigen. On association of this antigen-enzyme complex with the antibody the activity of the enzyme is completely eliminated, but on competition with the free antigen in the sample, enzyme linked antigen is released, restoring enzyme activity. The increase of enzyme activity is thus proportional to the amount of antigen in the sample under investigation, i.e.

$$\underset{\text{active}}{\text{B}'} + \text{A} \rightleftharpoons \underset{\text{inactive}}{\text{B}'\text{A}} \qquad (2.66)$$

$$\text{B} + \text{B}'\text{A} \rightleftharpoons \text{BA} + \underset{\text{active}}{\text{B}'} \qquad (2.67)$$

Therefore, the concentration of B is proportional to the released enzymic activity of B'.

Table 2.3: Enzyme Analysis of Food Constituents

Substance	Source	Enzyme	Reactions
Cholesterol esters	Meat	1. Cholesterol esterase 2. Cholesterol oxidase 3. Catalase	Cholesterol ester $\rightarrow$ Cholesterol + Fatty acid Cholesterol $\xrightarrow{O_2}$ Δ^4 - Cholesterol + H_2O_2 $H_2O_2 \rightarrow H_2O + \frac{1}{2}O_2$
Citric acid	Wine, diabetic foods	1. Citrate lyase 2. Malate DH	Citrate $\rightarrow$ Oxaloacetate Oxaloacetate $\xrightarrow{NADH}$ Malate
Fructose	Diabetic food	1. Hexokinase 2. Phosphoglucoisomerase 3. Glucose-6-P DH	Fructose $\xrightarrow{ATP}$ Fructose-6-P Fructose-6-P $\rightarrow$ Glucose-6-P Glucose-6-P $\xrightarrow{NADP}$ Gluconate-6-P
Glucose	Diabetic food	1. Hexokinase 2. Glucose-6-P DH	Glucose $\xrightarrow{ATP}$ Glucose-6-P
Lactose	Milk chocolate	1. β-Galactosidase 2. β-Galactose DH	Lactose $\rightarrow$ Glucose + β-Galactose β-Galactose $\xrightarrow{NAD}$ Galactonic acid
Maltose	Glucose syrups, germinating barley	1. α-Glucosidase 2. Hexokinase ... glucose-6-P DH	Maltose $\rightarrow$ 2 × Glucose
Starch	Dietetics	1. Amyloglucosidase 2. Hexokinase/glucose-6-P-DH or glucose oxidase/peroxidase	Starch $\rightarrow$ Glucose
Sucrose	Molasses	1. Invertase 2. Hexokinase/glucose-6-P DH	Sucrose $\rightarrow$ Glucose + Fructose

The inhibition of enzyme activity is probably due to steric hindrance by the antibody, preventing access of the substrate to the enzyme active site, although conformational changes in the enzyme induced by the large antibody molecule cannot be ruled out. The choice of enzyme and its attachment to the antigen are important. In particular enzymes have been chosen that do not adversely affect the antigen-antibody association, and are not inhibited by constituents of biological fluids or found in these fluids. Enzymes such as lysozyme, malate dehydrogenase, peroxidase, alkaline phosphatase, β-galactosidase and α-glucosidase fulfil these criteria. They are also easily assayed, fairly cheap and active at the pH of maximum antibody-antigen binding.

Using these enzymes, drugs such as the amphetamines, barbiturates, methodone, opiates, morphine, morphine glucuronide, and addictive drugs, such as cocaine have been monitored in human urine; digoxin, phenobarbital and other drugs for the treatment of epilepsy have been assayed in plasma. In addition, hormones such as progesterone and insulin have been measured.

2.3.3 Enzyme Assays in Metabolite Estimation

The concentration of many hundreds of substances can be estimated enzymically. These compounds and their relevant enzymes have been extensively catalogued[21] (see also Chapter 7). Only a few typical examples of analyses performed on foodstuffs are selected to illustrate the principle (Table 2.3).

One of the most useful applications of enzymes is the determination of pesticide concentrations in food. The non-competitive inhibition of cholinesterases has been used to estimate the organophosphates paraoxon, sarin and TEPP; and the DDT content of food has been quantitated by its inhibition of the lipase catalysed hydrolysis of the fluorogenic substrate 4-methyl umbelliferone heptonoate.

2.4 REFERENCES

1. *Techniques of Chemistry*, ed. A. Weissberger, Vol. IV, ed. K. W. Bentley and G. W. Kirby (Wiley, New York, 1972).
2. D. E. Koshland and S. S. Stein, *Fed. Proc.*, vol. 12 (1953), p. 233.
3. A. A. Frost and R. G. Pearson, *Kinetics and Mechanism*, 2nd edn. (Wiley, New York, 1961).
4. *Enzyme Nomenclature (1972)* (Elsevier, Amsterdam, 1973).

5. M. Eigen and L. M. Maeyer, 'Investigation of Rates and Mechanisms of Reactions', in G. G. Hammes (ed.), *Techniques of Chemistry*, Vol. VI, ed. A. Weissberger, (Wiley, New York, 1974).

6. S. Udenfriend, *Fluorescence Assay in Biology and Medicine* (Academic Press, New York, 1969).

7. M. Kessler (ed.), *Ion Selective and Enzyme Electrodes in Biology and Medicine* (Urban and Schwarzenberg, Berlin, 1975).

8. J. Charlton and P. Read, *J. Appl. Physiol.*, vol. 18 (1963), p. 1247.

9. K. Oldham, *Meth. Biochem. Anal.*, vol. 21 (1973), p. 191.

10a. V. Henri, *Lois Générales de l'Action des Diastases* (Hermann, Paris, 1903).

10b. L. Michaelis and M. Menten, *Biochem. Zeit*, vol. 49 (1913), p. 333.

11. G. E. Briggs and J. B. S. Haldane, *Biochem. J.*, vol. 19 (1925), p. 338.

12. H. Lineweaver and D. Burk, *J. Amer. Chem. Soc.*, vol. 56 (1934), p. 658.

13. W. W. Cleland, *Advances in Enzymology*, vol. 29 (1967), p. 1.

14. G. N. Wilkinson, *Biochem. J.*, vol. 80 (1961), p. 325.

15. W. W. Cleland, in P. Boyer (ed.), *Enzymes*, 3rd edn, (Academic Press, New York, 1970), vol. 4, p. 659.

16. K. Dalziel, *Biochem. J.*, vol. 114 (1969), p. 547.

17. L. Landon, *Am. Clin. Biochem.*, vol. 14 (1977), p. 90.

18. M. Bender, *J. Amer. Chem. Soc.*, vol. 88 (1966), p. 5890.

19. O. H. Lowry, J. V. Passonneau, D. Schutz and M. K. Rock, *J. Biol. Chem.*, vol. 236 (1961), p. 2746.

20. See the review by S. L. Scharoe, W. M. Cooreman, W. J. Blomme and G. M. Lackeman, *Clin. Chem.* vol. 22 (1976), p. 733.

21. H. V. Bergmeyer (ed.), *Methods of Enzymatic Analysis* (Academic Press, New York, 1974).

3 Modification of Enzyme Activity

In the previous chapter procedures were described which yield the basic information about an enzymic reaction. Although for many purposes this is often sufficient, considerably more detail is required for a real understanding of its chemical and physiological action. Some of this detail is provided by studies of the effects of temperature, pH, inhibitors, activators and regulators on the kinetics and thermodynamics of the enzymic reaction.

3.1 INFLUENCE OF TEMPERATURE

The general effect of temperature on an enzyme catalysed reaction is shown in Figure 3.1. The optimum, T_{max}, is the result of two processes. Initially, as the temperature is increased the rate also increases as for a typical chemical reaction, doubling or trebling every ten degrees but as the temperature is raised higher an increasing number of protein molecules become thermally inactivated.

Figure 3.1: Effect of Temperature on the Rate of an Enzyme Catalysed Reaction

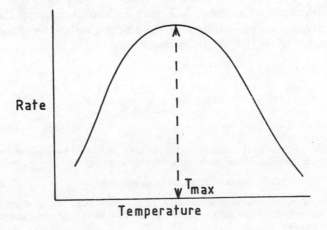

Since the energy required for denaturation is usually greater than that for catalysis, most enzymes exhibit a curve similar to Figure 3.1.

Thermal denaturation of the enzyme protein is studied separately by measuring, at a standard temperature, loss of activity as a function of temperature and time. The information obtained however is fairly limited and whether a conformational change, aggregation, disaggregation or other rearrangement causing loss of activity occurs during heating can only be resolved by simultaneous measurement of viscosity, molecular weight or optical rotatory changes.*

Enzyme catalysed conversions of substrates proceed via a series of steps each characterised by a rate constant. For example, those in the Michaelis-Menten mechanism are k_1, k_{-1} and k_2. Each of these constants will be affected differently by changes in temperature.

An expression for the temperature dependence of the rate constant has been derived by Arrhenius,

$$k = Z \cdot \exp\left(-\frac{E_a}{RT}\right) \qquad (3.2)$$

in which E_a is the Arrhenius activation energy (in J mol^{-1}), Z the collision frequency constant, R the gas constant (8.314 J mol^{-1} K^{-1}) and T the absolute temperature (Kelvin).

Descriptively, the Arrhenius activation energy is the size of the barrier which the reactants must overcome before products are formed. It is calculated as shown in Figure 3.2 from the slope of the $\log_{10}$ (rate constant) plotted against reciprocal absolute temperature, according to eq. 3.3, the linear form of eq. 3.2.

$$\log_{10} k = \log_{10} Z - \frac{E_a}{2.303\ R} \cdot \frac{1}{T} \qquad (3.3)$$

* For some enzymes heat inactivation is reversible, eq. 3.1, in which case the system can be kinetically and thermodynamically characterised.

$$\text{Native enzyme} \; \rightleftharpoons \; \text{Denatured enzyme} \qquad (3.1)$$

One interesting conclusion from such studies is the extremely large entropy change on denaturation, indicating that the native state possesses a relatively high degree of structural organisation.

Figure 3.2: Determination of the Arrhenius Activation Energy of a Reaction

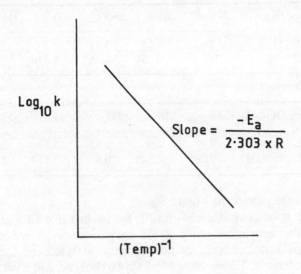

Example 3.1.
Both the enzyme alkaline phosphatase and hydroxyl ions catalyse the hydrolysis of *p*-nitrophenyl phosphate to yield *p*-nitrophenol and phosphate ion. At pH 9.5 and the temperatures listed below the following data were obtained.

1. Enzyme Catalysed

Temperature (°C)	0	10	20	30	40
V_{max} (mol min^{-1})	0.63	1.12	1.84	3.6	0.50

2. Base Catalysed

Temperature (°C)	60	65	70	75	80
Pseudo first order rate constant for basic hydrolysis $\cdot$(min^{-1})	1.84	3.36	6.31	13.6	24.8

Calculate the Arrhenius activation energies for both catalysts.

1. Enzyme Catalysed

Temperature (K)	273	283	293	303	313
T^{-1} (10^3 K^{-1})	3.66	3.53	3.41	3.3	3.2
$\log_{10}$ (V_{max})	−0.2	0.05	0.265	0.556	0.7

2. Base Catalysed

Temperature (K)	333	338	343	348	353
T^{-1} (10^3 K^{-1})	3.00	2.96	2.92	2.87	2.83
$\log_{10}$ (rate constant)	0.265	0.526	0.8	1.13	1.39

These data are plotted in Figure 3.3.

From the above calculations and the figures in Table 3.1 it can be seen that the activation energies for enzyme catalysis are much smaller than those for both the non-enzyme catalysed and the un-catalysed reactions. This is generally found to be true, and a catalyst can be considered as decreasing the energy barrier to reaction, with the most efficient catalyst decreasing it to the greatest extent.

The marked lowering of the activation energy by an enzyme is achieved by changing the mechanism of reaction to one with a different energy profile, providing an 'easier' route to products. This observation has prompted the definition of an enzyme as 'a protein with catalytic properties due to its power of specific activation',[1] or more precisely as involving 'the specific binding of the substrate to a catalytic site and a chemical interaction with this site which directly utilises the binding forces to decrease the free energy of activation of the catalysed reaction'.[2]

Basically, the Arrhenius equation is an empirical expression, valuable for the representation of experimental data but limiting in

Table 3.1: Activation Energies for Hydrogen Peroxide Decomposition

Catalyst	None	Fe^{2+}	Catalase
E_a (kJ mol^{-1})	70–76	42	7

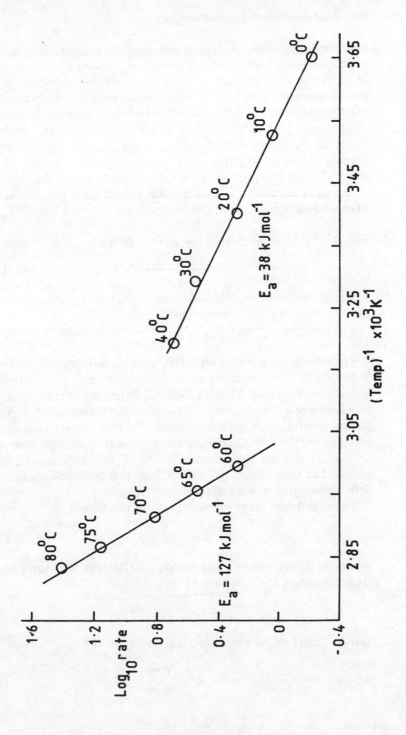

Figure 3.3: Example of an Arrhenius Activation Energy Calculation

interpretations of the numerical values of the parameters Z and E_a which it allows. More specific inference is possible when the energy changes in a chemical reaction are considered from the viewpoint of the 'transition state' or 'absolute reaction-rate' theory.[3] This theory postulates that the rate of a chemical reaction is proportional to the concentration of an 'activated complex' or 'transition state' in equilibrium with the reactants. A chemically reactive species of energy approximately coincident with the apex of the energy barrier (Figure 3.4), the transition state lies on the pathway to product formation. In line with this theory the Michaelis-Menten-Henri mechanism can be rewritten

$$E + S \rightleftharpoons E^*S \rightleftharpoons ES \rightleftharpoons E^*X$$

$$EP \rightleftharpoons E^*P \rightleftharpoons E + P \quad (3.4)$$

where ES and EP are Michaelis complexes and the starred species are the transition state complexes.

The energy diagram describing this pathway is shown in Figure 3.5.

The rate limiting (rate controlling, slow or rate determining) step is defined as the earliest step in the reaction with a rate equal to the overall reaction rate. Thus in Figure 3.5 the step characterised by rate constant k_2 would control the rate in the forward (left to right) direction while that with k_{-2} would be the slowest step in the reverse direction. An alternative description attributes the slow step to that possessing the transition state of the highest activation energy. For the reaction in Figure 3.5, as for most reactions, the two definitions agree in their designation.

The expression for the forward rate of reaction is:

$$v = k_0 \, E \cdot S. \quad (3.5)$$

where k_0 is the overall rate constant. The first step involving a transition state in this direction,

$$E + S \rightleftharpoons E^*S \quad (3.6)$$

can be described by an equilibrium constant, K^*

where
$$K^* = \frac{E^*S}{E \cdot S} \quad (3.7)$$

Figure 3.4: Energy Profile for Catalysed and Uncatalysed Reactions

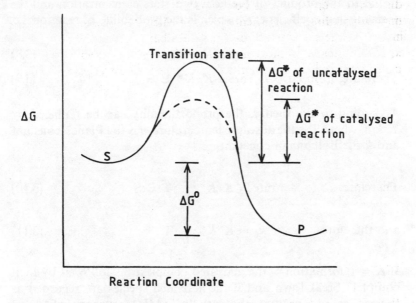

Figure 3.5: Energy Diagram for a One Substrate-One Product Enzyme Catalysed Reaction

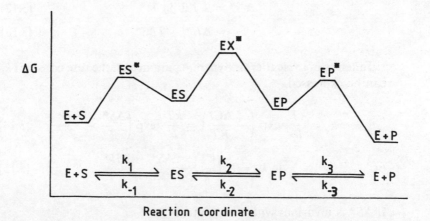

From transition-state theory the rate of the reaction is proportional to the product of the activated state concentration and the transmission coefficient, K, which is the probability of reaction, i.e.

$$\text{rate } \alpha \; K \cdot E^* S \tag{3.8}$$

Therefore, from eq. 3.7,   rate $\alpha \; K \cdot K^* \cdot E \cdot S$ (3.9)

According to the theory, the proportionality can be replaced by $T \cdot k/h$, where T is the absolute temperature, h is the Planck constant and k the Boltzmann constant,

Therefore, $$\text{rate} = K \cdot K^* \cdot \frac{k}{h} \cdot T \cdot E \cdot S \tag{3.10}$$

and therefore $$k_0 = K \cdot K^* \cdot \frac{k}{h} \cdot T \tag{3.11}$$

If $K = 1$, formation of the activated complexes is followed by their complete breakdown and if the activated species are regarded as thermodynamic entities, the enthalpy (ΔH^*), entropy (ΔS^*) and energy (ΔG^*) of activation can be calculated in the same way as the corresponding equilibrium parameters ΔH_0, ΔS_0 and ΔG_0; thus

$$\Delta G^* = -RT \ln K^* \tag{3.12}$$

$$= \Delta H^* - T\Delta S^* \tag{3.13}$$

In terms of these thermodynamic parameters, the rate constant k_0 can be expressed,

$$k_0 = \frac{kT}{h} \cdot \exp\left(-\frac{\Delta G^*}{RT}\right) = \frac{kT}{h} \cdot \exp\left(\frac{\Delta S^*}{R} - \frac{\Delta H^*}{RT}\right) \tag{3.14}$$

or $$\ln k_0 = \ln\left(\frac{kT}{h}\right) + \frac{\Delta S^*}{R} - \frac{\Delta H^*}{RT} \tag{3.15}$$

If ΔS^* is invariant with temperature

$$\frac{d}{dT}(\ln k_0) = \frac{\Delta H^*}{RT^2} + \frac{1}{T} = \frac{\Delta H^* + RT}{RT^2} \tag{3.16}$$

Comparing this with the differential of eq. 3.3,

$$\frac{d}{dT}(\ln k_0) = \frac{E_a}{RT^2} \qquad (3.17)$$

a relationship between Arrhenius activation energy and the enthalpy of activation can be derived,

$$E_a = \Delta H^* + RT \qquad (3.18)$$

or $\qquad\qquad \Delta H^* = E_a - RT \qquad\qquad\qquad (3.19)$

The enthalpy of activation is then easily calculated from the Arrhenius energy. At 30° the difference is 2.5 kJ mol^{-1}, a quantity which is within the precision of most experimental methods. From eq. 3.14,

$$\ln \frac{k_0}{T} = \ln \frac{k}{h} - \frac{\Delta G^*}{RT} = \ln \frac{k}{h} - \frac{\Delta H^*}{RT} + \frac{\Delta S^*}{R} \qquad (3.20)$$

thus the enthalpy of activation is obtained from a graph of $\ln k_0/T$ against $(RT)^{-1}$. Finally the energy of activation can be determined from k_0 and eq. 3.14, and the entropy of activation from the dependence of ΔG^* on absolute temperature, or if ΔH^* is known, eq. 3.13.

Interpretation in molecular and hence in mechanistic terms of these activation parameters is fraught with difficulty. Two major, complex effects suggested as contributing to their numerical values are those arising from the surrounding water molecules—a solvent effect, and from conformational changes occurring in the enzyme molecule during reaction—a structural effect.[4] But the contribution of each to the observed values is quantitatively uncertain. Often recorded are negative entropies of reaction which are consistent with a compactation of the reacting enzyme molecule, but equally such changes could arise from the formation of charged particles and the associated gain and ordering of solvent molecules.

It may be stated however that the function of the enzyme protein like that of other catalysts is to make the transition state for the reaction easier to reach. This is reflected in an increased rate and as has been seen, for a given reaction, an enzyme is more effective in this respect than other catalysts. An additional notion of its

efficiency may be obtained by contrasting the rates of enzyme catalysed reactions with the maxima theoretically attainable.
Consider the first order reaction,

$$S \xrightarrow{k_{cat}} P \tag{3.21}$$

The negative exponential relationship between the rate constant (k_{cat}) and the energy of activation (ΔG^*), eq. 3.14, expresses quantitatively that the smaller the ΔG^* the larger is the rate constant. In the lowest limit, the barrier to reaction will be negligible, thus ΔG^* will approach zero and the exponential term will approach unity; therefore

$$k_{cat} \longrightarrow kT/h \tag{3.22}$$

Insertion of the numerical values for the Boltzmann and Planck's constants, 1.381×10^{-23} J K^{-1} and 6.625×10^{-34} J s respectively, gives at physiological temperatures k_{cat} approximately 10^{-12} s^{-1}.

As can be seen from Table 3.2, this value is several orders of magnitude greater than the rates of dissociation of the enzyme-substrate complexes.

The first step in an enzyme catalysed reaction is not unimolecular but involves the association of enzyme with substrate molecules,

Table 3.2: Rates of Enzyme-Substrate Association and Dissociation

Enzyme	Substrate	Second order rate for complex formation (mol^{-1} s^{-1})	First order rate for complex dissociation (s^{-1})
Fumarase	Fumarate	10^9	4.5×10^4
Hexokinase	Glucose	3.7×10^6	1.5×10^3
Glutamate-	Glutamate	3.3×10^7	2.8×10^3
aspartate	Aspartate	10^7	5×10^3
transaminase	2-Oxoglutarate	5×10^8	5×10^4
	Oxaloacetate	7×10^7	1.4×10^2
Alcohol dehydrogenase	NADH	1.1×10^7	3.1

Source: Reference 5.

and hence is bimolecular. Before these react and form products two events must take place. Firstly the reactants must approach and invade each other's solvation shells—they must 'encounter'. Secondly, within their mutual solvation shell, the reactants must collide, and several collisions may take place before transformation occurs. If reaction occurs at every encounter the velocity of the reaction will depend only on the rate at which they initially come together—the rate of diffusion. The reaction is then said to be diffusion controlled, and the rate will be the upper limit of a bimolecular reaction.

The rate of diffusion is given[6] by

$$k = 4\pi D \rho N / 1000 \; \text{mol}^{-1} \cdot \text{s}^{-1} \qquad (3.23)$$

where D is the sum of the reactant diffusion coefficients (units $\text{cm}^2 \; \text{s}^{-1}$), ρ the sum of the radii of the reactants and N is Avogadro's number.

Insertion of the appropriate quantities into this equation yields values for the second order rate constant between 10^9 and $10^{11} \; \text{mol}^{-1} \; \text{s}^{-1}$. The upper part of this range approximates the limit for reactions involving protons or favourable charge-charge interactions, whereas the lower value is close to the rates of enzyme-substrate associations (Table 3.2).

3.2 EFFECT OF pH

The hydrogen ion concentration affects the affinity of an enzyme for its substrate, the maximum rate of reaction and the protein stability. The resultant of these effects is generally a narrow pH range of activity with an optimum, Figure 3.6. The contribution of each of these can be separated and experimentally evaluated relatively easily. Construction of Michaelis-Menten-Henri profiles and derived plots such as those of Lineweaver and Burk at each pH enable the individual pH-dependencies of K_m and V_{max} to be ascertained. Such plots also allow the analysis of other phenomena which may arise at certain hydrogen ion concentrations, such as substrate inhibition or activation.

Effects of pH on the macromolecular stability of the enzyme can be quantitated by exposing it to different hydrogen ion concentrations for varying times, followed by assay at a fixed, intermediate

Figure 3.6: Typical pH-Rate Profile of an Enzyme Catalysed Reaction

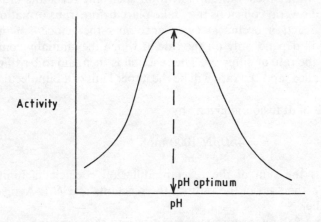

value. For the majority, inactivation and denaturation become significant only at pH extremes, where the catalytic activity would be very low anyway. At intermediate values, the environmental hydrogen ion concentration determines the state of ionisation of the amino acid side chains in the protein, the enzyme-substrate ionisations and the charged state of the substrate. Any effects on the last can be determined independently by titration and then taken into account when the complete pH dependence of the enzyme catalysed reaction is studied. In order that this study is rationally based, the reversible pH-effects on the enzyme (and enzyme-substrate complex), as exhibited in its catalytic parameters, are usually considered to occur at or near the active site, and in addition, since the enzyme is catalytically operational over a very limited range, to depend on one functional ionisation state of this site.

As an introductory basis for interpreting the observed pH-rate profiles in terms of side chain ionisations at the active site, the bell shaped graph (Figure 3.6) should be treated as the composite of two sigmoidal titration curves. Consider the left hand 'S' shape ascending from the more acid pH values to the optimum pH. This could be taken as corresponding to the ionisation of an acidic group, for example a carboxylic acid $(R-CO_2H)$, in the active site. As the pH is raised this ionises to $R-CO_2^-$. If catalytic activity depends on this carboxylate anion and not on the unionised acid, the rate will also

Figure 3.7: Effect of Hydrogen Ion Concentration on Active Site Group Ionisations

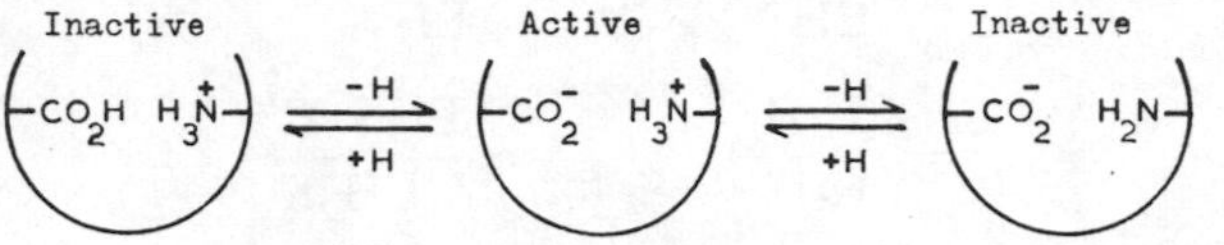

increase in the same sigmoidal manner and exhibit a pK_a value ($K_a = H^+ \cdot R-CO_2^-/R-CO_2H$) equal to the pH of half maximal activity. Similarly the second sigmoid could be considered due to the titration of the conjugate acid (H^+B) of a basic species (B) at the active site. As the pH is increased, the concentration of the unprotonated base increases, which if this is inactive will result in a progressive decrease in activity. These two effects can then be combined as shown in Figure 3.7.

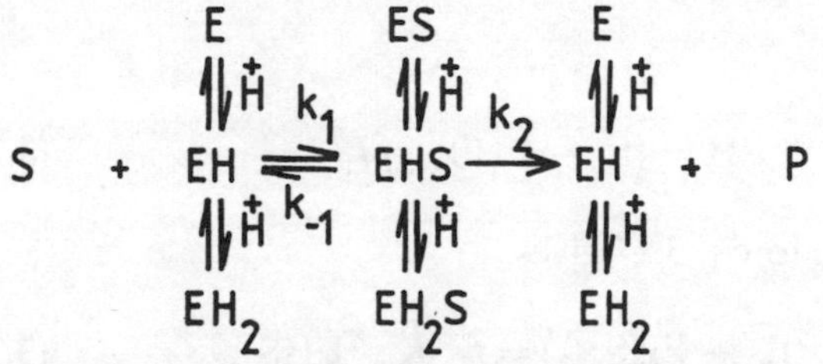

$$S + EH \underset{k_{-1}}{\overset{k_1}{\rightleftharpoons}} EHS \overset{k_2}{\longrightarrow} EH + P \qquad (3.24)$$

The schematic representation of eq. 3.24 allows a more quantitative analysis of the effects of pH on the reaction parameters to be made if it is assumed that (1) a basic Michaelis-Menten-Henri mechanism is followed in which breakdown of the enzyme-substrate complex (k_2) is rate limiting, (2) catalysis depends only on two ionisations at the active site with their rates of ionisation rapid compared to the rates of the catalytic process and finally (3) the rates of formation and breakdown of EHS are faster than those of either ES or EH_2S, i.e.

$$v = k_2 \cdot EHS. \qquad (3.25)$$

Based on these assumptions, the scheme can be subjected to steady-state analysis:[7]

$$\frac{d\,EHS}{dt} = 0 = k_1 \cdot EH \cdot S_0 - k_{-1} \cdot EHS + k_2 \cdot EHS \qquad (3.26)$$

$$\therefore EH \cdot S_0 = K_m \cdot EHS \qquad (3.27)$$

since $E_0 = E + EH + EH_2 + ES + EHS + EH_2S$ (3.28)

and if $K_{E1} = \dfrac{E \cdot H^+}{EH}$; $K_{E2} = \dfrac{EH \cdot H^+}{EH_2}$;

$$K_{ES1} = \frac{ES \cdot H^+}{EHS} ; \quad K_{ES2} = \frac{EHS \cdot H^+}{EH_2S}$$

then $E_0 = EH \cdot \dfrac{K_{E1}}{H^+} + EH + EH \cdot \dfrac{H^+}{K_{E2}} + EHS \cdot \dfrac{K_{ES1}}{H^+}$

$$+ EHS + EHS \cdot \frac{H^+}{K_{ES2}}$$ (3.29)

or $E_0 = EH \cdot f_1 + EHS \cdot f_2$ (3.30)

where $f_1 = 1 + \dfrac{K_{E1}}{H^+} + \dfrac{H^+}{K_{E2}}$ and $f_2 = 1 + \dfrac{K_{ES1}}{H^+} + \dfrac{H^+}{K_{ES2}}$

From eq. 3.30,

$$EH = (E_0 - EHS \cdot f_2)/f_1$$ (3.31)

thus substitution into eq. 3.27 gives,

$$S_0(E_0 - EHS \cdot f_2) = f_1 \cdot K_m \cdot EHS$$ (3.32)

$$\therefore EHS = \frac{E_0 S_0}{f_2 S_0 + f_1 K_m}$$ (3.33)

$$\therefore v = \frac{k_2 \cdot E_0 \cdot S_0}{f_2 S_0 + f_1 K_m}$$ (3.34)

or $$v = \frac{V' \cdot S_0}{S_0 + K_m'}$$ (3.35)

(where $V' = V_{max}/f_2$ and $K_m' = K_m \cdot f_1/f_2$)   (3.35)

$$\therefore V'/K_m' = V_{max}/(K_m \cdot f_1)$$ (3.36)

3.2.1 Effect of pH on the Maximum Rate

At saturating substrate concentrations all enzyme molecules will be complexed. Since the observed rate is governed by that for breakdown of this complex, pH effects on the maximum rate will reflect their influence on its ionised state.

From eq. 3.35, $$V' = V_{max}/f_2 = k_2^0 E_0/f_2 \qquad (3.37)$$

$$\therefore\ k_2' = k_2^0/f_2 \qquad (3.38)$$

where k_2^0 is a pH independent rate constant and f_2 is as before; eq. 3.38 then describes the variation of the maximum rate with pH. Expansion of eq. 3.37 yields,

$$\log V' = \log (k_2^0 \cdot E) + \log f_2^{-1} \qquad (3.39)$$

or $$\log V' = \log (k_2^0 \cdot E) + pf_2$$

(where p is the negative logarithm) $\qquad (3.40)$

The values obtained for the maximum rate at each pH, plotted according to this equation then enable numerical determination of the pK_a values of the ionising groups which control the breakdown of the enzyme-substrate complex, Figure 3.8.

3.2.2 Effect of pH on V_{max}/K_m

The pK_a values of the ionisations in the free enzyme pK_{EH1} and pK_{EH2} are given by $\log V'/K_m'$ versus pH plotted according to eq. 3.41, the expanded form of eq. 3.36, in Figure 3.9, in which K_m^0 is a pH-independent Michaelis constant.

$$\log (V'/K_m') = \log k_2^0 E/K_m^0 + pf_1 \qquad (3.41)$$

As an alternative to (V'/K_m') the first order rates at constant, small concentrations of substrate (which fulfil the condition $K_m' > S_0$) may be used.

3.2.3 Effect of pH on K_m
From eq. 3.35

$$K_m = K_m^0 \cdot f_1/f_2 \qquad (3.42)$$

or $$pK_m = pK_m^0 + pf_1 - pf_2 \qquad (3.43)$$

Dixon[7] has shown that the graphical representation of the dependence of pK_s or pK_m on pH, could be readily interpreted by application of a few simple rules. He recommended that straight lines taking gradients of 0, 1 or 2 be drawn to the experimental data. The intersection point of two lines then indicates the pK_a of a group

Figure 3.8: Effect of pH *on* $\log_{10} V_{max}$

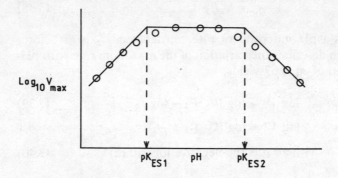

Figure 3.9: Effect of pH *on* $\log (V_{max}/K_m)$

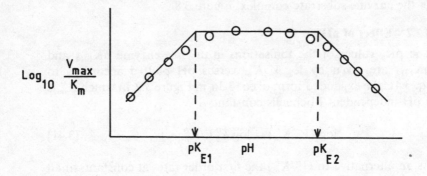

Figure 3.10: Effect of pH *on* pK_m

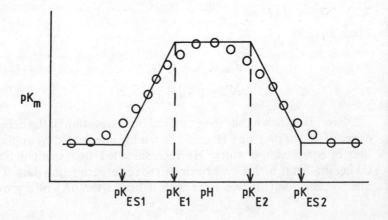

ionisation. If the pK_a is accompanied by an upwards change of slope then this corresponds to a group in the enzyme-substrate complex, if downwards then to the free enzyme or substrate (Figure 3.10).

3.2.4 Interpretation of pH Effects

The attraction of determining active site pK_a values lies in the possibility of thereby determining the identity of the groups responsible for catalysis and binding. Table 3.3 gives a list of the amino acid chains and their pK_a values when present in small molecular weight derivatives.

Table 3.3: pK_a Values for Amino Acid Side Chains

	pK_a (25°)
Aspartic acid (β-CO$_2$H)	3.0–4.5
Glutamic acid (γ-CO$_2$H)	4.4
Histidine (imidazole)	5.6–7.0
Cysteine (sulphydryl)	8.3
Lysine (ε-amino)	9.5–10.5
Tyrosine (phenolic hydroxyl)	10.0
Arginine (guanidine)	11.5–12.5

Source: Reference 8.

Such comparisons can only be tentative however and without other experimental evidence their interpretation should be deferred. One problem is to assign the observed ionisations to particular groups. Several side chains have overlapping pK_a values, those of tyrosine and lysine, and aspartate and glutamate for example. This problem of identification becomes additionally difficult if the side chains are located in micro-environments with properties different from aqueous solution, since these perturb the ionisations and give rise to anomalous pK_a values. The pK_a value of 6.5 found for glutamate 35 in lysozyme, for example, arises from its situation in a micro-environment of reduced polarity. Decreasing the local dielectric constant would tend to destabilise the carboxylate form relative to the unionised species and thus raise the pK_a. Similarly the observed pK_a 5.9 found for the ε-amino group of a lysine in acetoacetate decarboxylase can also be rationalised by assuming the neighbouring presence of non-polar side chains which would tend to depress its protonation. In contradistinction the value of 10.0 expressed by the α-amino group of isoleucine 16 in chymotrypsin (see Chapter 4) arises from stabilisation of the cation by salt link formation with aspartate 194.

3.2.5   Titration of Amino Acid Side Chains in Enzymes

Although kinetic tools are indispensable to the analysis of active site reactivity, they are not the only means whereby the pK_a values of groups in enzymes may be estimated. Several other techniques have been developed of which nuclear magnetic resonance, ultraviolet difference spectroscopy and chemical modification are among the most frequently applied. NMR measures the chemical shifts on protonation of histidine, tyrosine and cysteine which are usually well downfield from the protons in the rest of the molecule. The use of NMR, in conjunction with other physical data, in assigning pK_a values to the four histidines in ribonuclease stands as a model for the application of this technique.[9] Determination of pK_a values by chemical modification utilises the different rates of reaction of protonated and unprotonated forms of the side chains towards the same molecule (such as amino and ammonium groups towards dinitrofluorobenzene). The extent of reaction and the identity of the reactive groups are determined by degradation of the enzyme after modification.[10] Ultraviolet difference spectroscopy measures the ionisations of tyrosine, histidine and cysteine side chains.[11]

3.3 ENZYME INHIBITION

The normal state of living matter is a delicately balanced, spatially and temporally co-ordinated organisation. If a substance causes an adverse effect on this balance it is usually termed a poison, alternatively if it redresses a pathophysiological imbalance it is regarded as a drug. Both may be enzyme inhibitors.

An enzyme inhibitor decreases the activity of an enzyme without significantly disrupting its three-dimensional macromolecular structure. Inhibition is therefore distinct from denaturation and is the result of a specific action by a reagent directed or transmitted to the active site region. Studies of enzyme inhibition can yield much information on the mechanism of enzyme catalysis. The topographical and energy requirements of the active site can often be inferred from the pattern and extent of inhibition by a series of compounds chemically related to the substrate, information often not as readily afforded by the substrate. A distinct type of inhibitor, the transition state analogue, exerts its specificity through mimicking the transition state of the enzyme catalysed reaction, thus insights into this highly energetic barrier to reaction and a more selective type of inhibitor may be derived.

The study of enzyme inhibition in fact encompasses much of enzymology and biochemistry on one hand and pharmacology, physiology, toxicology and pathology on the other, progress in one area often having important ramifications on the others. One underlying feature common to these areas is the specificity of inhibitor action. Elucidation of the steps in the Krebs cycle relied partly on the accumulation of succinate resulting from malonate inhibition of succinate DH and the irreversible action of iodoacetate on glyceraldehyde-3-phosphate DH, preventing triose phosphate metabolism, was crucial to the unravelling of glycolysis. The realisation that selectivity of inhibitor action can extend to yield a physiological response has particularly influenced the synthesis and pharmacological testing of potential active site directed enzyme inhibitors. Many insecticides are enzyme inhibitors; the development and investigation of these are improving agricultural efficiency and the treatment of accidentally poisoned farm workers. Several natural and synthetic antibiotics, e.g. penicillin and sulphonamide exert their action at the enzyme level, abnormal mental states are induced by monoamine oxidase inhibitors, and neuromuscular blockers such as neostigmine act by modifying cholinesterase activity.

The first detailed kinetic analysis of enzyme inhibition was presented by Michaelis and Menten in their papers on the concept of the enzyme-substrate complex. These appeared contemporaneously with Warburg's, now classical, experimentation on cyanide ion inhibition of respiration. The subject came of age however with Haldane's work and rationalisations in 1930–32 after an intervening period during which the mechanisms of action of several previously well known inhibitors, such as the arsenicals, were determined and two modes of inhibition, competitive and non-competitive, were distinguished.

3.3.1 Reversible and Irreversible Inhibition

Reaction of an enzyme with an inhibitor involves initial reversible association to form an inactive enzyme-inhibitor complex. If in this complex covalent bond formation takes place, dissociation will be prevented and the enzyme will become irreversibly inhibited (eq. 3.44).

$$E + I{-}X \underset{k_{-1}}{\overset{k_1}{\rightleftharpoons}} E{-}{-}{-}I{-}X \overset{k_2}{\longrightarrow} E{-}I + X \quad (3.44)$$

Active   Non-covalent   Covalent
                           inactive complex   inactive complex

Enzyme inhibitors can therefore be classified into two types, *reversible* and *irreversible*. In the presence of an irreversible inhibitor, loss of enzyme activity is proportional to the rate of covalent bond formation, and thus a progressive decrease in activity will be observed, with rate constant k_2 (see for example Figure 3.13). Since leaving group (X) is displaced by a reactive group in the enzyme, resulting in the formation of a stable covalent bond, the inhibitor cannot be removed by dialysis or gel chromatography. Irreversible inhibitors are particularly useful therefore for labelling and determining the chemical nature of binding sites.

On the other hand, *reversible* inhibition is not progressive and is characterised by an equilibrium constant, $K_I = E \cdot I / EI = k_{-1}/k_1$. The inhibitor is not covalently bound in the inactive complex and so can be removed by dialysis or gel chromatography to restore full enzymic activity. A reversible inhibitor can fall into any one of five categories: (i) competitive, (ii) non-competitive, (iii) uncompetitive, (iv) mixed or (v) allosteric.

3.3.1.1   Competitive Inhibition

As the name implies, inhibitors of this type compete with the substrate for the same active site. Usually the inhibitors have molecular structures similar to those of the substrate, therefore both cannot associate with the site at the same time.

Enzyme catalysis in the presence of a competitive inhibitor can be represented,

$$E + S \underset{k_{-1}}{\overset{k_1}{\rightleftharpoons}} ES \overset{k_2}{\longrightarrow} E + P \quad (3.45)$$

$$E + I \rightleftharpoons EI \quad (3.46)$$

Equations 3.45 and 3.46 indicate that the substrate and inhibitor cannot bind simultaneously to the enzyme and only the enzyme-substrate complex forms products. The scheme can be analysed by the steady-state procedure used in the derivation of the Briggs-Haldane equation:

$$\frac{d(ES)}{dt} = 0 = k_1 \cdot E \cdot S - k_{-1} \cdot E - k_2 \cdot ES \qquad (3.47)$$

If E_0 is the total enzyme concentration $= E + ES + EI$

then

$$(E_0 - ES - EI)S = K_m \cdot ES \qquad (3.48)$$

and if

$$K_I = \frac{E \cdot I}{EI} = (E_0 - ES - EI) \cdot \frac{I}{EI} \qquad (3.49)$$

then

$$EI = \frac{E_0 - ES}{1 + \dfrac{K_I}{I}} \qquad (3.50)$$

and substitution of eq. 3.50 into eq. 3.48 gives,

$$ES = \frac{E_0}{1 + \dfrac{K_m}{S}\left(1 + \dfrac{I}{K_I}\right)} \qquad (3.51)$$

$$\therefore v = k_2 \cdot ES = \frac{k_2 \cdot E_0}{1 + \dfrac{K_m}{S}\left(1 + \dfrac{I}{K_I}\right)} \qquad (3.52)$$

or $\quad v = \dfrac{V^c}{1 + \dfrac{K_m^c}{S}}$, where $K_m^c = K_m\left(1 + \dfrac{I}{K_I}\right)$ and $V^c = V_{max}$

$$(3.53)$$

Equation 3.53 is, like the Briggs-Haldane equation, also that for a hyperbola, and shows a maximum velocity of reaction unaltered by the presence of the competitive inhibitor. When the substrate concentration is increased while that of the inhibitor is held constant, the probability that a substrate molecule will bind increases. Thus at high substrate concentrations, V_{max} will be approached. However since substrate binding is impaired, the measured apparent K_m will be increased in proportion to the concentration of inhibitor and its dissociation constant. Therefore if the concentration of inhibitor is known, its dissociation constant can be estimated.

The comments regarding the application of the Michaelis profile for determining enzyme parameters outlined in Chapter 2 apply equally to eq. 3.53. Thus, for a more accurate evaluation of V^c and K_m^c, the experimental results are usually plotted according to the reciprocal (eq. 3.54) as in Figure 3.11a. When $I = 0$ this simplifies to that for the uninhibited reaction.

$$\frac{1}{v} = \frac{1}{V_{max}} + \frac{K_m}{S \cdot V_{max}}\left(1 + \frac{I}{K_I}\right) \tag{3.54}$$

Dixon[8] expanded eq. 3.54 to yield,

$$\frac{1}{v} = \frac{1}{V_{max}} + \frac{K_m}{S} + \frac{1}{S} \cdot \frac{K_m}{V_{max}} \cdot \frac{I}{K_I} \tag{3.55}$$

which predicts a linear relationship between reciprocal rate and inhibitor concentration at constant substrate concentration. This method of plotting the experimental data (Figure 3.11g) has the advantage of directly outputting K_I.

(Both the Lineweaver-Burk and Dixon plots are of diagnostic value for the differentiation of the type of reversible inhibition, Figure 3.11a–1.)

3.3.1.2   *Non-competitive Inhibition*

In strict non-competitive inhibition the inhibitor does not affect the binding of the substrate to enzyme (K_S) but when bound renders the enzyme-substrate complex inactive. Separate binding sites for inhibitor and substrate can then be envisaged, but which interact with each other. This type of inhibition differs from the competitive case, involving the formation of an inactive ternary enzyme-substrate-inhibitor complex and a maximum velocity reduced as a consequence of a decreased proportion of freely available enzyme molecules.

The system may be represented,

$$E + S \underset{k_{-1}}{\overset{k_1}{\rightleftharpoons}} ES \xrightarrow{k_2} E + P \tag{3.56}$$

$$E + I \rightleftharpoons EI; \; EI + S \rightleftharpoons EIS; \; ES + I \rightleftharpoons EIS \tag{3.57}$$

If $K_S = \dfrac{\text{EI}\cdot\text{S}}{\text{EIS}}$ and $K_I = \dfrac{\text{E}\cdot\text{I}}{\text{EI}} = \dfrac{\text{ES}\cdot\text{I}}{\text{EIS}}$

then by the steady state procedure,

$$v = \frac{k_2\,E_0}{\left(1 + \dfrac{K_m}{S_0}\right)\left(1 + \dfrac{I}{K_I}\right)} \quad \text{or} \quad v = \frac{V^{nc}}{1 + K_m/S_0} \qquad (3.58)$$

where $V^{nc} = V_{max}/(1 + I/K_I)$

The maximum velocity is reduced by the factor $(1 + I/K_I)$, but the equation is still that of a hyperbola.

Reciprocation of eq. 3.58 gives

$$\frac{1}{v} = \frac{1}{V_{max}}\left(1 + \frac{K_m}{S_0}\right)\left(1 + \frac{I}{K_I}\right) \qquad (3.59)$$

$$\therefore \quad \frac{1}{v} = \frac{1}{V_{max}}\left(1 + \frac{I}{K_I}\right) + \frac{1}{S_0}\cdot\frac{K_m}{V_{max}}\cdot\left(1 + \frac{I}{K_I}\right) \qquad (3.60)$$

the Lineweaver-Burk representation, Figure 3.11b,

or $\qquad \dfrac{1}{v} = \dfrac{1}{V_{max}}\left(1 + \dfrac{K_m}{S_0}\right) + I\cdot\dfrac{1}{V_{max}\cdot K_I}\left(1 + \dfrac{K_m}{S_0}\right) \qquad (3.61)$

the Dixon representation, Figure 3.11h.

3.3.1.3 Uncompetitive Inhibition

True examples of this type of inhibition are extremely rare although other catalytic mechanisms, such as product inhibition can give the same kinetic behaviour. The accepted action of an uncompetitive inhibitor is its association with the enzyme-substrate complex:

$$\text{E} + \text{S} \rightleftharpoons \text{ES} \longrightarrow \text{E} + \text{P}; \quad \text{ES} + \text{I} \rightleftharpoons \text{ESI} \qquad (3.62)$$

for which $\qquad v = \dfrac{V_{max}}{1 + \dfrac{I}{K_I} + \dfrac{K_m}{S_0}} = \dfrac{V^u}{1 + \dfrac{K_m^u}{S_0}} \qquad (3.63)$

Figure 3.11: Graphical Schemes for the Representation of Enzyme Inhibition

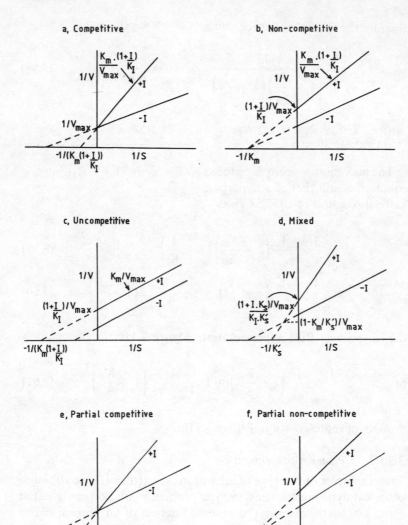

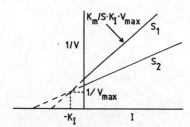

g, Competitive

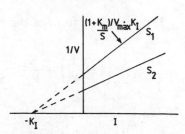

h, Non-competitive

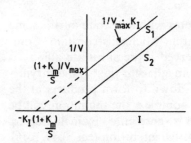

i, Uncompetitive

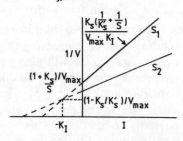

j, Mixed

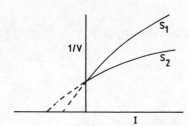

k, Partial competitive

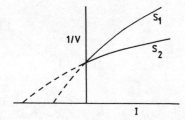

l, Partial non-competitive

where $K_I = \dfrac{ES \cdot I}{ESI}$; $V^u = \dfrac{V_{max}}{1 + \dfrac{I}{K_I}}$ and $K_m^u = \dfrac{K_m}{1 + \dfrac{I}{K_I}}$.

The reciprocal form of the rate equation:

$$\frac{1}{v} = \frac{1}{V_{max}}\left(1 + \frac{I}{K_I}\right) + \frac{K_m}{V_{max}} \cdot \frac{1}{S_0} = \frac{1}{V_{max}} + \frac{K_m}{V_{max}} \cdot \frac{1}{S_0} + \frac{1}{V_{max}} \cdot \frac{I}{K_I}$$

$$(3.64)$$

indicates that in both the Lineweaver-Burk, Figure 3.11c and Dixon plots, Figure 3.11i, parallel lines of constant slope (K_m/V_{max} and $1/V_{max} \cdot K_I$ respectively) will be observed.

3.3.1.4 Partial Competitive Inhibition

A pure competitive inhibitor competes with substrate for the same binding site, so increasing its apparent Michaelis constant. Competitive inhibition will also be observed if the substrate affinity is reduced by the inhibitor binding at an adjacent site.* If the enzyme-substrate-inhibitor complex breaks down to form products at the same rate as the enzyme-substrate complex, the inhibition will be partially competitive. This situation is represented by including eq. 3.65 in addition to those for competitive inhibition (eqs. 3.45, 3.46).

$$
\begin{array}{c}
EI + S \\
\qquad\qquad\searrow \\
\qquad\qquad\qquad EIS \xrightarrow{\ k_2\ } EI + P \qquad (3.65)\\
\qquad\qquad\nearrow \\
ES + I
\end{array}
$$

if $K_I = \dfrac{E \cdot I}{EI}$, $K_S' = \dfrac{EI \cdot S}{EIS}$ and $K_I' = \dfrac{ES \cdot I}{EIS}$

then $v = k_2(ES + EIS)$ (3.66)

---

* In non-competitive inhibition not the binding but the transformation of substrate is inhibited.

and
$$v = \frac{V_{max}}{1 + \frac{K_m^{pc}}{S_0}}$$

where
$$K_m^{pc} = K_m\left(1 + \frac{I}{K_I}\right) \cdot \left(1 + \frac{I}{K_I'} \cdot \frac{K_S}{K_S'}\right)^{-1} \quad (3.67)$$

When no inhibitor is present, $K_m^{pc} = K_m$ and the equation becomes that for the uninhibited reaction and when $K_I' = K_S' = 0$ the equation reduces to that for full competitive inhibition.

It is clear from inspection of the reciprocal of this equation that partial competition cannot be differentiated from complete competition by means of the Lineweaver-Burk plot, Figure 3.11e; they may be distinguished however by application of the Dixon plot. For partial competitive inhibition curves are obtained, Figure 3.11k.

3.3.1.5 *Partial Non-competitive Inhibition*

In this type of reversible inhibition, the enzyme-inhibitor-substrate complex, which is totally inactive in pure non-competitive inhibition, has partial activity. Equations 3.56 and 3.57 are therefore also applicable but the system is described more fully by the addition of an equation describing the breakdown of the ternary complex, eq. 3.68.

$$EIS \xrightarrow{k_2'} E + I + P \quad (3.68)$$

where $k_2' < k_2$

$$v = k_2 \cdot ES + k_2' \cdot EIS \quad (3.69)$$

$$\therefore v = \frac{V^{pn}}{1 + \frac{K_m}{S_0}} \quad \text{where} \quad V^{pn} = \frac{E_0(k_2 + k_2' \cdot I/K_I)}{1 + \frac{I}{K_I}} \quad (3.70)$$

When $k_2' = 0$ the equation is that for full non-competitive inhibition. If $k_2' \gg k_2$ the ternary complex breaks down to form products faster than the binary complex, and I then acts as an activator. Dixon and Webb[8] have given procedures whereby the various inhibition constants may be determined (Figure 3.11j).

3.3.1.6   Mixed Inhibition

Strict competitive inhibition can be readily distinguished from the other types since this is the only one in which a tertiary enzyme-substrate-inhibitor complex cannot be formed. For the other types, in which such complexes are involved, the complexity of the binding processes and the concomitant co-operative movements in the protein molecule mean that the inhibition behaviour exhibited by a particular enzyme may not fall into one or other of the above well defined classes, but such conformational alterations may cause the inhibitory patterns to be mixed. These can include partially competitive/non-competitive and partially competitive/partially non-competitive inhibitions. Both will result in the numerical values of K_m and V_{max} being affected.

Partial competitive/non-competitive inhibition can be represented by

$$E + S \rightleftharpoons ES \longrightarrow E + P \qquad (3.71)$$

$$ES + I \rightleftharpoons EIS; \quad E + I \rightleftharpoons EI;$$

$$EI + S \rightleftharpoons EIS \qquad (3.72)$$

with
$$K_I = \frac{E \cdot I}{EI}; \quad K'_S = \frac{EI \cdot S}{EIS}; \quad K'_I = \frac{ES \cdot I^*}{EIS} \qquad (3.73)$$

$$\therefore \; v = k_2 \cdot ES = \frac{V^m}{1 + \dfrac{K_m^m}{S}} \qquad (3.74)$$

where
$$V^m = \frac{k_2 \cdot E_0}{1 + \dfrac{I}{K_I} \cdot \dfrac{K_S}{K'_S}} \quad \text{and} \quad K_m^m = K_m \frac{1 + \dfrac{I}{K_I}}{1 + \dfrac{I}{K_I} \cdot \dfrac{K_S}{K'_S}}$$

When $K_S = K'_S$ these equations reduce to those corresponding to non-competitive inhibition.

3.3.1.7   Other

Although the graphs may indicate that a certain pattern is exhibited by the enzyme inhibitor, this is not proof that a mechanism of the

---

* In partial competitive inhibition both ES and EIS can form products and in non-competitive inhibition $K_S = K'_S$ and $K_I = K'_I$.

suggested type is followed. For example if $K_m \neq K_S$ (i.e. if $k_2 \simeq k_{-1}$) then a non-competitive inhibitor will affect K_m in addition to V_{max} and in the limit $k_2 \gg k_1$ ($K_m \simeq k_2/k_{-1}$) parallel Lineweaver-Burk plots will result and suggest uncompetitive inhibition instead.

Borate ions have been found to inhibit alcohol dehydrogenase by a mechanism kinetically manifested as competitive inhibition. However the effect does not result from direct borate interaction with the enzyme but from its association with the coenzyme NADH. The active concentration of the latter is reduced leading to a decrease in reaction rate. Kinetic analysis[12] of the system $(E + S \rightleftharpoons ES \rightarrow E + P; S + I \rightleftharpoons SI)$ leads to a rate equation identical in form to eq. 3.52, but in this case K_I is the dissociation constant of the inhibitor—NADH complex. In this mechanism, enzyme and inhibitor compete for substrate whereas in the more usual type substrate and inhibitor compete for enzyme.

3.3.2 Examples of Reversible Enzyme Inhibition

The classical example of a competitive inhibitor is the malonate $(CH_2(CO_2^-)_2)$, inhibition of the Krebs' cycle enzyme succinate dehydrogenase, eq. 3.75.

$$^-O_2C \cdot CH_2 \cdot CH_2 \cdot CO_2^- \xrightarrow{\text{Succinate DH}} {}^-O_2C \cdot CH{=}CH \cdot CO_2^-$$

$$\text{Succinate} \qquad\qquad\qquad\qquad \text{Fumarate}$$

$$(3.75)$$

The similarity between the structures of succinate and malonate in the two carboxylate groupings, suggests that both would bind to the same site and hence compete with one another. However, since malonate lacks the second methylene group, removal of its two hydrogens would be unfeasible, so it behaves as an inhibitor. Malonate is not alone in producing this effect on succinate dehydrogenase; many molecules containing two anionic groups separated by a suitable interatomic distance will also inhibit, for example, pyrophosphate and meta-phthalate.

Competitive inhibition, in which the compound binds to the same centre as the substrate but remains intact is extremely common. For example the amide ketone in oxamate $(NH_2 \cdot CO \cdot CO_2^-)$ requires a lower redox potential for reduction than the α-ketone of pyruvate $(CH_3 \cdot CO \cdot CO_2^-)$, thus although it binds with a similar affinity as the substrate to the same site of lactate dehydrogenase, oxamate is not reduced by NADH.

Table 3.4: Energy of Lysozyme-Oligosaccharide Interactions

Oligosaccharide*	K_I (mM)	$-\Delta G_0$ ($RT \ln K_I^{-1}$) (kJ mol^{-1})
NAG	50	6.8
NAG$_2$	0.18	21.8
NAG$_3$	0.01	29.0
NAG$_4$	0.005	30.6

* NAG, N-acetyl glucosamine.

A different mode of competitive inhibition is exhibited by short oligosaccharides binding to the active cleft of lysozyme. This enzyme catalyses the lysis of cells by first binding the polysaccharide component of the cell wall in a channel containing six sub-sites. Di- or tri-saccharides containing N-acetylglucosamine rings, can also bind in these sub-sites, but without traversing the catalytic region and so are able to prevent substrate binding. From the dissociation constants for the various oligosaccharide-lysozyme complexes the Gibbs free energy for association at each sub-site can be estimated (Table 3.4).

Elucidation of the precise modes of many enzyme-inhibitor associations will have to await advances made in the crystallisation of complexes suitable for X-ray crystallography, as has elegantly been achieved, for example, with lysozyme. Even so, the fidelity of the results and estimates of the magnitude of the forces involved in enzymatic catalysis require independent validation and quantitation. In these respects, competitive inhibitors are valuable tools. Bergman, Nieman and others,[13] from a study of the effect of different substituents on the reactivity of a series of amides with trypsin and chymotrypsin devised a concept of complementary binding localities (I).

$$P_3$$
$$\boxed{CO \cdot R_3}$$
$$P_1 \ \left\langle R_1 \cdot CO \cdot NH - C - H \right] P_H$$
$$\boxed{R_2}$$
$$P_2$$

I

Table 3.5: Binding of Alkylammonium Ion Inhibitors to Trypsin

	K_I (mM)	$-\Delta G_0$ (kJ mol^{-1})
$CH_3 \cdot \overset{+}{N}H_3$	340	2.9
$CH_3 \cdot CH_2 \cdot \overset{+}{N}H_3$	62	6.7
$CH_3 \cdot (CH_2)_2 \cdot \overset{+}{N}H_3$	8.7	11.8
$CH_3 \cdot (CH_2)_3 \cdot \overset{+}{N}H_3$	1.7	16.0
$CH_3 \cdot (CH_2)_4 \cdot \overset{+}{N}H_3$	12	10.9

Source: Reference 14.

The p_2 area confers specificity on the enzyme. For example if R_2 is aromatic the resultant is a good substrate or inhibitor of chymotrypsin but a poor one for trypsin. Conversely, if R_2 is an alkylammonium or an alkyl guanidinium ion the resulting compound is trypsin specific. Systematic approaches of this kind can afford considerable information concerning the spatial and energetic requirements of the active site. For example trypsin is inhibited by straight chain alkylammonium ions but, as Table 3.5 shows, the binding energy is maximised at the C_4 homologue.

These results are consistent with the insertion of the alkylammonium chain into a hole of a specific size in the enzyme. Such a 'hole' was later verified by X-ray crystallographic experiments and was shown to have buried at one extremity an ionised carboxylic acid group. If it is assumed that the methyl group contributes very little to binding the electrostatic force would be of the order 2.9 kJ mol^{-1}. Each methylene group can then be calculated to contribute approximately 4.2 kJ mol^{-1} hydrophobic plus van der Waals binding energy.[14]

3.3.2.1 *Enzyme Inhibition and Chemotherapy*

Enzyme inhibitors are essentially poisons, and one of the basic tasks of chemotherapy is to ensure their selectivity of action. This has proven relatively easy in those diseases caused by an infecting agent using metabolic pathways different from those of the host, but more difficult when the targets are enzymes in abnormal cells. Pathological cells may contain fewer enzymes or differ only quantitatively from normal cells. One main difference between neoplastic and normal cells however is the more rapid proliferation of the former.

For these reasons, a target for chemotherapeutic attack of both parasitic and neoplastic diseases is nucleic acid biosynthesis. Nucleic acids, the genetic determinants, are required by all cells for division and protein synthesis. Interference with their formation, replication or transcription could therefore adversely affect cell growth. All ribonucleic acids and deoxyribonucleic acids are synthesised in the cell nucleus from their constituent purine and pyrimidine bases through the action of specific polymerases. The precursors are in turn constructed *de novo* from glycine, glutamine and ribofuranose units. Each step in these anabolic pathways is catalysed by a specific enzyme. At least twelve are required for a recognisable purine to be formed from ribose-5-phosphate, and hence in principle each is a possible target for inhibition.

The biosynthetic routes from ribose-5-phosphate to the nucleic acids including the sites of inhibition by several nautral and synthetic antibiotics and antitumour agents are outlined in Figure 3.12. The antileukemic compound, 6-mercaptopurine has been extensively studied at the molecular level and found to inhibit several enzymes involved in purine biosynthesis. *In vivo* it is first activated to thioinosinate (II, $R_1 = SH$, $R_2 = H$) by combination with 5-phosphoribosyl-1-pyrophosphate under the direction of inosinate pyrophosphorylase. Normally this enzyme converts hypoxanthine into inosinate (II, $R_1 = OH$, $R_2 = H$), thus the physiologically active inhibitor is formed *in situ* by the cell itself. Thioinosinate

II

III

Figure 3.12: Enzyme Mediated Route to the Nucleic Acids (PR, phosphoribosyl-; MP, monophosphate. Inhibitors are capitalised in boxes: azaserine, O-diazoacetyl-L-serine; DON, 6-diazo-5-oxo-L-norleucine; 6MP, 6-mercaptopurine, 5FDU, 5-fluorodeoxyuridine)

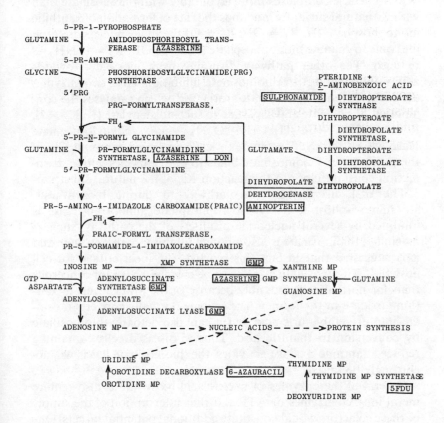

inhibits at least four enzymes involved in the conversion of inosinate into nucleic acid precursors, inosinate dehydrogenase, adenylosuccinate synthetase, adenylosuccinate lyase and glutamine-5-phosphoribose-1-pyrophosphate amidotransferase. Inosinate dehydrogenase complexes more favourably with thioinosinate than with its natural substrate, and thus the rate of formation of xanthine monophosphate (II, R_1 = OH, R_2 = O) the main intermediate on the route to guanine monophosphate (II, R_1 = OH, R_2 = $-NH_2$) is reduced. The other pathway diverging from inosinate, towards adenylate, is subject to thioinosinate inhibition. Here two enzymes are involved, adenylosuccinate synthetase, which catalyses the conjunction of inosinate and aspartate to succinoadenylate (II, R_2 = H, R_1 = $-NH \cdot CH(CO_2H) \cdot CH_2 \cdot CO_2H$) and adenylosuccinate lyase, which removes succinate to leave an amino group at C-6 of the purine ring. Thioinosinate binds to both enzymes with a dissociation constant close to the reaction K_m of the natural substrate.

The first enzyme on the pathway to purine biosynthesis, glutamine-5-phosphoribosyl-1-pyrophosphate amidotransferase is inhibited by several nucleotide products of the pathway, such as inosinate; thioinosinate is also a potent inhibitor of this enzyme and it is suggested that in fact this block is the most critical for cell growth.[14] Thioinosinate appears to be more selective for cancerous than for normal cells, possibly because of the lower levels of xanthine oxidase in the former. The enzyme normally converts inosinate into excretable uric acid, but it can also detoxify thioinosinate by conversion to thiouric acid. Thus in diseased cells containing reduced xanthine oxidase activities, the thioinosinate levels will accordingly be raised.

Several of the enzymes of nucleic acid biosynthesis also require metal ions, coenzymes or ATP and thus interruption of the supply of these cofactors would constitute additional potential targets. One coenzyme necessary for the functioning of these pathways is tetrahydrofolate (III). In mammals a deficiency of this coenzyme (derived from the vitamin folic acid) results in a failure to grow, clinical anaemia and at the molecular level, impaired purine and pyrimidine biosynthesis. Thus the biosynthesis or utilisation of tetrahydrofolate would also be a potential target.

The discovery of Prontosil in 1935 marked the beginning of bacterial chemotherapy and a rational approach to the design of enzyme inhibitor drugs. The molecular similarity between sulphonamides such as Prontosil (IV) and *p*-aminobenzoic acid (V), provides a basis for rationalising the competitive inhibition of folate biosynthesis observed *in vitro*. For a host cell, folic acid is provided in the diet, whereas bacterial and many other parasitic organisms synthesise the coenzyme *de novo*. Thus the probable target in the sulphonamide treatment of bacterial infections is the 'condensing enzyme', dihydropteroate synthetase, which directs the formation of pteroate from *p*-aminobenzoic acid and pteridine.

Folic acid itself inhibits dihydrofolate dehydrogenase but replacement of its C-4 hydroxyl with an amino group decreases its enzyme dissociation constant by about 10^3 fold; the numerical value of the constant, 5×10^{-10}M, can be compared to a Michaelis constant of 10^{-6}M. The compound, aminopterin, and its *N*-10-methyl derivative amethopterin (methopterin) which binds just as strongly to the enzyme are thus powerful anti-malignancy drugs. Being less specific they are of limited use as antimalarial or antiparasitic agents. Malaria has been treated by the potent antagonist pyrimethamine (VI) which competes with the substrate for dihydrofolate dehydrogenase, binding approximately ten fold better.

IV   V   VI

3.3.2.2 *Naturally Occurring Protein Protease Inhibitors*

Many mammalian and plant tissues elaborate proteins which specifically inhibit proteolytic enzymes. These naturally occurring inhibitors, whose normal physiological function is probably protection, may be divided into three categories:

(a) Extracellular pancreatic antiproteases.
(b) Soybean anti-trypsin, also designated Kunitz inhibitor in honour of one of the earlier workers.
(c) Serum α-1-antitrypsin. Found in blood plasma, the action of this antiprotease is not restricted to trypsin but elastase, collagenase, plasmin, thrombin, fungal and leukocyte proteases are also inhibited.

Protein protease inhibitors bind stoichiometrically to their cognate enzymes and competitively inhibit their catalytic functions; they are therefore substrate analogues. All have molecular weights in the range $1-100 \times 10^3$ daltons, and are highly stable to high temperature and proteolytic attack. Internally they appear to be very rigid. Their enzyme association is maximal around physiological pH and the association constants, of the order 10^{10}M, are at least two orders of magnitude greater than those of low molecular weight amide substrates, which reflects the high affinity between the two. The second order rate constants for association are 10^5-10^6 mol^{-1} s^{-1}.

Association was originally thought to be essentially that of the non-covalent classical competitive type but with the observation of a two stage process on mixing trypsin and soybean inhibitor, the mechanism was judged to be more complex. To rationalise the experimental observations[15] a reversible enzyme catalysed reaction has been proposed in which after initial complex formation the enzyme cleaves a susceptible peptide bond in the inhibitor to form an acyl-enzyme complex. Similar complexes have been observed in trypsin catalysed hydrolyses of small substrates, but with the antiproteolytic proteins it was further proposed that the released α-amino group remains held in close juxtaposition to the acyl derivative. In such a position it would facilitate transpeptidation and hence reversibility.

Selective chemical modification and radioactive labelling of amino acid side chains before and after dissociation of the enzyme-inhibitor complexes have enabled two types of anti-protease to be distinguished, those such as the pancreatic inhibitors which possess a lysyl residue in the binding site and the serum and soybean antitrypsins with arginine at this site. A schematised picture of the reaction between trypsin and soybean inhibitor is shown below.

$$\mathrm{Arg}_{64}.\mathrm{CO}.\mathrm{NH}.\mathrm{Ile}_{65} \xrightarrow{\text{Trypsin}} \mathrm{Arg}_{64}.\overset{\text{O-Trypsin}}{\mathrm{CO}} \quad \mathrm{H_2N}.\mathrm{Ile}_{65} \qquad (3.76)$$

3.3.2.3   Transition State Analogues[16]

The inhibitors described in the previous sections have, in the main, structures similar to those of the corresponding substrates. Thus their interactions with the enzymes will have been much the same as those in the Michaelis complex. However, products can only be formed by the system first passing through a transition-state configuration, that species on the reaction pathway possessing maximum energy. Based on Pauling's[16] proposition that an enzyme possesses a configuration complementary to the activated complex, a different approach to the design of enzyme inhibitors has been developed.

Pauling's hypothesis states that the active site shows greater complementarity to the transition state than to the ground state of the substrate. Accordingly an extra degree of specificity, that for the reaction mechanism, is envisaged for the enzyme, and thus a stable molecule which resembles the transition state should be more tightly bound than a substrate analogue, possibly by several orders of magnitude.

The enzymic reaction including transition state complexes may be represented by the complexus,

$$
\begin{array}{ccc}
\text{ES} & \rightleftharpoons & \overline{\text{ES}}{}^* \longrightarrow \text{E} + \text{P} \\[2pt]
+\text{E} \updownarrow & & +\text{E} \updownarrow \\[2pt]
\text{S} & \rightleftharpoons & \text{S}^* \longrightarrow \text{P}
\end{array}
\tag{3.77}
$$

where S^* is the transition state concentration of the uncatalysed reaction, $\overline{ES}^*$ represents the transition state of the enzyme catalysed reaction and S is the ground state concentration. If k_0 is the first order rate constant of the uncatalysed reaction, k_e the first order rate constant for conversion of ES to EP, and if $K^* = S^*/S$, $K_s = ES/E \cdot S$, $K_T = \overline{ES}^*/E \cdot S^*$ and $K_e^* = \overline{ES}^*/ES$ then

$$
K_T/K_s = K_e^*/K^* = k_e/k_0
\tag{3.78}
$$

Equation 3.78 predicts that the ratio of association constants for transition state-enzyme binding to substrate-enzyme binding equals the ratio of the enzymic to non-enzymic rates of catalysis. Enzymic rate enhancements of the order $10^6–10^{12}$ have been variously calculated and thus from the equation these are the magnitudes by which the enzyme could be expected to bind the transition state over and above the substrate. The tighter the association (and hence the more specific) the greater the reactivity.

The transition state analogue is a molecule with a structure similar to the activated state of the substrate. Therefore before a potential enzyme inhibitor can be synthesised, a reasonable mechanism for its catalytic action must first have been advanced. In many cases this is approached by comparing the enzymic reaction with its non-enzymic counterpart, and is often aided by the similarities in mechanisms exhibited by enzymes catalysing analogous reaction types. Thereafter the main criterion by which a compound is judged a transition state analogue is the magnitude of its enzyme association constant, which must be several orders of magnitude greater than that of either the substrate or substrate analogue.

Under certain conditions, non-enzymic ester and amide alcoholyses proceed via intermediate tetrahedral addition complexes;

$$R_1 \cdot O^- + R_2 \cdot CO \cdot X \rightleftharpoons R_1 -O-\overset{\overset{\displaystyle R_2}{|}}{\underset{\underset{\displaystyle X}{|}}{C}}-O^- \rightleftharpoons R_1 \cdot O \cdot CO \cdot R_2 + X^-$$

$$(3.79)$$

This mechanism is also most probably adopted by several proteolytic enzymes. The 'serine' proteases, e.g. trypsin, chymotrypsin, elastase and subtilisin, employ a serine side chain as attacking alcoholate ion $(R_1 \cdot O^-)$ whose nucleophilicity is increased by a charge relay system. These enzymes are potently inhibited by two chemically different types of molecule, boronic acids and aldehydes, both of which form relatively stable tetrahedral compounds. Phenylethane boronic acid, $C_6H_5(CH_2)_2 \cdot B(OH)_2$, ionises in alkaline solution (where the enzymes are active) to give a tetrahedral anionic structure (VII), isosteric and isoelectronic with the postulated transition state of the proteases. Chymotrypsin, which catalyses the hydrolysis of aromatic amides and peptides, binds phenylethane boronic acid approximately 150 fold more strongly than hydrocinnamic acid chloride, $C_6H_5 \cdot (CH_2)_2 \cdot CO \cdot NH_2$.

$$C_6H_5 \cdot (CH_2)_2 -\overset{\overset{\displaystyle OH}{|}}{\underset{\underset{\displaystyle OH}{|}}{B}}-O-Enz$$

(VII)

Intermediates with the tetrahedral structure known to form between aldehydes and alcohols, $R_1 \cdot O \cdot CH(OH)R_2$, have also been proposed as the transition state analogue inhibitors formed when peptides such as $CH_3 \cdot CO \cdot$pro-ala-pro-ala-CHO bind to elastase.[16] This tetrapeptide $(K_I = 8 \times 10^{-7})$ binds approximately 5,000 fold more tightly than $CH_3 \cdot CO \cdot$pro-ala-pro-ala-NH_2 $(K_s = 4 \times 10^{-3}M)$ and 1,000 fold better than $CH_3 \cdot CO \cdot$pro-ala-pro-alaninol $(K_I = 6 \times 10^{-4}M)$. The oligopeptide aldehyde was chemically synthesised, but potent inhibitors of this type do in fact exist in nature. Streptomyces species produce leupeptins of general structure acetyl-x-x-arginine-CHO where x is L-leucine (mainly), iso-leucine or valine. These inhibit the proteolytic enzymes plasmin, papain, thrombokinase and trypsin with dissociation constants close to those of the elastase inhibitor above. They are substrate competitive and probably inhibit in a similar manner.

Considerable attention has been directed to the inhibition of nucleic acid biosynthesis as a means of controlling parasitic infections and the rapid division of cells. It is in this area that the specific application of transition state analogs has potential value. One step necessary for nucleic acid production is the interconversion of cytidine and uridine, catalysed by cytidine deaminase. This enzyme is rapidly and competitively inhibited 10^4 fold more effectively by tetrahydrouridine (VIII) $(K_I = 2.4 \times 10^{-7}M)$ than by either dihydrouridine $(K_I = 3.4 \text{ mM})$ or uridine $(K_I = 2.4 \text{ mM})$. Tetrahydrouridine also binds about 10^3 fold better than the substrate $(K_m = 0.2 \text{ mM})$ a result attributed to the similarity between its structure and the probable tetrahedral intermediate in the cytidine deaminase reaction (IX).

VIII   IX

Specific acid catalysed hydrolyses of glycosides are usually discussed in terms of the cyclic mechanism in which initial protonation of the glycoside to the conjugate acid is followed by fission of the glycoside-oxygen bond to yield a carbonium ion (X).

During the formation of this reactive intermediate the monosaccharide ring experiences a change in conformation from a chair to a
half-chair in which the ring oxygen and carbon atoms C1, C2 and
C5 occupy the same plane (X). From crystallographic and kinetic
investigations there is evidence that several glycosidases, notably
lysozyme, glycogen phosphorylase, β-N-acetylglucosaminidase, α-
and β-glucosidase and β-galactosidase also catalyse glycoside
hydrolyses and glycosyl transfers through the intermediacy of such
glycosyl carbonium ions. It might be expected therefore that more
stable molecules with similar planar conformations, such as glycon-
olactones (XI) would also bind more tightly than the substrate
chairs.

The 1-5 gluconolactone derived from the reducing monosaccharide of an N-acetylglucosamine tetramer (XI, R = $-$NH$\cdot$CO$\cdot$CH$_3$)
is calculated to bind 6×10^3 fold more strongly to lysozyme than
the chair form. This, it has been suggested, is the extent to which
structural distortions in the transition state contribute to the enhancement of catalytic rate.[17]

A group of naturally occurring antibiotics are the penicillins
(XII) which block the final step in bacterial cell wall biosynthesis,
the transpeptidase catalysed crosslinking of the peptidoglycan
chains. The enzyme removes a D-alanine residue from the end of
one peptidoglycan strand and replaces it with the N-terminus of a
second. Because of the similarity between the β-lactam function in
penicillin and the C-terminus of the glycan strand (XIII) and argu-
ing that the angle of the β-lactam dihedral (135°) is also subtended
by the planar peptide bond as it approaches the tetrahedral state

(109°), Lee[18] has suggested that penicillin first binds strongly to the transpeptidase as a transition state analogue and then its lactam ring is ruptured, simultaneously acylating the active site sulphydryl group. An alternative explanation for the potent action of penicillin is that the enzyme binds the inhibitor in a mode which strains the β-lactam ring into a planar conformation so increasing its acylating reactivity towards the active site residue.[19]

3.3.3 Irreversible Enzyme Inhibition

Chemical modification is particularly valuable for probing the physico-chemical character of an enzyme and for determining the nature and reactivity of its constituent amino acids. It is therefore an important part of the investigation of proteins that have yet to be crystallised and for membrane bound proteins. Even for those that have been subjected to X-ray analysis, data on their structures and behaviours in solution are needed to complement those obtained in the crystal. Modification of the enzyme can indicate the position of the active site and which of several possible amino acids are essential to its function. In addition the preparation of derivatives for peptide sequencing, the production of isomorphous heavy atom derivatives for X-ray analysis and crosslinking stabilisation of the enzyme all depend on the judicious application of modifying reagents. These reagents can be divided into group and site selective types.

3.3.3.1 Group Specific Reagents

A plethora of reagents designed to specifically modify most of the reactive groups in proteins has emerged over many years, and their chemical reactivities and applications are frequently, thoroughly reviewed.[20] Table 3.6 gives a very short, representative list of the more selective reagents.

Experimental determination of the groups essential to enzymic activity may be made in several ways. One widely used method is differential labelling. Group selective modification of the enzyme is first performed in the presence of a substrate analog, to protect the active site. The protector is then removed and the enzyme again exposed to the modifier. If activity is retained by the substrate protected enzyme but lost when deprotected, then the group tested for essentiality is presumed present at or near the active site (Figure 3.13). If prior to the second treatment, both inhibitor and excess reagent are removed and the latter replaced by radioactively

Table 3.6: Chemical Modification of Amino Acid Side Chains

Amino acid	Reagent
Arginine	Nitromalondialdehyde, phenylglyoxal
Aspartic acid/ glutamic acid	Triethyloxonium fluoroborate, water soluble carbodiimide plus glycine methyl ester
Cysteine/ cystine	Phosphorothioate, performic acid, 5,5'-dithio*bis*(2-nitrobenzoic acid), *p*-chloromercurybenzoate
Histidine	Iodoacetamide, diazonium-1·H-tetrazole
Lysine	Methyl acetimidate, maleic anhydride, 1-fluoro-2,4-dinitrobenzene
Methionine	Hydrogen peroxide, β-propiolactone, α-haloketones
Tryptophan	Iodine, *N*-bromosuccinimide, sulphenyl halides
Tyrosine	Tetranitromethane, *N*-acetylimidazole, *p*-diazoarsanilic acid

labelled modificant, the active site will be specifically tagged. Radio-actively labelling the group facilitates the isolation of active site fragments and their sequencing by the usual protein degradative procedures. Table 3.7 shows the amino acid sequences around the active site seryl group in several esterases. Each was radioactively tagged by a specific organohalophosphate.

One drawback of these methods is the lack of absolute specificity shown by a modifier for a given functional group. Alkyl halides for example, although preferentially reactive with lysine and α-amino groups will also modify cysteine and threonine. This multiplicity of

Table 3.7: Active Site Fragments of Several Esterases

Chymotrypsin	-ser-cys-met-gly-asp-Ser*-gly-gly-pro-leu-
Trypsin	-ser-cys-gly-gly-gly-asp-Ser*-gly-pro-val-cys-
Cholinesterase	-phe-gly-glu-Ser*-ala-gly-
Alkaline phosphatase	-pro-asp-tyr-val-thr-asp-Ser*-ala-ala-ser-ala-

* Group modified.

Figure 3.13: Modification of Calf-intestinal Alkaline Phosphatase with Phenylglyoxal (Specific for the guanidine side chain of arginine) in the Presence (×) *and Absence* (O) *of the Competitive Inhibitor Potassium Phosphate.*

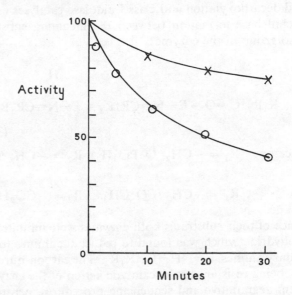

reaction leads to considerable 'noise', complicating the isolation procedures and the interpretation of labelling patterns. A second source of ambiguity is that a lack of reactivity cannot be taken as evidence for the absence of a functional group. Neighbouring residues may either sterically hinder reagent approach to a 'buried' group or restrict formation of the transition state for reaction even though the residue may have been initially accessible.

To avoid these difficulties and to aid interpretation of the acquired experimental data, modification techniques have been designed to attack the active site directly.

3.3.3.2 Site Specific Modification

Four main types of compound have been found which specifically modify active sites: (a) substrates, (b) pseudosubstrates, (c) affinity labels, and (d) 'suicide' or 'k_{cat}' inhibitors.

Substrates. The substrate is the almost perfect site-specific reagent; unfortunately the bonds formed in the Michaelis complex are often labile and transitory making isolation of an enzyme bound species almost impossible. In certain circumstances however it has proven possible to trap an intermediate in the enzyme reaction.

Reasonable mechanisms proposed for acetoacetate decarboxylase catalysed decarboxylation and class I aldolase catalyses envisage initial schiff base formation between the incoming substrate and an amino group in the enzyme:

$$\text{E—NH}_2 + \text{R}_1\text{R}_2\cdot\text{C}{=}\text{O} \rightarrow \text{E—N}{=}\text{CR}_1\text{R}_2 \rightarrow \text{E—}\overset{\text{H}}{\underset{+}{\text{N}}}{=}\text{CR}_1\text{R}_2$$

(3.80)

Dihydroxyacetone phosphate　　$R_1 = -CH_2 \cdot O \cdot PO_3H_2$　　$R_2 = -CH_2 \cdot OH$

Acetoacetate　　$R_1 = -CH_2 \cdot CO \cdot CH_3$　　$R_2 = -CO_2H$

In the presence of their substrates both enzymes were inhibited by sodium borohydride which was found to reduce the imines to the stable secondary amines, $\text{E—NH}{-}\text{CHR}_1\text{R}_2$. The carbon-nitrogen single bonds, being resistant to the catalytic action of the enzymes and to protein degradative and sequencing procedures, permitted the isolation and identification of active site sequences[23] (Table 3.8). By an analogous procedure lysine groups have been identified as the sites of attachment of the prosthetic pyridoxal-5-phosphate groups to glycogen phosphorylase and to the lysine and arginine decarboxylases of *E. coli*. The catalytic consequences of this prior imine formation are discussed in Chapter 4.

Pseudosubstrates. The distinction between pseudosubstrates, which possess certain characteristics in common with the actual

Table 3.8: **Active Site Fragments from Acetoacetate Decarboxylase and Aldolase Labelled by Substrate plus Sodium Borohydride**

Acetoacetate decarboxylase	-glu-leu-ser-ala-tyr-pro-Lys*-lys-leu-
Aldolase	-glu-gly-thr-leu-leu-Lys*-pro-asn-met-val-

* Group modified.

enzyme substrates, and affinity labels is partly semantic in origin, both can be included under Baker's[13] classification of 'active-site directed irreversible inhibitors'. However the designation pseudo-substrates has been applied particularly to diisopropylfluorophosphate (DFP, XIV) and its analogues. These organohalophosphates react rapidly and irreversibly with the enzymes trypsin, chymotrypsin, thrombin and acetylcholinesterase. In each case the reaction is stoichiometric, resulting in the loss of a single active site seryl residue.

$$((CH_3)_2CHO)_2-P(=O)F$$

XIV

DFP has been the progenitor of many similar organophosphate inhibitors. Many different aryl, alkyl, phosphoryl, sulphonyl and carbonyl halides have been synthesised, most of which inhibit the 'serine' enzymes. Not unexpectedly the greater the apparent resemblance to the substrate the greater is the effectiveness of the inhibitor. For example phenylmethanesulphonyl fluoride reacts 10^4 fold faster than methanesulphonyl fluoride with chymotrypsin, presumably due to the additional phenyl ring fitting into the hydrophobic cavity of the enzyme active site.

The mechanism of inactivation by these organohalophosphates involves initial reversible association with the enzyme, followed by rapid covalent bond formation, eq. 3.81.

$$E \cdot CH_2OH + (RO)_2P(O)F \rightleftharpoons \underset{(RO)_2P(O)F}{E \cdot CH_2OH} \longrightarrow E \cdot CH_2OP(O)(OR)_2 \quad (3.81)$$

That reaction takes place at the active site is shown by several pieces of evidence, including loss of activity and the protection afforded by competitive inhibitors. The covalent bond formed between the phosphate and serine is stable to mild protein degradation and sequencing methods, enabling precise location of the activated residue in the primary sequence (Table 3.7). Such information is valuable in assigning possible roles for particular residues in the reaction mechanism and in aiding the interpretation of other physico-chemical data, such as locating the active site in models constructed from X-ray crystallographic data.

Studies on the mechanism of inhibitor action at the molecular level have proved of enormous value to the design of tools not only for elucidating enzyme function but also to understanding the control of physiological processes. This latter aspect is graphically illustrated by the work on acetylcholinesterase (AChE).

Acetylcholinesterase inactivates the neurotransmitter acetylcholine by catalysing its hydrolysis;

$$(CH_3)_3 \overset{+}{N} \cdot CH_2 \cdot CH_2 \cdot O \cdot CO \cdot CH_3 \xrightarrow{H_2O} CH_3 \cdot CO_2^-$$

$$+ (CH_3)_3 \cdot \overset{+}{N} \cdot CH_2 \cdot CH_2 \cdot OH \quad (3.82)$$

Thus the enzyme plays a key role in the control of nerve excitability at post-synaptic sites. It is also important as a site of action of therapeutically useful drugs, and toxic agents.

The Wilson model[24] of the enzyme active site envisages two subsites, a fairly unspecific 'anionic' site to which the trimethylammonium cation binds through a combination of the coulombic attraction of the protonated nitrogen and hydrophobic interactions of the methyl groups, and at 5 Å distance an 'esteratic' site containing an organohalophosphate modifiable serine. The nucleophilicity of this 'essential' serine is probably augmented through electronic interaction with the imidazole ring of a neighbouring histidyl residue.

Directed modification of the esteratic site serine by phosphate esters and carbamates is the basis of many applied aspects of AChE inhibition. Some of the more important inhibitors are listed in Table 3.9. Sarin is an extremely potent nerve gas and parathion is at present a commonly used and effective insecticide. The latter accounts for the majority of accidental poisonings of agriculture workers. Both sarin and parathion react with the AChE serine group by a similar mechanism to eq. 3.81. Under normal conditions this reaction is virtually irreversible, however based on his active site model, Wilson[24] synthesised the heterocyclic derivatives pyridine-2'-aldoxime methiodide (PAM) and trimethylene-*bis*-4-PAM (TMB-4). Both of these have proved successful in reversing parathion inactivation of AChE, probably via the mechanism ttfrfffrffffffffff

Table 3.9: Acetylcholinesterase Inhibitors

Malaoxon	$(CH_3O)_2\text{-}P(=O)S\text{-}CH(CO_2C_2H_5)\cdot CH_2\text{-}CO_2C_2H_5$
Parathion	$(C_2H_5O)_2\text{-}P(=S)OC_6H_4NO_2$
Sarin	$(CH_3)_2CH\cdot O\cdot P(=O)(CH_3)F$
Prostigmine	$(CH_3)_2\cdot N\cdot CO\cdot OC_6H_4\overset{+}{N}(CH_3)_3$
Physostigmine (Eserine)	

shown in eq. 3.83.

$$(3.83)$$

Stigmines, the carbamate drugs, are possibly the most useful anti-cholinesterase compounds because of their behaviour as poor substrates. The intermediates, $E\cdot CH_2\cdot O\cdot CO\cdot N(CH_3)_2$, are less stable than the corresponding phosphates and are readily hydrolysed by water or oximes to return enzyme activity.

Affinity Labels. Reversible binding is very limited in the information it can provide concerning the active site and a more fruitful approach has been devised by Schoellman and Shaw.[25] They suggested the synthesis of substrate analogues possessing the molecular requisites complementary to the active site but which in addition have incorporated chemically reactive groupings. By mimicking the substrate, they argued, such molecules would be held at high concentrations in the sites and once in position the reactive group(s) would then form irreversible covalent attachments to amino acid side chains in their vicinities. Hence such compounds are designated *affinity labels*. There are several criteria that the compound should also possess. For greater discrimination the chemical groupings should be relatively inert outside the site and the affinity reagent should be designed with regard to the mechanism of binding and catalysis.

As an example the chloromethyl ketone of tosylphenylalanine (TPCK, XV) was synthesised as a possible affinity reagent for chymotrypsin.

$$CH_3 \cdot C_6H_4 \cdot SO_2 \cdot NH \cdot CH \cdot CO \cdot CH_2Cl$$
$$|$$
$$CH_2 \cdot C_6H_5$$

XV

The specificity requirement was fulfilled by incorporation of the large aromatic benzyl group (chymotrypsin preferentially cleaves peptide bonds adjacent to phenylalanine, tryptophan and tyrosine), and the enzyme-inhibitor complex was most probably also stabilised by hydrogen bond formation between the ketone and the active site serine. The protease was irreversibly and rapidly inactivated by TPCK, and amino acid analysis indicated that one histidine had been modified, subsequently shown to be at position 57. Modification of the active site was verified as follows. Inactivation was reduced by the additional incorporation of a reversible competitive inhibitor, a 1 : 1 stoichiometric amount of labelled TPCK was incorporated by the protein and no label was incorporated by urea denatured or DFP inactivated enzyme. TPCK also had no effect on trypsin.

Trypsin is however inactivated by TLCK, in which the benzyl group of TPCK is replaced by the tetramethylene ammonium ion, $-(CH_2)_4 \cdot \overset{+}{N}H_3$, but conversely not by TPCK. TLCK also causes irreversible modification of other 'trypsin-like' enzymes, notably thrombin, plasmin, kallikrein, papain, ficin and clostripapain.

Compounds of similar molecular structures which preserve their affinities for the enzyme but which differ in their chemical reactivities can be used to explore the topology and side chain reactivities of the active site. For example, the three different affinity labels for chymotrypsin listed in Table 3.10 all employ a large phenyl ring as the specificity determinant but their different structures and electro-philicities have identified dissimilar groups at the same site.

Chymotrypsin catalyses the hydrolysis of esters and amides via a pathway involving the intermediate formation of a covalent acylser-ine 195 complex. Deacylation of this active site residue is slow under certain conditions. Thus if a second reactive group is built into the substrate, advantage of the first enzyme directed covalent attach-ment may be taken. Chymotrypsin releases p-nitrophenol from N-bromoacetyl-α-amino-isobutyrate, p-nitrophenyl ester (XVI, R = p-nitrophenyl) to form the acyl-enzyme (XVI, R = $-O-CH_2$-enzyme) in which the activated substrate analogue is fixed into the active site. Consequently the attached bromoacetamide group is in a favourable position to be attacked by any suitably orientated nucleophile, in this case the thioether side chain of methionine 192.

$$Br \cdot CH_2 \cdot CO \cdot NH \cdot C(CH_3)_2 \cdot CO \cdot R$$
$$XVI$$

Table 3.10 gives a short list of specific affinity labels. As this table indicates, acyl halides and α-haloketones have been by far the fav-ourite activating combinations because of their ready electrophilici-ties; however such chemically reactive species suffer from several drawbacks. Nucleophilic attack only is possible, thus reactions with other types of amino acid residues such as tryptophan, valine, leuc-ine, etc. are precluded. Nucleophiles are also present elsewhere in the protein molecule besides the active site, and reaction at these secondary sites will reduce the effective concentration of the affinity label and confuse the inhibition pattern. In addition the nucleo-philes in highest concentration are water molecules. Reagents which react with a wider variety of different hydrophobic and hydrophilic groups and peptide bonds, and which in addition become reactive only when suitably positioned, are therefore called for.

Two chemical species fulfilling the above requirements are car-benes and nitrenes. Carbenes are formed by thermolysis and pho-tolysis from diazoalkanes, diazirines or α-ketodiazo precursors and nitrenes in the same way from alkyl, acyl and aryl azides. As thermal generation is of limited practicality in biological systems, photogen-esis is generally preferred. The precursors themselves must of course be chemically unreactive and activatable at wavelengths remote from enzyme absorption.

Table 3.10: Affinity Labels

Enzyme	Reagent	Residue modified
Aspartate aminotransferase	β-Bromopyruvate β-Chloroalanine	Cystine Lysine
Carboxypeptidase B	α-N-Bromoacetyl-D-arginine Bromoacetyl-p-aminobenzyl-succinate	Glutamate Methionine
α-Chymotrypsin	Tosyl-L-phenylalanine chloromethyl ketone (TPCK) Glycidol phenyl ether Phenylmethanesulphonyl fluoride	Histidine 57 Methionine 192 Serine 195
Formylglycinamide-ribotide amidotransferase	Azaserine	Cysteine
Fumarase	Bromomesaconate	Methionine, histidine
β-Galactosidase	N-Bromoacetyl-β-D-galactosylamine	Methionine
20-β-Hydroxysteroid dehydrogenase	Cortisone 21-iodoacetate	Histidine
Lactate dehydrogenase	3-Bromoacetylpyridine	Cysteine, histidine
Lysozyme	2',3'-Epoxypropyl-β-D-(N-acetyl glucosamine)$_2$	Aspartate 52
Methionyl-tRNA synthetase	p-Nitrophenyl-carbamyl-methionyl-tRNA	Lysine
RNA polymerase	5-Formyl-uridine-5'-triphosphate	Lysine
Staphylococcal nuclease	3'(N-Bromoacetyl-p-amino phenylphosphonyl)deoxy-thymidine-5'-phosphate	Lysine 48 Lysine 49 Tyrosine 115
Triose phosphate isomerase	Glycidol phosphate	Glutamate

The potential utility of carbene photogenesis for enzyme labelling was confirmed by Westheimer *et al.*[26] by work on diazomalonyl ethyl ester modification of trypsin (eq. 3.84). After digestion of the inactivated enzyme, *S*-carboxymethyl cysteine was identified as a major product and this could only have arisen from carbene attack on a sulphide bond; in addition a small amount of ^{14}C-glutamic acid was obtained, indicating that carbene insertion into the methyl group of a neighbouring alanine had also taken place; eq. 3.84.

$$E \cdot CH_2 \cdot OH + C_2H_5 \cdot O \cdot CO \cdot C(N_2) \cdot CO \cdot O \cdot pNP$$

$$\longrightarrow E \cdot CH_2 \cdot O \cdot CO \cdot C(N_2) \cdot CO_2C_2H_5$$

$$\Big\downarrow h\nu$$

$$E \cdot CH_2 \cdot O \cdot CO \cdot \ddot{C} \cdot CO \cdot O \cdot C_2H_5$$

$$\overset{HR \cdot X}{\diagdown}$$

$$\underset{H}{\overset{RX}{\underset{|}{E \cdot CH_2 \cdot O \cdot CO \cdot C \cdot CO \cdot O \cdot C_2H_5}}} \qquad\qquad (3.84)$$

$$\Big\downarrow \overset{+}{H}$$

$$HO_2C \cdot CH(NH_2)CH_2 \cdot R \cdot CO_2H$$

$$pNP = p\text{-nitrophenyl-}$$
$$R = -CH_2- \quad \text{and} \quad -CH_2-S-$$
$$X = \overset{\displaystyle -CH}{\underset{\displaystyle \cdots NH \quad CO \cdots}{\diagup \ \diagdown}}$$

Nitrenes are generally more reactive than carbenes but the azide precursors, especially aryl azides are more stable. Aryl azides can be photolysed at longer wavelengths than acyl or alkyl azides, thus precluding enzyme inactivation and are less subject to rearrangement, both before and after activation. Like carbenes however, nitrenes are capable of insertion into C—H bonds. The first investigations of nitrenes as indiscriminate enzyme labelling agents were carried out by Foster.[27] ^{14}C-labelled p-azidocinnamic acid, p-nitrophenyl ester was used to acylate serine 195 of chymotrypsin.

Photolysis of the intermediate complex and digestion of the inactivated enzyme enabled location of the radioactivity in that part of the primary sequence forming the aromatic binding locus of the chymotrypsin active site.

'*Suicide*' or '*k-cat*' *Inhibitors.*   The first of these expressions was coined by Abeles and Maycock[28] and the second by Rando.[19] The two main characteristics of this type of irreversible inhibitor are (i) non-reactivity of the precursor before it interacts with the enzyme active site and (ii) activation and hence subsequent inhibition catalysed by the same active site residues as those responsible for substrate transformation. This second characteristic thus adds an extra dimension to the specificity of inhibition, in that the catalytic reactivity of the enzyme is employed to effect its modification. Such compounds can therefore be regarded as catalytic inhibitors as opposed to affinity labels or transition-state analogs whose selectivities reside in their binding powers. Suicide inhibitors can then provide additional evidence on the mechanism of an enzymic reaction.

β-Hydroxydecanoyl thioester dehydrase is the prime example of an enzyme so inhibited. Required for fatty acid biosynthesis it catalyses the interconversions

$$R \cdot CH_2 \cdot CH(OH) \cdot CH_2 \cdot CO \cdot NAC \rightleftharpoons R \cdot CH_2 \cdot CH \overset{trans}{=\!=\!=\!=}$$

$$CH \cdot CO \cdot NAC \rightleftharpoons R \cdot CH \overset{cis}{=\!=\!=\!=} CH \cdot CH_2 \cdot CO \cdot NAC \quad (3.85)$$

where $R = -(CH_2)_5 \cdot CH_3$;

$NAC = -S \cdot CH_2 \cdot CH_2 \cdot NH \cdot CO \cdot CH_3$ (*N*-acetyl-cysteamine)

$\Delta^{3,4}$ decynoyl-NAC, which contains an acetylenic grouping instead of the *cis*-double bond of the substrate has been described as an irreversible, 'k_{cat}' inhibitor of this enzyme, as its reaction was found to be stoichiometric and complete within one turnover. The dehydrogenase was also selectively inhibited *in vivo*. The most likely mechanism for its inhibitory action is enzymic conversion to the highly reactive conjugated allene (XVII), eq. 3.85, followed by irreversible covalent attachment to a histidine in the active site.

$$R \cdot C \equiv C \cdot CH_2 \cdot CO \cdot NAC \rightleftharpoons$$

$$R \cdot CH = C = CH \cdot CO \cdot NAC \xrightarrow{\text{his}} R \cdot CH = \overset{\displaystyle \text{his}}{\underset{\displaystyle |}{C}} \cdot CH_2 \cdot CO \cdot NAC$$

$$\text{XVII} \hspace{5cm} (3.86)$$

The postulated intermediary catalytic formation of activated double bonds causing the observed inhibitions is a common feature of suicide inhibitors, and the selectivity this implies opens up a fascinating approach to the design of pharmacologically active agents. For example, pargyline,

$$C_6H_5 \cdot CH_2 \cdot N(CH_3) \cdot CH_2 \cdot C \vdots CH$$

currently employed clinically in antihypertension therapy inhibits monoamine oxidase stoichiometrically, probably via enzymic formation of a relatively stable flavin-allene adduct (XVIII). Once formed the electrophilic allene is in a favourable position to attack any adjacent nucleophile and inactivate the enzyme, and the natural antibiotic, rhizobitoxine

$$NH_2 \cdot CH(CH_2OH) \cdot CH_2 \cdot O \cdot CH \vdots CH \cdot CH(NH_2)CO_2H$$

which resembles the substrate cystathionine

$$NH_2 \cdot CH(CO_2H) \cdot CH_2 \cdot S \cdot CH_2 \cdot CH(NH_2)CO_2H$$

of bacterial β-cystathionase, irreversibly prevents its cleavage into homocysteine and pyruvate probably also by acting as a suicide inhibitor.

XVIII

3.4   REGULATION OF ENZYME ACTIVITY

Living cells and organisms require strict organisation and co-ordination of their function in order to balance energy supply and demand. The prime agents in energy metabolism are of course the enzymes and thus a multitude of controlling devices have evolved through which their activities may be regulated in response to changing local conditions. Varying concentrations of substrates, cofactors, activators, inhibitors, inducers, repressors, and degradative systems, intramolecular conformational changes, and enzyme compartmentation and multiplicities all serve to modulate the activity of an enzyme. And gradually we are gaining an understanding of the contributions each of these various factors makes to cellular homeostasis. Although the details are far from complete a simplified scheme such as that in Figure 3.14 may be visualised, in which an

Figure 3.14: Regulation of Enzyme Activity

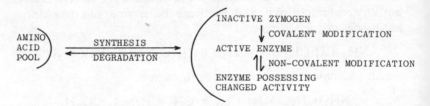

enzyme is subject to control at three separate and distinct levels: synthesis *de novo* from its constituent amino acids, regulation of its post-synthetic activity and control of its degradation, inactivation or catabolism back to its components.

3.4.1   Control of Enzyme Concentration

Enzyme activity is a function of the concentration of its intact molecules and is thus a resultant of a balance between its synthetic and degradative pathways, themselves subject to different and specific regulatory influences.

In catabolic pathways two general types of enzyme may be distinguished, 'constitutive' enzymes which are always present in fairly constant concentrations and 'inducible' enzymes whose DNA operons are usually repressed but which in the presence of activators are de-repressed and transcribed. Numerous examples of induction and repression exist. In certain prokaryotic cases, more than one enzyme in a pathway may be simultaneously induced by the same substance. Thus lactose in the medium will induce in *E. coli* the synthesis of β-galactosidase, galactose permease and galactoside transacetylase. If an alternative energy source, such as glucose, is made available, these enzymes are repressed.

In lower organisms investigations on enzyme induction and repression were stimulated by the Jacob and Monod hypothesis concerning the LAC operon, the DNA stretch on the *E. coli* gene that codes for the three enzymes induced by lactose. Figure 3.15 schematically represents the LAC operon for the three enzymes and its

Figure 3.15: The LAC Operon

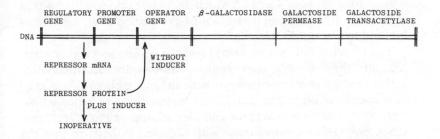

three associated genes, the regulatory gene, the promotor gene which contains an RNA polymerase recognition site, and the operator gene.

An RNA polymerase molecule is envisaged to interact with the regulator gene and transcribe a repressor mRNA molecule, which then translates a repressor protein. In the absence of lactose, or inducer, this protein binds to the operator gene so preventing transcription of the structural genes. However the inducer is proposed to be capable of binding and causing a conformational change in the protein and so to prevent its association with the operator site. Transcription of the three structural genes by the RNA polymerases will now be unhindered. Constitutive organisms possess regulator genes unable to synthesise the repressor protein.

As would be expected, in multicellular organisms more complex and elaborate communication and regulation systems have evolved; the stimuli for enzyme-mediated physiological effects are effected by endocrine hormonal messages to the target cells. A considerable amount of information has accumulated from both steroid and thyroid hormone studies to support Karlson's[29] proposal that these hormones act by regulating gene activity.* In addition they appear to exhibit many similarities in their patterns of interaction at the molecular level although their physiological effects are often quite different.

Each endocrine hormone is secreted into the circulating blood stream from which, in a free form, it passes through the plasma membrane and into the cytoplasm of the target cell. Once in the cytosol, it complexes with a specific hormone 'receptor'. The complex is then envisaged to undergo a conformational transition to a form suitable for breaching the nuclear membrane. Specific target cell receptors for oestradiol, progesterone, aldosterone and testosterone have been isolated and purified. All are protein complexes sedimenting between 4 and 8 S which dissociate into smaller units at high ionic strength. Importantly all have been shown capable of binding to nuclear DNA.

The direct result of introduction of the hormone-receptor complex into the nucleus is transcription, RNA synthesis. Three interactive mechanisms are consistent with the available evidence as to how transcription is initiated. Direct interaction and hence activation by the hormone-receptor complex of one or more precursor RNA polymerases, interference with protein synthesis so that the concentration of RNA polymerase molecules is increased, or derepression of the appropriate DNA operon by removal of a repressor substance—in essence, similar to the LAC operon model. Some evidence points to the last alternative, at least for oestradiol stimulated RNA synthesis where increases in the length of DNA chains have been observed. Some of the newly synthesised RNA molecules then move into the cytoplasm for translation into proteins. Among the enzymes induced by the glucocorticosteroid hormones are tyrosine aminotransferase, alanine aminotransferase, phosphoenolpyruvate carboxylase and tryptophan oxidase; glutamate DH, acid phosphatase, ribonuclease, β-glucuronidase, cytochrome oxidase and carbamoylphosphate synthetase have been observed induced by the thyroid hormones.

Although this intracellular hormonal regulation of enzymic activity is very efficient, it does not allow rapid adaptation to changing

* The polypeptide hormones, such as glucagon, stimulate adenylate cyclase activity in the cell membrane (Section 5.2.1).

metabolic conditions. There has also evolved provision for faster responses. These may be regarded as 'fine' controls affecting enzyme activity, superimposed on the former 'coarse' or 'stop-go' control of enzyme synthesis.

3.4.2 Modulation of Enzyme Activity through Physical Interactions

3.4.2.1 Negative Feedback Inhibition

One of the fundamental concepts of metabolic regulation—the regulation of the metabolite flux in a biochemical pathway through the control of the participating enzyme concentrations and activities—was discovered by Umbarger, Yates and Pardee (described in reference 30). Investigating the pathways of branched chain amino acid biosynthesis in *E. coli*, they found that the first directionally committed enzymes were feedback inhibited by the eventual products. The principle of this 'negative feedback' inhibition is illustrated in Figure 3.16. Feedback inhibition by a later

Figure 3.16: Negative Feedback Inhibition. (Capitals, except E, represent metabolites, E$_i$ represent the corresponding enzymes)

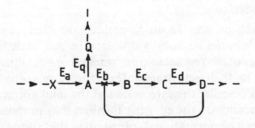

product on the first committed enzyme is tactically very effective since the flux of all the intermediates depends on the availability and rate of production of the first. If this is inhibited through a build-up in the end product concentration, no wasteful accumulation of intermediates can occur but as soon as the product is utilised its inhibitory influence is lessened, the first enzymic activity is de-inhibited and the pathway again becomes functional.

The same principle has been observed to operate in the control of branched and parallel pathways where simultaneous regulation is needed to prevent a product imbalance from occurring.[30] For

Figure 3.17: Sequential Feedback Inhibition of Aromatic Amino Acid
Biosynthesis

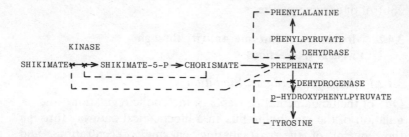

example in the formation of the aromatic amino acids phenylalanine and tyrosine from shikimate via prephenate in *Bacillus subtilis*,
sequential feedback inhibition is observed as illustrated in Figure
3.17. The end-product of each branch inhibits its branching enzyme,
thus phenylalanine inhibits prephenate dehydrogenase and tyrosine
inhibits prephenate dehydrase but an accumulation of the intermediate prephenate is avoided by its main stem feedback inhibition
of flow at shikimate kinase.

3.4.2.2   Allosterism

Monod, Changeux and Jacob[31a] pointed out that whereas many
traditional inhibitors resemble substrates, i.e. are isosteric, the feedback modifiers of the regulatable enzymes often bear little structural
resemblance to their substrates. They therefore coined the term
allosterism to describe the interaction of the inhibitor at a separate
'allosteric' locality on the enzyme. Further, they proposed that the
ligand induces a conformational transition in the 'allosteric' enzyme
which alters its catalytic behaviour. The result of this change is then
either a positive or negative modulation of its kinetic properties.
This implies that there must be an interaction between the catalytic
site and the regulatory site.

Support for the presence of at least two different sites on regulatory enzymes is plentiful. Many such enzymes are 'desensitised'
towards their allosteric effectors by heat, cold or by chemical
modification while sustaining no concomitant effect on their catalytic activities. Allosteric enzymes are also polymeric with each subunit possessing a defined geometry. Aspartate transcarbamylase, for
example, which converts aspartate into carbamylaspartate and is

Figure 3.18: Kinetic Behaviour of a Regulatable Enzyme

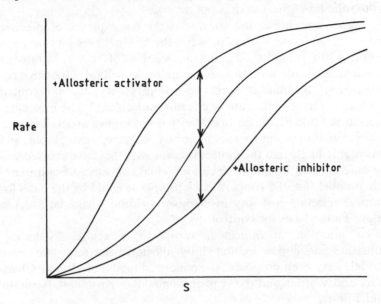

feedback inhibited by cytidine triphosphate (CTP), possesses two different types of subunit, binding CTP and aspartate respectively.

Kinetic studies of many regulatable enzymes have shown that they depart considerably from the Michaelis-Menten model. The variation of velocity with substrate concentration they frequently exhibit is sigmoidal (S-shaped),* Figure 3.18. The modifier (effector) does not alter the maximum rate of catalysis but at non-saturating substrate levels it exercises considerable control, increasing (activating) or decreasing (inhibiting) the activity of the enzyme (Figure 3.18). At constant effector concentration the enzymic activity is relatively insensitive to low substrate concentrations, but above a certain substrate level small increases produce marked rises in activity. In the mid-range allosteric modifiers produce significant effects, affording higher degrees of control than is possible on enzymes obeying Michaelis-Menten kinetics; in Figure 3.18 the arrows indicate the relative extents of rate modulation produced by allosteric activators and inhibitors on a regulatable enzyme. The large changes in rate produced by the allosteric modifiers effectively

* Sigmoidicity has tended to become synonymous with allosterism even though the latter was introduced as a structural concept.

'switch-on' and 'switch-off' enzymic activity* and so regulate the metabolic flow (flux) at that point.

This malleability of the enzyme is the consequence of effective 'co-operative' interactions between the binding sites and between subunits. In other words, the association of one substrate or modifier molecule with an enzyme molecule facilitates (positive co-operativity) or inhibits (negative co-operativity) the subsequent binding of further substrate or effector molecules.** The molecular mechanism underlying co-operativity is not known in detail for any enzyme as yet (except for the 'honorary' enzyme haemoglobin) and it is highly likely that the molecular processes that have evolved will be different and unique to each individual enzyme. As Stadtman[30] has pointed out, the cooperative effect *per se* could be the basis for natural selection and any mechanism yielding sigmoidal kinetics might be the basis for control.

To interpret in molecular terms the observed 'S-shaped' substrate-rate curves exhibited by allosteric enzymes, two main models have been proposed, the concerted model of Monod, Changeux and Wyman and the sequential model of Koshland, Nemethy and Filmer.

3.4.2.3   The Monod-Wyman-Changeux (MWC) Concerted Model[31b]

This simple, elegant model interprets the sigmoidal binding curve by postulating that (a) allosteric enzymes are oligomeric, composed of definite numbers of protomeric subunits, (b) each protomer exists in two conformational states in equilibrium, a relaxed (R) form with a higher affinity for substrate and a tense (T) form with a lower affinity which is the predominant form when unliganded, and (c) the conformational change from T to R occurs with conservation of symmetry, i.e. all subunits in each state have the same conformation and the same intrinsic ligand affinity.

According to this model therefore the binding of one substrate molecule will facilitate the association of further molecules as a result of binding site interactions in which conversion from one

---

* This can be interpreted as the enzyme having a high elasticity towards its modifier, see Section 5.2.7.

** Interactions induced by the binding of identical substrate molecules are termed 'homotropic' and the interactions between allosteric effectors and substrates are termed 'heterotropic'.

state to another is concerted with all subunits undergoing a simultaneous transition. For the simplest case of an allosteric dimer, if $\square$ represents the tense form and $\bigcirc$ the relaxed state of each subunit, substrate binding can be represented

$$\square\square \;\rightleftharpoons\; \bigcirc\bigcirc \;\overset{+S}{\rightleftharpoons}\; \bigcirc\text{\textcircled{s}} \;\overset{+S}{\rightleftharpoons}\; \text{\textcircled{s}}\text{\textcircled{s}} \qquad (3.87)$$

Substrate molecules are proposed to bind with higher affinity to the relaxed state than to the tense state and so formation of RS will cause the concentration of enzyme molecules in the free R-state to be decreased. But since this state is in equilibrium with the T-form some of this will be relaxed. If initially very little of the R-state is present, i.e. if the ratio T/R of the unliganded species is large, the relative amount of the T-state converted to the R-state on substrate binding will also be large but if the enzyme already exists predominantly in the relaxed state the relative amount of conformational change will be small. In other words the greater the numerical value of T/R, which is called the *allosteric constant*, L, the greater is the degree of cooperation between subunits; alternatively if L is small so that essentially all the enzyme is in the R-state, then there will be no effective co-operative subunit interaction and the substrate binding curve will be hyperbolic.

The model can also be extended to explain the control exercised by allosteric effectors. If the effector preferentially binds to the R-state and so increases the concentration of the more active form (and decreases L) it will activate, and by the argument above, decrease the substrate co-operativity; alternatively, if the effector increases the relative concentration of the T-state it will act as an inhibitor and increase the co-operative effect, eq. 3.88.

$$\square\square \;\overset{+I}{\rightleftharpoons}\; \square\boxed{I} \;\overset{+I}{\rightleftharpoons}\; \boxed{I}\boxed{I}$$

$$\big\updownarrow \qquad\qquad\qquad\qquad\qquad\qquad (3.88)$$

$$\bigcirc\bigcirc \;\overset{+A}{\rightleftharpoons}\; \bigcirc\text{\textcircled{A}} \;\overset{+A}{\rightleftharpoons}\; \text{\textcircled{A}}\text{\textcircled{A}}$$

The success of the MWC concerted model, besides providing a simple explanation of control, is its prediction of theoretical curves based on the parameters L, K_R the dissociation constant for substrate binding to the relaxed state and K_T the dissociation constant for substrate binding to the tense state, which fit the haemoglobin-oxygen binding data and the substrate association profiles of enzymes such as pyruvate kinase and isocitrate DH. However, only positive and not negative co-operativity is predicted by the model. It assumes also that the enzyme is always symmetrical. Therefore before the model can be applied to a particular enzyme this would have to be experimentally demonstrated. It also postulates an equilibrium between two forms of the enzyme which may be difficult to establish. For example, the T state of haemoglobin has been estimated to be many thousand fold in excess, so that verification of the R form would require experimental demonstration of less than 0.01 percent of the molecules in this conformational form.

3.4.2.4   The Koskland-Nemethy-Filmer (KNF) Sequential Model

An alternative proposal put forward by Koshland *et al.*[32] to account for the experimental observations on regulatory enzymes differs from the above by including the possibility of hybrid conformational states (such as RT in an allosteric dimer) and that different binding sites for a ligand in a given state may be dissimilar. They proposed that (a) only one unliganded conformation exists for an enzyme, (b) the association of a ligand molecule with one subunit induces a conformational change in that subunit which (c) is transmitted to adjacent unfilled subunits so altering their ligand association.

For an allosteric dimer, this 'induced fit' model (see Section 1.1.3) can be represented by,

$$\square\square \; \underset{\rightleftharpoons}{K_1} \; \square \text{\textcircled{S}} \; \underset{\rightleftharpoons}{K_2} \; \text{\textcircled{S}}\text{\textcircled{S}} \qquad (3.89)$$

where K_1 and K_2 are association constants.

Each subunit can exist in two conformational states, $\square$ and $\bigcirc$; binding of the substrate to a subunit changes its conformation, in this case from $\square$ to $\bigcirc$. The result of this conformational change is an alteration in the specificity and strength of the intersubunit contacts. They could facilitate subsequent substrate binding, i.e. elicit positive co-operativity ($K_2 > K_1$), hinder association of the second substrate molecule, negative co-operativity ($K_1 > K_2$), or have no apparent effect on substrate binding ($K_1 = K_2$). The main differences between this model and the concerted one of Monod *et al.* are firstly, the involvement of mixed intermediates formed as a consequence of the subunits undergoing conformational changes in sequence and secondly, negative cooperativity, as found for example in NAD binding to rabbit muscle glyceraldehyde-3-phosphate DH, can be accounted for since the model assumes that symmetry is not conserved on ligand binding, the conformational transitions each subunit undergoes depends on the energy of interaction.

It can be extended to the association of inhibitors and activators, eq. 3.90.

$$\bowtie \underset{+I}{\rightleftharpoons} \bowtie\square \underset{+I}{\rightleftharpoons} \square\square \overset{+A}{\rightleftharpoons} \square\langle A \rangle \overset{+A}{\rightleftharpoons} \langle A \rangle\langle A \rangle \qquad (3.90)$$

3.4.3 Control of Enzyme Activity by Covalent Modification

3.4.3.1 Energy Linked Chemical Modification

Certain enzymes are predisposed to co-operativity by ATP linked post-translational chemical modification. These enzyme linked reactions thus constitute additional elements of control. A fascinating example is gradually emerging from studies of glutamine synthetase (GS), important in maintaining the nitrogen balance in cells.[33]

Glutamate : ammonia ligase (ADP forming) (EC 6.3.1.2) catalyses the energy linked formation of glutamine:

$$\text{Glutamate} + \overset{+}{\text{NH}_3} + \text{ATP} = \text{Glutamine} + \text{ADP} + \text{P}_i \qquad (3.91)$$

The *E. coli* enzyme has MW 6×10^5 daltons and comprises twelve identical subunits arranged into two hexagonal layers. Its activity is controlled in two ways, by feedback inhibition and by covalent modification.

Glycine, CTP, try, ala, his, AMP and carbamyl phosphate, products of biosynthetic pathways emanating from glutamine all feedback inhibit GS. At the physiological concentrations of the compounds however mixtures are needed for effective inhibition. Stadtman[30] has termed this effect 'cumulative feedback inhibition'. It is probable that ala and gly bind at the active site but the others are thought to act at separate regulatory sites within close proximity of the catalytic region.

Covalent modulation of GS activity is superimposed on this feedback effect. The subunits can each exist in two interconvertible forms which differ by the presence of an AMP group attached to the hydroxyl of a tyrosyl residue:

$$GS(C_6H_4OH)_{12} \underset{12\,ADP \quad 12\,P_i}{\overset{12\,ATP \quad 12\,PP_i}{\rightleftharpoons}} GS(C_6H_4OAMP)_{12} \qquad (3.92)$$

The adenylylated form is less active, exhibits a requirement for Mn(II) cations and is more susceptible to inhibition by his, AMP, CTP and try. The deadenylylated form requires Mg(II).

In viable *E. coli* cells the extent of adenylylation depends on the nitrogen balance, effectively the ammonia : glutamine ratio. When the ammonia level is high, deadenylylation is stimulated but when the glutamine concentrations are high and those of ammonia low the adenylylated form predominates. Thus a mechanism has evolved by which the substrates limit their own formation. There is good evidence to support a regulation mechanism in which the substrates allosterically interact with a GS interconverting system.

Originally it was thought that adenylylation and deadenylylation were catalysed by two different enzymes, an adenylyltransferase (AT) and a deadenylylating enzyme (DA). However the latter preparation is resolvable into two protein fractions, P_I and P_{II}. P_I is able to catalyse both directions of eq. 3.92 but its activity in the two opposing directions responds differently to small molecules. Adenylylation is inhibited by 2-oxoglutarate* but activated by glutamine whereas deadenylylation is inhibited by glutamine and stimulated by 2-oxoglutarate. Fraction P_{II} is not catalytic and can exist in two interconvertible forms, P_{IIA} and P_{IID} which stimulate respectively the adenylylation and deadenylylation reactions. P_{II} therefore controls P_I activity since without it a futile cycle would be set up only accomplishing the wasteful hydrolysis $ATP + P_i \rightleftharpoons ADP + PP_i$.

The P_I fraction has been further differentiated into two fractions with uridylyltransferase (UT) and deuridylylating (DU) activities, which catalyse respectively the addition and removal of UMP to

---

* Glutamate is formed enzymically from 2-oxoglutarate, ammonia and NADPH by glutamate dehydrogenase.

P_{IIA}. P_{IIA} has in fact been shown to be a tetramer of MW 44×10^3 binding four molecules of UMP. The overall GS control system can then be represented as in eq. 3.93.

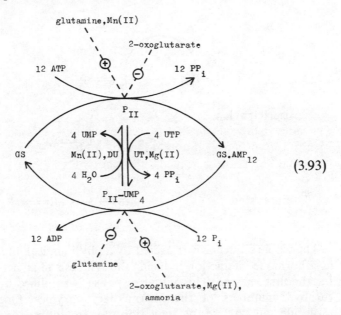

$$(3.93)$$

Glutamine synthetase activity in *E. coli* is controlled by variations in the degree of covalent adenylylation which in turn is allosterically modulated by the levels of glutamine, 2-oxoglutarate, the energy content (ATP plus UTP) and the concentration of the metal ions Mg(II) and Mn(II). High levels of glutamine activate the adenylylation reaction and inhibit the deadenylylation and uridylyltransferase activities thus decreasing the GS activity; 2-oxoglutarate has the opposite effect, stimulating DA and inhibiting AT.

Three other well characterised and important examples of control of enzyme activity by covalent modification, phosphorylase, glycogen synthetase and protein kinase are discussed in Section 5.2.6.2.

3.4.3.2 *Zymogen Activation by Limited Proteolysis*

The series of consecutive activation processes preceding glucose mobilisation (section 5.2) is a common feature of many biological control processes, blood coagulation (Figure 6.5) and zymogen activation being examples. The basic principles involved in these 'cascade' processes are described by Figure 3.19.

Figure 3.19: An Enzyme 'Cascade'

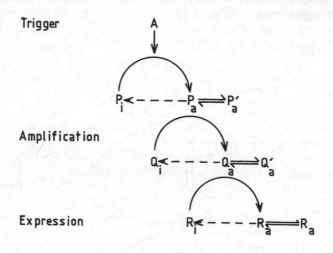

The whole sequence is triggered by a physiological impulse, such as a hormone interaction inducing the synthesis of an activator (A) for the first enzyme. This initial response is amplified by the high turnover numbers of the intermediary enzymes. One enzyme molecule will catalyse the activation of many thousands of its pro-enzyme substrate molecules so an avalanche effect is produced, the degree of which depends on the number of activating enzymes, their catalytic efficiencies and specificities. Finally the resultant is enhanced stimulation of a biochemical pathway leading to an observed physiological expression. Each of the components may then be inactivated by an allosteric interaction or by enzyme catalysed modification back to its precursor.

Zymogen activation describes the cascade process whereby inactive or partially active precursors are converted into proteases of full biological activity. Originally the term was applied to the activation of digestive enzymes, but as Neurath and Walsh[34] have pointed out, many other proteolytic processes, blood coagulation, fibrinolysis, and complement fixation amongst them, proceed by a cascade series of limited proteolytic reactions. That protease zymogens are synthesised ready for delayed activation *in situ* is an element of control important to the survival of the cell. Each protein is synthesised on a ribosome where the amino acids are linked together, commencing at the N-terminus. It is also an N-terminal polypeptide which is usually split off, in the correct location, to confer enzymic activity on the zymogen. Thus an inactive molecule and not the protein degrading agent is present during ribosomal synthesis and transport to the proper site of action. Table 3.11 gives a short list of enzyme precursors.

Table 3.11: Enzyme Precursors

Physiological process	Zymogen
Development	Prococoonase, procollagenase
Blood coagulation	Prothrombin, plasminogen
Digestion	Pepsinogen, trypsinogen, chymotrypsinogen, procarboxypeptidase, prophospholipase

A group of proteolytic enzymes—the trypsins—are formed from trypsinogen in the alimentary canal by the action of enterokinase.* Trypsinogen is synthesised by the acinor cells of the pancreas, and although initially found to be 229 amino acids in length, the first six of which are removed by enterokinase, a longer precursor, containing eighteen additional residues has been isolated.[35] The extra sequence is probably responsible for correct addressing of the zymogen molecule. Pro-trypsinogen is synthesised in the pancreas but enterokinase is produced by the intestinal brush border and its site of action is probably the villous epithelial cells. An extra element of control is thus present in which trypsinogen is activated as a consequence of the convergence of two different processes, Figure 3.20. By this means trypsin is not activated in the pancreatic juice so disorders such as pancreatitis are normally obviated.

Another zymogen of similar size and three-dimensional structure to trypsinogen and which is also carried by the pancreatic juice into the small intestine is chymotrypsinogen. Here its activity is induced by reaction with trypsin and further reaction products arise autocatalytically (Figure 3.21). The three-dimensional forms of α, β, δ, and π chymotrypsins appear to be identical, with an overall folding

---

* The importance of enterokinase in the activation processes is seen by the gross nitrogen malabsorption and disordered protein digestion in patients with congenital enterokinase deficiency.

Figure 3.20: Activation of Trypsinogen

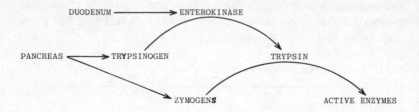

similar to that of the zymogen. There is a significant difference between the two however. The α-amino group of isoleucine16, exposed on activation, moves to form an intramolecular ionic bond with the ionised β-carboxylic acid group of aspartate194, and this

Figure 3.21: The Chymotrypsin Family. (CT*g, chymotrypsinogen;* CT, *chymotrypsin;* T, *trypsin*)

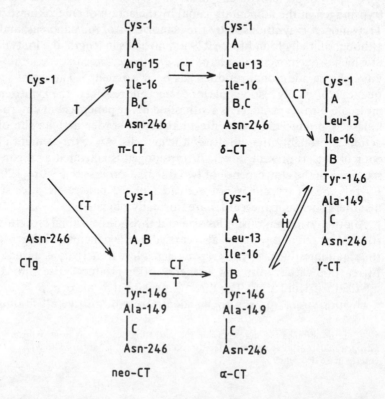

movement causes a conformational movement in methionine192 to form a flexible lid for the binding site, which is absent in the zymogen (see Chapter 4).

3.5   REFERENCES

1. H. Gutfreund, *Essays in Biochem.*, vol. 3 (1967), p. 19.
2. W. P. Jencks, *Catalysis in Chemistry and Enzymology* (McGraw-Hill, New York, 1969).
3. A. A. Frost and R. G. Pearson, *Kinetics and Mechanism* (Wiley, New York, 1961).
4. K. Laidler, *Disc. Faraday. Soc.*, vol. 20 (1955), p. 83.
5. M. Eigen and G. G. Hammes, *Adv. Enzym.*, vol. 25 (1963), p. 1.
6. R. M. Noyes, *Prog. React. Kinetics*, vol. 1 (1961), p. 129.
7. M. Dixon and E. C. Webb, *Enzymes* (Longmans, London, 1964).
8. R. M. C. Dawson, D. C. Elliott, W. H. Elliott and K. M. Jones, *Data for Biochemical Research* (Oxford University Press, Oxford, 1969).
9. J. L. Markley, *Acc. Chem. Res.*, vol. 8 (1975), p. 70.
10. L. A. Cohen, *The Enzymes*, vol. 1 (1970), p. 149.
11. S. N. Timasheff, *The Enzymes*, vol. 2 (1970), p. 371.
12. A. Bannister and R. L. Foster, *Analyt. Biochem.*, vol. 82 (1977), p. 184.
13. J. B. Jones, C. Niemann and G. E. Hein, *Biochem.*, vol. 4 (1965), p. 1735.
14. B. R. Baker, *Design of Active-Site-Directed Irreversible Enzyme Inhibitors* (Wiley, New York, 1967).
15. M. Laskowski and R. W. Sealock, *The Enzymes*, vol. 3 (1971), p. 375.
16. R. Wolfenden, *Acc. Chem. Res.*, vol. 5 (1972), p. 10; R. N. Lindquist, *Drug Design*, vol. 5 (1975), p. 24.
17. I. I. Secemski, S. S. Lehrer and G. E. Lienhard, *J. Biol. Chem.*, vol. 247 (1972), p. 4740.
18. B. Lee, *J. Mol. Biol.*, vol. 61 (1971), p. 463.
19. R. R. Rando, *Science*, vol. 185 (1974), p. 320; *Biochem. Pharmacol.*, vol. 24 (1975), p. 1153.
20. See the *Annual Reviews in Biochemistry* and *Advances in Protein Chemistry*.
21. G. E. Means and H. Feeney, *Chemical Modification of Proteins* (Holden-Day, San Francisco, 1971).
22. R. L. Foster and M. S. Patel, Unpublished Observations (1977).
23. R. A. Lauson and F. H. Westheimer, *J. Am. Chem. Soc.*, vol. 88 (1966), p. 3426.
24. I. B. Wilson, *Drug Design*, vol. 2 (1971), p. 213.
25. G. Schoellman and E. Shaw, *J. Biol. Chem.*, vol. 240 (1965), p. 694.
26. F. H. Westheimer, *J. Biol. Chem.*, vol. 241 (1966), p. 421.
27. R. L. Foster, Unpublished Observations quoted in J. Knowles, *Acc. Chem. Res.*, vol. 5 (1972), p. 155.
28. R. H. Abeles and A. L. Maycock, *Acc. Chem. Res.*, vol. 9 (1976), p. 313.
29. P. Karlson, D. Doenecke and E. Sekeris, *Comprehensive Biochem.*, vol. 25 (1975), p. 1.
30. E. R. Stadtman, *The Enzymes*, vol. 1 (1969), p. 458.
31a. J. Monod, J.-P. Changeux and F. Jacob, *J. Mol. Biol.*, vol. 6 (1963), p. 306.
31b. J. Monod, J. Wyman and J.-P. Changeux, *J. Mol. Biol.*, vol. 12 (1965), p. 88.
32. D. E. Koshland, G. Nemethy and D. Filmer, *Biochem.*, vol. 5 (1966), p. 365.
33. E. R. Stadtman and A. Ginsberg, *The Enzymes*, vol. 10 (1974), p. 755.
34. H. Neurath and K. A. Walsh, *Proc. Natl. Acad. Sci. U.S.*, vol. 73 (1978), p. 3825.
35. A. Devilliers-Thierg, *Proc. Natl. Acad. Sci.*, vol. 72 (1975), p. 5016.

4 Mechanisms of Catalysis

An understanding of the mechanisms by which enzymes exert their catalytic effects is one of the aims of much current research. These mechanisms are all the more intriguing when on initial examination of the three-dimensional structure of a protein there are often few obvious clues as to its biological function, notwithstanding its chemical mode of action. Like all proteins, enzymes appear to be composed of random sequences of some twenty amino acid residues, divided almost equally between hydrophilic and hydrophobic side chains; but from a study of those whose structures and actions are known, two inferences can be made. First, the physiological behaviour of an enzyme (as of any protein) is a function of its primary structure and secondly, its mechanism of catalysis can be explained in the same molecular terms as those of less complex molecules.

These realisations have prompted an approach to their mechanisms through observations on 'enzyme-models', less complex non-enzymic systems which incorporate some of the features of enzymes. Of course, such systems can only roughly mimic enzymic reactions and thus extrapolation from these to the observed behaviour of enzymes will only be tenuous. Nevertheless, the information provided by a study of non-enzymatic reactions has proved valuable, especially regarding the behaviour of molecules in homogeneous solution and possible mechanisms of rate enhancement.

4.1 CATALYSIS IN SOLUTION

4.1.1 Acid-Base Catalysis

Most of the amino acid side chains listed in Table 1.2 have been found or implicated in the catalytic activity of enzymes. For example two carboxylic acid groups have been observed in the binding cleft of lysozyme, and serine 195 together with histidine 57 are undoubtedly involved in chymotrypsin catalysed hydrolysis of esters. These residues and the other side chains listed in Table 1.2 most obviously chemically reactive, e.g. lysine, arginine and tyrosine, can behave as acids and bases.

Brönsted defined an acid as a substance with a tendency to provide or donate a proton to a suitable acceptor and a base as a proton acceptor, eqs. 4.1, 4.2.

$$HX + H_2O \rightleftharpoons H_3\overset{+}{O} + X^- \tag{4.1}$$

$$X^- + H_2O \rightleftharpoons HX + HO^- \tag{4.2}$$

In aqueous solutions the highly reactive protons and hydroxyl ions are able to catalyse the transformation of susceptible molecules, such reactions are then said to be *specific acid* or *specific base* catalysed. Many examples of these types of catalysis have been observed, for example the hydrolysis of glycosides and the alcoholysis of most esters, eq. 4.3, are hydrogen ion catalysed and the aldol condensation of ketones and aldehydes are catalysed by hydroxide ions, eq. 4.4.

$$R_1 \cdot CO \cdot OR_2 \overset{\overset{+}{H}}{\rightleftharpoons} R_1 \cdot CO \cdot \overset{\overset{H}{\bullet}}{\underset{+}{O}}R_2 \overset{R_3OH}{\longrightarrow}$$

$$R_1 \cdot CO \cdot OR_3 + R_2 \cdot OH \tag{4.3}$$

$$CH_3 \cdot CO \cdot CH_3 + \bar{O}H \rightleftharpoons CH_3 \cdot CO \cdot \bar{C}H_2 \overset{CH_3 \cdot CO \cdot CH_3}{\longrightarrow}$$

$$CH_3 \cdot CO \cdot CH_2 \cdot \underset{\underset{CH_3}{|}}{\overset{\overset{CH_3}{|}}{C}} - O^- \tag{4.4}$$

Protons and hydroxyl ions are, however, powerfully reactive and hence rather indiscriminate species. More significant, in this case, are the weaker, more discriminating acids and bases such as acetic acid, methylamine, etc. Reactions catalysed by undissociated acids and bases in solution are described as subject to *general acid* or *general base* catalysis. For example the rate of glucose mutarotation (eq. 4.5), depends not only on the pH but also on the concentration of the buffering components, i.e. the reaction is susceptible to both specific and general catalysis. One mechanism consistent with this observation involves rapid donation of a proton from the buffer acid to the glucose followed by slower reaction of the cation with the conjugate base (B) of the buffer (eq. 4.6).

$$\text{(4.5)}$$

$$G + \overset{+}{H} \; \rightleftharpoons \; \overset{+}{GH} \; \xrightarrow[\text{slow}]{+\,B^-} \; BH + G \qquad (4.6)$$

for which

$$v = k \cdot \overset{+}{GH} \cdot B^- = k \cdot K_e \cdot B^- \cdot \overset{+}{H} \cdot G = k \cdot K_e \cdot K_B \cdot BH \cdot G \quad (4.7)$$

where $\qquad K_e = \overset{+}{GH}/G \cdot \overset{+}{H} \quad$ and $\quad K_B = B^- \cdot \overset{+}{H}/BH$

$$\therefore \; v = k_0 \cdot BH \cdot G \qquad (4.8)$$

Mechanism 4.6 predicts that the anomerisation rate is dependent on the total acid concentration (BH) and not just on the pH. Kinetically, however this mechanism cannot be distinguished from one in which the weak acid first associates with the substrate, i.e. where in the first step, the proton is shared by both the catalyst and substrate, instead of being completely donated before reaction with the conjugate base (eq. 4.9). A mechanism of this type has been proposed for the mutarotation of tetramethyl glucose in non-aqueous solvents:

$$G + HA \; \rightleftharpoons \; GHA \; \xrightarrow[+\,B]{\text{slow}} \; P \qquad (4.9)$$

$$\therefore \; v = k \cdot GHA \cdot B = k \cdot K' \cdot G \cdot HA \cdot B \qquad (4.10)$$

where $\qquad\qquad K' = \dfrac{GHA}{G \cdot HA}$

Prior catalyst-substrate association followed by slow disproportionation of the complex also accounts for the general base catalysis of activated esters by weak bases such as acetate anions and amines.

$$R \cdot CO \cdot OX \underset{B}{\overset{B}{\rightleftharpoons}} \overset{\displaystyle O^-}{\underset{\displaystyle B}{R \cdot C \cdot OX}} \xrightarrow{\text{slow}} R \cdot CO \cdot B + X \quad (4.11)$$

$$S \qquad\qquad SB$$

$$v = k \cdot SB = k \cdot K_e \cdot B \cdot S \quad (4.12)$$

i.e. the rate is proportional to the base concentration.

The imidazole catalysed hydrolysis of acetyl-imidazole is an example of a base catalysed reaction with some parallels to the chymotrypsin catalysed hydrolysis of esters[1] (Section 4.3.2). Unlike the general catalysis mechanisms described above, in which transition state proton transfer between the catalyst and substrate is the first significant step, in its catalysed hydrolysis of acetyl-imidazole, the base first removes a proton from a water molecule. By this means the nucleophilicity of the water molecule is increased so facilitating its attack at the carboxyl carbon atom, but without involving the intermediate formation of a highly charged hydroxide ion:

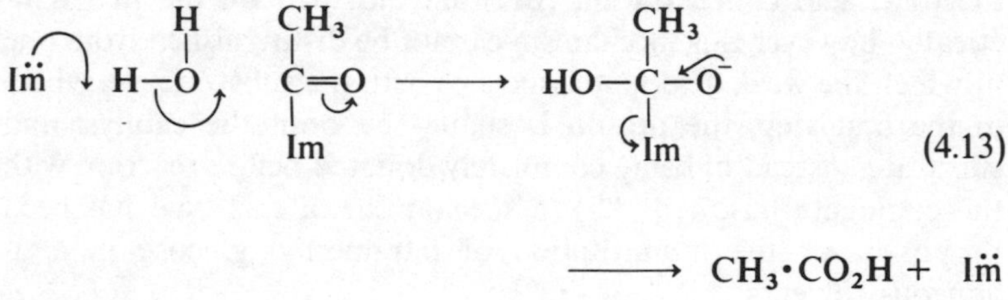

$$\text{Im} \downarrow \text{H--O} \quad \underset{\text{Im}}{\overset{\displaystyle CH_3}{C{=}O}} \longrightarrow \text{HO--}\underset{\displaystyle \zeta\text{Im}}{\overset{\displaystyle CH_3}{C \cdot O^-}} \quad (4.13)$$

$$\longrightarrow CH_3 \cdot CO_2H + \text{Im}$$

A similar mechanism has been proposed for the deacylation of acyl-chymotrypsin $(E \cdot CH_2 \cdot O \cdot CO \cdot R)$ the intermediate postulated in the chymotrypsin catalysed hydrolysis of esters and amides (Section 2.2.2.4). Acyl-chymotrypsin can be considered as an activated ester and there is considerable evidence to implicate a reactive histidyl residue in its active site (see Section 4.3.2). Its deacylation rate depends only on the concentration but not the basicity of acceptor. Thus the pathway shown in Figure 4.10 has been conceived for the deacylation of acyl-chymotrypsin.

From the foregoing it is apparent that a given reaction may be susceptible to catalysis by many different species in solution. For example, the mutarotation of glucose is catalysed by water molecules, hydrogen ions, hydroxyl ions, carboxylic acids and carboxylate anions. For such reactions the total rate of substrate depletion will depend on the concentration of each solute species and its individual efficacy. In the simplest case, where each species contributes additively, the general rate equation for acid-base catalysis is,

$$k_{obs} = k_S \cdot H_2O + k_H \cdot \overset{+}{H} + k_{OH} \cdot \overset{-}{OH} + k_{HA} \cdot HA + k_A \cdot A^-$$

(4.14)

where k_S, k_H, k_{OH}, k_{HA} and k_A are the rate constants for catalysis by the corresponding species, solvent, protons, hydroxide ions, undissociated acid and conjugate base molecules in solution.

The first task in the mechanistic elucidation of a solute species catalysed reaction is therefore to separate out and evaluate the individual rate constants. This is performed by experimentally determining the rate dependence on buffer concentration. Increasing the buffer concentration while maintaining a constant ratio of buffering species will not alter the hydrogen ion concentration, thus if the rate remains unchanged no general contributions to catalysis can be inferred. On the contrary, if a buffer concentration dependency is observed, general acid or base catalysis may be present (Figure 4.1). Extrapolation of the observed rate constants to zero buffer concentration gives numerical values on the ordinate equal to $(k_S + k_H \cdot [\overset{+}{H}])$ for acid catalysis and $(k_S + k_{OH} \cdot [\overset{-}{OH}])$ for

Figure 4.1: Dependence of General Acid Catalysis on Buffer Concentration

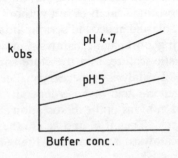

base catalysis: from these values at each pH therefore, the contributions from specific catalysis and spontaneous hydrolysis can be evaluated. Catalysis arising from the buffering species is obtained from the slopes of Figure 4.1. Each gradient is the apparent second order constant (k') of the buffer catalysed reaction, i.e.

$$k' = \frac{k_{obs} - k_S}{\text{Buffer concentration}} = \frac{k_{obs} - k_S}{HA + A^-} \qquad (4.15)$$

$$\therefore \quad k' = \frac{k_{HA} \cdot HA}{HA + A^-} + \frac{k_A \cdot A^-}{HA + A^-} \qquad (4.16)$$

The quantitative values of k_A and k_{HA} are then easily derived from a plot of k' against mole fraction free acid or base, Figure 4.2.

Figure 4.2: Dependence of Buffer Catalysis on Mole Fraction of Buffering Components

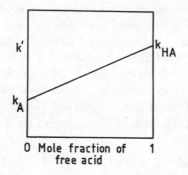

Catalysis by weak acids and bases means that the same end result can be achieved as by protonic catalysis but without the formation of highly reactive, unstable and hence indiscriminate species in solution. This has a bearing on enzyme catalysis in which reaction control is a characteristic feature, and therefore some idea of the efficiency of this type of catalysis is necessary. The capacity of a molecule to act as a general acid or base depends on its strength. This factor is expressed in terms of the dissociation constant (K_a) of the acid or conjugate acid. Thus if k_a and k_b are the rate constants for acid and base catalysis respectively then their catalytic efficiencies can be expressed,

$$k_a = G_a \cdot K_a^{\alpha} \qquad \text{or} \qquad \log k_a = \log G_a - \alpha p K_a \qquad (4.17)$$

$$k_b = G_b \cdot K_a^{-\beta} \qquad \text{or} \qquad \log k_b = \log G_b + \beta p K_a \qquad (4.18)$$

where G_a, G_b, α and β are constants whose values depend on the reaction type, temperature and solvent type.

These equations, known as the 'Brönsted catalysis laws',[2] predict a linear relationship between the logarithm of the catalytic constant and pK_a (Figure 4.3). The slopes (α and β) of the lines can vary numerically between zero and unity, reflecting the sensitivity of the reaction in question to general acid or general base catalysis respectively. A low value signifies a relative insensitivity, e.g. when $\alpha = 0$ no transfer of proton occurs and the reaction is independent of acid strength, whereas a value approaching unity indicates the opposite,

Figure 4.3: Brönsted Plot for Base Catalysis

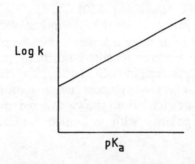

e.g. when $\alpha = 1$ the proton is completely transferred and the reaction is sensitive to the acid strength. (The mutarotation of glucose for example, shows a moderate susceptibility to catalysis by both general acids, $\alpha = 0.27$ and general bases $\beta = 0.34$). Strictly, in general acid or base catalysis the catalyst participates in proton transfer in the transition state, and the value of the Brönsted constant indicates to what extent it is transferred compared to specific catalysis.

By means of the Brönsted relationships an idea of the rate enhancements afforded by general acid-base catalysis can be calculated. For example at pH 7 the rate accelerations of one molar concentration of an acid of pK_a 7 are, with $\alpha = 0$, 1 fold, $\alpha = 0.5$, 200 fold and $\alpha = 1$, 5×10^6 fold over the water catalysed reaction. It can be inferred therefore that several general acid-base pairs acting concertedly on different parts of the substrate molecule may contribute more effectively to its catalytic rearrangement.

Possibly the clearest example of concerted general acid-general base catalysis is the mutarotation of tetramethylglucose in aprotic solvents.[3] The reaction proceeds very slowly in benzene unless traces of water or pyridine plus phenol are added. In the presence of the latter combination, a third order rate law is obeyed, first order with respect to tetramethylglucose, phenol and pyridine, suggesting that the acidic and basic components act together to promote mutarotation. This conclusion was supported by the considerable rate enhancement found on addition of 2-pyridone which contains both a phenolic and amino functional group. Although combination in 2-pyridone weakens the individual strengths of the two groups compared to phenol and pyridine, the rate was increased by 10^2 fold. Compatible with these observations a 'push-pull' mechanism has been proposed, in which the acidic and basic groups simultaneously and concertedly remove or add protons to effect hemiacetal fission and synthesis (eq. 4.19).

As a model for enzyme catalysis the system has obvious attractions but the extent of its analogy is lessened because of the hydrophobic medium used. Enzyme reactions normally take place in highly polar, aqueous systems; in such systems, 2-pyridone has no appreciable catalytic effect on glucose mutarotation. However many enzyme active sites have been shown to have microenvironments decreased in polarity with a limited number of water molecules present.

(4.19)

4.1.2 Covalent Catalysis

Covalent catalysis, characterised by the formation of a covalently bonded substrate-catalyst intermediate, is of direct relevance to enzymic catalysis. Many enzymic reactions have been shown to

Table 4.1: Covalent Enzyme Bound Intermediates

Enzyme	Type of complex	Residue modified
Acetyl-CoA acetyltransferase	Acyl-E	cys
Alkaline phosphatase	Phosphoryl-E	ser
D-Amino acid oxidase	Schiff base	lys
ATP-citrate (pro-3S)-lyase	Phosphoryl-E	glu
Fructose-bisP aldolase	Schiff base	lys
Glyceraldehyde-3-P DH	Acyl-E	cys
Glycine amidinotransferase	Amidino-E	—
Papain	Acyl-E	ser
Phosphoglucomutase	Phosphoryl-E	ser
Phosphoribosylformyl-glycinamide synthetase	Glutamyl-E	cys
Succinyl-CoA synthetase (ADP)	Phosphoryl-E	his
Sucrose phosphorylase	Glucosyl-E	Carboxylic acid
Transaldolase	Schiff base	lys
Trypsin	Acyl-E	ser

proceed via covalent enzyme-substrate complexes (Table 4.1), including the protease and esterase catalysed hydrolysis of amides and esters, decarboxylation of acetoacetate and alkaline phosphatase catalysed transphosphorylation. One striking feature from this table is the widespread distribution of the many different types of complex, imine, acyl-enzyme, etc. throughout the enzyme classes, indicating that they are not limited to only a few reaction types.

Consideration must therefore be given to the factors which make intermediate formation of a covalent enzyme-substrate complex more favourable under certain circumstances than either direct chemical reaction or acid-base catalysis. In model systems this is translated into determining what types of reaction are subject to covalent catalysis and evaluating the factors that contribute to and limit its efficacy.

4.1.2.1 Nucleophilic Catalysis

Of particular biochemical relevance are reactions involving catalysed acyl transfer from activated substrates, such as anhydrides or esters, with good leaving groups, to suitable acceptors. For example, imidazole enhances acetyl transfer from p-nitrophenylacetate to other nucleophilic acceptors,

$$CH_3 \cdot CO \cdot OX + Im \xrightarrow{\overset{X\bar{O}}{\rightleftharpoons}} \overset{+}{Im} - CO \cdot CH_3$$

$$\xrightarrow{QH} Im + CH_3 \cdot CO \cdot Q$$

(4.20)

X = p-nitrophenyl

QH = H_2O, $R \cdot NH_2$ or other nucleophile.

For imidazole to act as a catalyst in reactions of this type it must have a higher nucleophilic reactivity than the terminal acceptor and the intermediate covalent complex should be more reactive than either the substrate or product. Since molecular nucleophilicity is determined essentially by basicity, imidazole, with pK_a 7, is a strong base at physiological pH values and hence highly nucleophilic. The imidazole side chain of histidine probably nucleophilically promotes phosphate transfer in the enzymes phosphoglycerate mutase and nucleoside diphosphokinase but for enzymes acting on unactivated esters and amides, such as the serine proteases (e.g. trypsin and chymotrypsin) and the thiol proteases (e.g. papain, ficin and bromelain), it behaves as an acid-base catalyst enhancing the nucleophilicity of the hydroxyl and sulphydryl groups respectively.

Other side chains probably acting as active site nucleophiles are the β-carboxylate ion of aspartate in pepsin which forms an acyl-enzyme, the ε-amino group of lysine in acetoacetate decarboxylase and twelve tyrosyl phenols in glutamine synthetase which are adenylylated by an equal number of ATP molecules; some other examples are listed in Table 4.1.

A major group of reactions involves nucleophilic attack at the carbonyl level of oxidation. Semicarbazone formation by aromatic aldehydes such as pyridoxal or pyridoxal phosphate, catalysed by primary and secondary amines, probably follows the sequence,

$$Ar \cdot CO \cdot H + HNR_2 \underset{slow}{\overset{\overset{+}{H}}{\rightleftharpoons}} Ar \cdot \overset{+}{C}H : NR_2 \xrightarrow[fast]{NH_2 \cdot NH \cdot CO \cdot NH_2}$$
$$\overset{}{} I$$

$$Ar \cdot CH : N \cdot NH \cdot CO \cdot NH_2 + R_2NH \quad (4.21)$$

in which intermediate Schiff base (I) formation between the amine and aldehyde is the rate determining first step, followed by rapid trans-imination of the highly reactive cationic intermediate with semicarbazine. The reactivity of the complex (I) to nucleophilic attack is attributed to its ease of protonation relative to the aldehyde, and the greater resonance stabilisation of the final product. The catalytic benefits enjoyed by enzymes such as aspartate transaminase, cystathionase and phosphorylase, all of which have the pyridoxal phosphate cofactor initially condensed in an imine linkage to a neighbouring main chain lysine ε-amino group can then be envisaged, Figure 4.4. Preformation of a potentially reactive imine will facilitate transimination with the incoming amino substrate by eliminating the potentially slow step of eq. 4.21.

Figure 4.4: Formation of Imine Linkage between Pyridoxal Phosphate and Lysine Amino Group

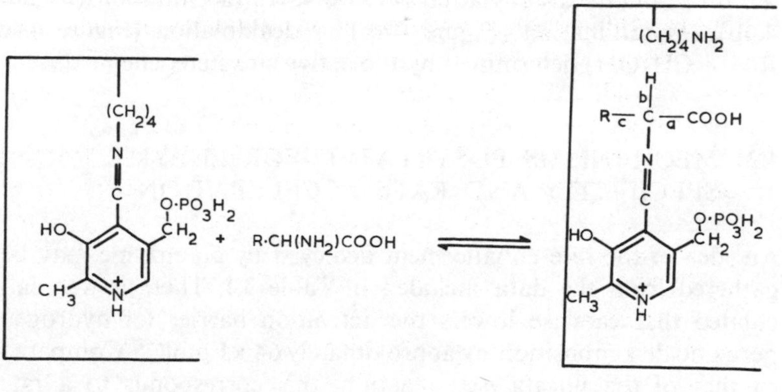

4.1.2.2 Electrophilic Catalysis

Electrophilic catalysis is described as the creation of an electron deficiency at the reaction centre by the catalyst. The latter, often an easily protonated entity acts as an 'electron-sink'. For example the decarboxylation of 2-oxo acids catalysed by primary amines probably follows a mechanism in which the amine first reacts with the carbonyl to form the imine (II), eq. 4.22. Protonation of this will yield an electrophilic centre to provide the driving force for decarboxylation.

$$R \cdot CO \cdot CR_2' \cdot CO_2^- \;\rightleftharpoons\; R \cdot C - CR_2' - C \overset{O}{\underset{O}{\cdots}} \;\rightarrow\; R \cdot C = CR_2' \rightarrow R \cdot CO \cdot CHR_2' \qquad (4.22)$$

II

A mechanism similar to that of eq. 4.22 is also consistent with the observed decarboxylase catalysed removal of carbon dioxide from acetoacetate $(CH_3 \cdot CO \cdot CH_2 \cdot CO_2^-)$ and with the aldolase catalysed condensation of glyceraldehyde-3-phosphate with dihydroxyacetone phosphate to form fructose-1,6-bisphosphate. An intermediate corresponding to (II, $R'' =$ enzyme) formed during the action of both enzymes has been stabilised by reduction of the imine with radioactively labelled sodium borohydride. After protein degradation most of the label was found associated with lysine residues (Section 3.3.3.2).

The conjugated pyridine ring of pyridoxal (Figure 4.4) provides a very effective electrophilic centre which by withdrawing electrons weakens the bonds to the amino acid α-carbon atom. Energetically, the reaction is then translated partway towards the products, with the direction, decarboxylation (Figure 4.4a), racemisation (Figure 4.4b), transamination (Figure 4.4b) or dealdolation (Figure 4.4c $R = -CH_2OH$) determined by the active site stereochemistry.

4.2   MECHANISMS POSTULATED FOR ENZYME SPECIFICITY AND RATE ACCELERATION

An idea of the rate enhancement achieved by an enzyme may be gathered from the data included in Table 3.1. There it was calculated that catalase lowers the activation barrier for hydrogen peroxide decomposition by approximately 64 kJ mol^{-1}. Compared to that of the uncatalysed reaction, this corresponds to a rate acceleration of about 10^{10} fold, a value which is also several orders of magnitude higher than those of non-enzymic catalyses. What features present in enzymes evoke such rate enhancements and what theories have been put forward to account for their outstanding catalytic properties?

4.2.1   Approximation and Orientation

The basic mechanism of enzyme catalysis involves encounter of the enzyme with its substrate(s), solvation changes in both, mutual orientation, electronic rearrangements, release of the products and rearrangement of their solvation shells. It follows that one function of the enzyme is to gather the reactants together within the same proximity and to orientate them with respect to the catalytic groups in the active site. It would be informative therefore to try and estimate the increase in encounter frequency and hence the rate increase that could arise from increasing the effective concentration of reacting groups, i.e. from their approximation.*

One approach to the proximity effect has been through the analysis of model systems in which the catalytic and leaving groups are made to adopt suitable orientations within the same molecule. The kinetic properties of these 'intramolecular' reactions can then be compared to those in which the groups are not favourably orientated or to the corresponding intermolecular reactions.[4]

An interesting model reaction for lysozyme catalysis is the hydrolysis of 2-carboxyphenyl-β-D-glucopyranoside (III, R = —OH). This glycoside undergoes cleavage of the glycosidic bond, assisted by the 2-carboxyl group, approximately 10^4 fold faster than the 4-carboxy derivative. To rationalise this result, the authors[5] favour a mechanism involving intramolecular general acid catalysis by the weak carboxylic acid (III). Replacement of glucose with N-acetylglucosamine (III, R = —NH·CO·CH$_3$) increases the rate by a further 20 fold, probably as a result of additional nucleophilic attack by the carboxyl of the N-acetyl group at C-1 of the pyranoside (see Figure 4.8, VI). Consistent with this mechanism is the observation that methanolysis proceeds with retention of the β-configuration, a result which parallels that found for lysozyme catalysed hydrolysis of N-acetylglucosamine oligosaccharides.[5]

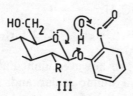

III

An extensive study of the effect of proximity has been pursued by Bruice and Pandit.[6] Hydrolyses of the p-bromophenol monoesters listed below proceed via intramolecular nucleophilic attack by the ionised carboxylate anions followed by departure of the phenolate

* Also called the 'proximity', 'propinquity' or 'probability and entropy' effect.

anions and cyclisation to the intermediate anhydrides. The anhydrides then rapidly hydrolyse to the dicarboxylic acids.

Relative
rates   1   20   230 10^4 5.3×10^4

As the probability increases of obtaining a population of conformers in which the catalytic group and reaction centre adopt orientations favourable for reaction, so do the observed rates. Finally, in the *exo*-3,6-*endoxo*-Δ^4-tetrahydrophthalic acid derivative in which the two groups are rigidly held, facilitation of the rate amounts to 5×10^4 fold over molecules possessing many degrees of rotational freedom.

It is of interest to compare the rates of intramolecular reactions with those of the corresponding bimolecular reactions in solution. The first order rate constant for cyclisation of N,N-dimethyl aminobutyric acid, phenyl ester (eq. 4.23) is 0.17 s^{-1}, while the aminolysis of phenyl acetate proceeds with a second order rate constant of $1.3 \times 10^{-4} \text{ mol}^{-1} \text{ s}^{-1}$, eq. 4.24.

$$(CH_3)_2\overset{..}{N} \cdot (CH_2)_3 \cdot CO \cdot OC_6H_5$$

$$\longrightarrow (CH_3)_2N{-}CO + \bar{O}C_6H_5 \qquad (4.23)$$

$$(CH_3)_3\overset{..}{N} + CH_3 \cdot CO \cdot OC_6H_5 \longrightarrow$$

$$(CH_3)_3 \cdot N \cdot CO \cdot CH_3 + \bar{O}C_6H_5 \qquad (4.24)$$

Any comparison of the two however immediately encounters the problem of contrasting a unimolecular and a bimolecular rate constant. The ratio, 1.25×10^3 M, is often regarded as the concentration of the catalyst required to promote the reaction with a pseudo first order rate constant equal to that of the intramolecular reaction. This amount is then the 'effective concentration' of the catalyst in the intramolecular reaction.

From geometric considerations, it has been calculated that the rate increases that accrue from just collecting the catalytic groups and substrates together within the same space on the enzyme are relatively insignificant.[7] This conclusion was reached mainly because of the low concentration of active sites in solution and because of the assumption that the concentration of intramolecular groups was the same as the molar concentration of water (55M). A value of this size is found for some intramolecular general acid-base catalysed reactions, but for intramolecular nucleophilic catalysis, with its more ordered transition state, the 'effective molarities' are much higher, as shown for reaction 4.23. In enzyme catalysed reactions the crucial component is the formation of an enzyme-substrate complex and Jencks and Page[8] have drawn attention to the large deficits in translational and rotational entropies* consequent upon its formation and to the fact that reaction takes place within this complex.** Correctly positioned within the active site little or no additional entropy is lost as the substrate forms the transition state whereas in a bimolecular reaction the entropy is lost on reaction of the catalyst and substrate. Because of this initial binding of enzyme and substrate the catalytic groups in the active site will then have high 'effective concentrations'. He obtained a numerical value for this 'effective molarity' and hence some idea of the rate accelerations to be expected from enzymic catalysis by calculating the decrease in entropy when two molecules combine. One set of translational plus rotational entropies is lost which together are of the order 60 kJ mol^{-1} at 30°C. This is therefore the amount by which the energy of reaction is reduced, and corresponds to a rate acceleration of about 10^{10}. From such a calculation it can be inferred that the greater the entropy loss, i.e. the greater the degree of approximation and interaction between the active site and the substrate, the higher the rate.

4.2.2 Binding Energy and Catalysis

The foregoing section has shown how considerations of the entropy changes, and mechanisms of acid-base and covalent catalysis can account for a fairly large part of the rate enhancements associated with enzyme catalysis. The next step is to understand how the binding energies between substrate and enzyme are employed to reduce the activation energy of reaction.[9]

---

* By entropy is meant the degree of randomness or disorder.

** Jencks and Page[8] have called the 'utilization of strong attractive forces to lure a substrate into a site in which it undergoes an extraordinary transformation of form and structure', the 'Circe effect'.

The observed rate of a reaction following the Michaelis-Menten equation is given by eq. 2.32. If the substrate concentration is low, so that few enzyme molecules are complexed, this reduces to,

$$v = \frac{k_{cat}}{K_m} \cdot E \cdot S \tag{4.25}$$

(where E and S are concentrations of free enzyme and free substrate and $k_{cat} = V_{max}/E_0$).

k_{cat}/K_m is then the apparent second order rate constant for reaction of the free species i.e.

$$E + S \; \underset{\phantom{k_{cat}/K_m}}{\overset{k_{cat}/K_m}{\rightleftharpoons}} \; \text{ES*} \tag{4.26}$$

From Figure 4.5 it is seen that the activation energy ΔG_A^* for eq. 4.26 is reduced compared to that (ΔG_B^*) for the chemical reaction ES $\rightleftharpoons$ ES* by the binding energy ΔG_C^*

$$\Delta G_A^* = \Delta G_B^* - \Delta G_C^* \tag{4.27}$$

i.e. the activation energy for the bimolecular reaction (k_{cat}/K_m) is lowered by the binding energy.

Much published experimental data supports this conclusion. Inouye and Fruton,[10] for example, found little difference in the exhibited K_m values for binding of a series of peptides to pepsin (Table 4.2), but a considerable range of maximum velocities for their hydrolyses. According to the foregoing, these observations can

Figure 4.5: *Energy Profile for Reaction* E + S $\rightleftharpoons$ ES*

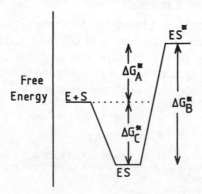

Table 4.2: Pepsin Catalysed Peptide Hydrolyses

Substrate*	K_m (mM)	k_{cat} (s^{-1})	k_{cat}/K_m (s^{-1} mol^{-1})
Cbz-His-Phe-Try-OEt	0.23	0.51	2220
Cbz-His-Phe-Phe-OEt	0.18	0.31	1720
Cbz-His-Phe-Tyr-OEt	0.23	0.16	690
Cbz-His-Tyr-Tyr-OEt	0.24	0.0094	39

* Cbz = carbobenzyloxy-
Source: Reference 10.

be rationalised by the additional energies of association being employed not to tighten the binding but to increase the rate.

It is difficult to envisage how one conformation of an active site can bind both the substrate and the product of reaction.[8] In the reversible reaction given in eq. 4.28,

$$\text{Glucose-1-phosphate} + \text{fructose} \rightleftharpoons \text{Sucrose} + \text{phosphate}$$
(4.28)

catalysed by sucrose phosphorylase, if the active site had a rigid structure complementary to the monosaccharides the result would be a destabilisation of the enzyme-disaccharide complex through production of strain in the disaccharide; conversely if the enzyme were rigidly specific for sucrose little of the reverse reaction would be observed. On the other hand, if the active site had a conformation complementary to the transition-state of the reaction some strain and distortion would have to be induced in both the enzyme-substrate and enzyme-product complexes. In this case the substrate would be energetically transported partway along the reaction pathway, in effect decreasing the activation barrier to reaction. This implies therefore that the enzyme would have a greater affinity for the transition state than either substrate or product. If this proposal holds, then compounds with structures analogous to the postulated transition state of an enzyme catalysed reaction would be held more tightly than either the substrate or product and be potent inhibitors. Such compounds, termed transition state analogues have been discussed in the previous chapter. For example cytidine deaminase is reported to bind tetrahydrouridine, which with a tetrahedral C-4 carbon atom mimics the proposed transition state for deamination of cytidine to uridine, 10^3 fold better than its substrate. This corresponds to approximately 16 kJ mol^{-1} of free energy and on the above argument this amount of energy would be available for specific substrate distortion and hence rate enhancement. More direct evidence for enzyme-transition state complementarity is provided by X-ray diffraction data some of which are discussed in Section 4.3.

Fersht[9] has shown that when the enzyme and transition state are complementary the quotient k_{cat}/K_m is maximised and that to optimise the rate, the enzyme 'should have' evolved to exhibit a K_m value greater than its physiological substrate level i.e. bind its substrate weakly.

4.2.3   Strain and Induced Fit

Both these theories have been put forward to explain, in chemical terms, how enzymes bring about enzyme-transition state correspondence.

The strain theory assumes that the active site has a fixed structure complementary to the substrate transition state and that the enzyme-substrate binding energy is used to strain or distort the bonds in the substrate towards their transition state configurations. The hypothesis was enunciated before hard physical evidence for distortion of the substrate was available but X-ray diffraction evidence has shown that a mechanism of this type may be, at least partly, involved for some enzymes. For example in Section 4.3.1 it will be seen that lysozyme appears to sterically force the monosaccharide ring binding to sub-site D towards the half chair configuration of a possible glycosidic carbonium ion intermediate. This would therefore be an example of a strain mechanism. (However Levitt and Warshal[11] have calculated that the ring can also bind in its unstrained chair form and thus it may be that the strain theory should be superseded in certain cases by a more modern equivalent, 'transition-state stabilisation' which proposes that the enzyme binds the transition-state more favourably than the substrate.)

There are numerous reactions which can act as models for the effect of strain on reaction rate. For example ethylene phosphate (IV) undergoes alkaline hydrolysis of the P—O bond with a rate increase close to that expected for enzyme catalysis, about 10^8 fold over the diester (V). This was attributed to prior distortion of the cyclic phosphate towards a configuration resembling the reaction transition state.[12] Presumably if (V) were sufficiently constrained by a catalyst, a similar rate enhancement would be effected.

$$\text{IV} \qquad\qquad \text{V}$$

According to the induced fit theory of Koshland,[13] the substrate is assumed to be rigid and the active site assumed to be flexible. So that on binding, a 'good' substrate will induce the correct catalytic conformation in the active site peptides, whereas a 'poor' one will not. Koshland[13] has summarised the evidence in support of his theory and there is no doubt that some flexibility is inherent in the active site residues of many enzymes. For example, try 62 moves about 1 Å when inhibitors bind to lysozyme and association of NAD with lactate dehydrogenase induces a movement of about 13 Å in the oligopeptide sequence 98–114 (Section 4.3.4). The theory was put forward originally to interpret the specificity exhibited by enzymes and not as a mechanism for utilisation of binding energy to reduce activation. It does in fact predict a decrease in k_{cat}/K_m because energy is needed to distort the enzyme.[9]

The two theories represent extremes and it is probable that in reality enzymic catalysis involves some contribution of both effects. This is indeed found to be so for many of the enzyme-inhibitor complexes studied crystallographically.

Although both, and particularly the induced fit theory, offer an explanation of specificity, Fersht[9] has argued that specificity, if defined as 'discrimination between competing substrates' is independent of both strain and induced fit since neither favourably alters the ratio k_{cat}/K_m and it is this that differentiates between competing substrates. For competing substrates

$$P + E \leftarrow ES_2 \overset{S_2}{\rightleftharpoons} E \overset{S_1}{\rightleftharpoons} ES_1 \rightarrow E + P_1 \qquad (4.29)$$

$$v_{S_1}/v_{S_2} = \frac{k_{cat}^{S_1}}{K_m^{S_1}} \cdot S_1 \bigg/ \frac{k_{cat}^{S_2}}{K_m^{S_2}} \cdot S_2 \qquad (4.30)$$

Therefore the ratio k_{cat}/K_m determines the relative rates.

The concept of substrate transition state-complex complementarity discussed in the previous section predicts that k_{cat}/K_m is maximised when the transition state binds optimally to its enzyme. Thus it has been suggested[8, 9] that maximisation of both rate and specificity depend on the same factors in the enzyme (see Section 5.2.6.1 for a specific example of enzyme specificity).

4.3   THE STRUCTURES AND CATALYTIC ACTIONS OF SELECTED ENZYMES

In this section the structures and chemical modes of action of three enzymes of known three-dimensional structure will be considered, lysozyme (the first to be successfully analysed crystallographically), chymotrypsin (arguably the most investigated enzyme) and lactate dehydrogenase (a polymeric protein).

4.3.1   Lysozyme (EC 3.2.1.27)

A place for hen's egg white lysozyme (HEWL) in the history of enzymology was secured with the elucidation of its three-dimensional structure, the first to be determined by X-ray crystallography, by Phillips *et al.*[14]. This work is further distinguished by acquisition of data from which models of lysozyme-inhibitor complexes could be constructed. These aided in the location of its active site and together with the known cleavage pattern of the enzyme the proposal of a reasonable mechanism for its catalytic action.

Discovered by Alexander Fleming in 1922 in tears and nasal mucus, lysozyme was hoped originally to possess the antibacterial capability later discovered in penicillin. However, it was found to be operative only against certain Gramme-positive bacteria and in particular *Micrococcus lysodeikticus*. The action of the enzyme is to break open the bacterial cell wall by catalysing rupture of the $\beta(1-4)$ glycosidic bonds between the *N*-acetylmuramic acid (NAM) and *N*-acetylglucosamine (NAG) monosaccharide rings of its peptidoglycan backbone (Figure 4.6). In the intact cell walls, the 3-lactyl groups are joined through amide bonds to crosslinking polypeptides, and consequently they are not involved in the catalytic action of lysozyme. Homo-oligosaccharides with degrees of polymerisation greater than five NAG units are effective substrates, being catalytically hydrolysed at rates close to those of the cell wall polysaccharides. Shorter oligosaccharides are able to behave as sub-

Figure 4.6: Site of Action of Lysozyme on Bacterial Cell Wall,
$R = -OCH(CH_3)COOH$

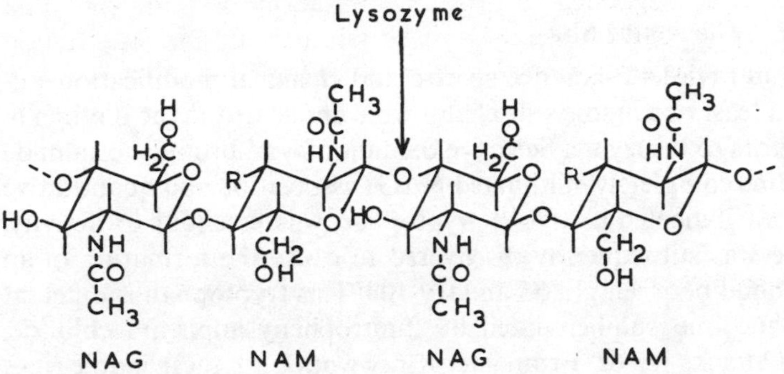

strates but they can also associate with non-catalytic sites and so inhibit lysozyme activity.

Lysozyme from hen's egg white is a single polypeptide chain of 129 amino acids, giving a molecular weight of 15,000 daltons (Figure 1.2). It contains four disulphide bridges, between half cystines, 6–127, 30–115, 64–80 and 76–94. Only the first, joining together the two ends of the molecule, can be cleaved with no loss of activity. Phillips *et al.*[14] showed the folded molecule to be roughly ellipsoidal in shape with dimensions approximately 45 × 30 × 30 Å. The three-dimensional conformation of its amino acid residue α-carbon atoms is shown in Figure 1.4. Amino acid residues 1 to 40 are seen to fold back on themselves to form a hydrophobic wing with a polar surface; in this wing are two sections of distorted α-helix, 5 to 15 and 24 to 34. The next fourteen residues fold into the recognisable secondary structure of an anti-parallel pleated sheet which buries the hydrophobic groups 55 and 56. A second wing is then provided by 57 to 87, also with a short α-helix formed from residues 80 to 85. The gap between the wings is partially filled by a helix of residues 88 to 96. This helix helps to define a cleft in the molecule which consequently has, on this side, a hydrophobic environment, conferred principally by the side chains try 28, 108, 111 and tyr 23 and on the other flank a more polar environment formed from residues 40 to 80 and which is in contact with solvent. The remaining amino acid residues, of which 119 to 122 also form a helix, then are seen to partly enfold the first wing before becoming attached to the N-terminus by the disulphide bridge.

The polar and basic side chains, including the terminal α-amino and α-carboxylate groups, are situated on the surface of the molecule, the only exceptions being ser 91 and glu 57.

4.3.1.1   The Active Site

From ultraviolet difference spectra and chemical modification studies, at least two indole side chains were implicated in the binding of inhibitors to lysozyme. Selective oxidation by N-bromosuccinimide or iodine completely eliminated NAG_3 association and quantitative analyses indicated that one tryptophan was essential to activity. Iodine was subsequently discovered to cause the formation of an ester bond between glu 35 and try 108. This tryptophan is different from the one sulphenylated by 2-nitrophenylsulphenyl chloride, which attacks try 62. From later X-ray work both these indole rings were shown to be intimately concerned with the stabilisation of lysozyme-oligosaccharide complexes.

Association with NAG_3 also perturbs two sets of ionisations with pK_a values 4 and 6 in the lysozyme molecule and kinetic experiments show that the catalysed hydrolyses of NAG_5 and NAG_6 are dependent on groups with dissociation constants in the pH ranges 3–4 and 5.5–6.5. The attribution of both classes to carboxylic acid ionisations was tested by Hoare and Koshland.[15] They esterified lysozyme with a glycine methyl ester-water soluble carbodiimide mixture. In the presence of NAG_3, to protect the active site, the enzyme retained 60 percent activity with an average 7.3 groups modified, whereas with no protection the enzyme was totally inactivated with 8.4 carboxylic acid groups esterified. Removal of the inhibitor and further treatment of both deprotected and unprotected enzyme with radioactively labelled glycine ester additionally incorporated 1.4 and 0.5 equivalents respectively. By reference to the known primary sequence, Koshland and Hoare identified the two groups as aspartate 52 and glutamate 35.

One of the successes of the X-ray investigations on lysozyme was the acquisition of high resolution electron density maps from isomorphous crystals containing inhibitor molecules. From these the location of the active site and the juxtaposition of the various groups discussed above could be ascertained. Of particular value were the data acquired at 2 Å resolution for the binding of the competitive inhibitor, NAG_3. The trisaccharide was found to bind along the cleft observed from crystals of the native molecule, with its free reducing end pointed inwards. If the three monosaccharide rings are designated ABC, where A is the non-reducing terminus, then the contacts it makes to residues in the cleft can be shown diagrammatically as in Figure 4.7. Ring A is hydrogen bonded via its acetamide NH to the ionised side chain of asp 101 and via its ring oxygen to the 3-hydroxyl of NAG ring B. Both NAG A and B make several van der Waals contacts to groups in their vicinities and in addition B is also hydrogen bonded via its C-6 oxygen to the same aspartate as ring A. Rings B and C are hydrogen bonded together in the same manner as A and B, but C makes by far the most numerous contacts of the three. Its acetamide group is fixed by two hydrogen bonds directed to opposite sides of the cleft, NH to the carbonyl of ala 107 and CO to the NH of residue 59, and by a non-polar interaction between the acetamido methyl group and tryptophan. The ring is positioned by two more hydrogen bonds between the indole NH's of tryptophans 62 and 63 and the hydroxymethyl and 3-hydroxyl groups respectively. These bonds and the many van der Waals contacts are responsible for the large free energy of stabilisation of the enzyme-inhibitor association. Binding induces several positional alterations in the side chains of the lysozyme molecule. In particular a movement of about 1 Å in the orientation of the hydrophobic side chain of tryp 62 was observed to sandwich the oligosaccharide.

In order to illustrate the specific closeness of fit between the inhibitor and its enzyme, the number of contacts and the relative contribution each ring makes to the overall binding energy are also listed in Figure 4.7.

Since NAG_3 acts as a competitive inhibitor it was assumed in the interpretation of the data, that although it bound to the active cleft, it did not traverse a catalytic region. The 2 Å resolution data indicated that in fact only half the cleft was filled, and by careful model building three additional monosaccharide rings could be incorporated. A fourth NAG unit, D, could however only be accommodated if first twisted into a half chair conformation. This was necessary in order to relieve steric interactions between its C-6 and O-6 atoms and residues 52, 108 and the acetamido group of ring C. In the half chair orientation hydrogen bonding was then feasible between O-1 and the carboxylic acid side chains of asp 52 and glu 35. Two further monosaccharide rings E, F could then be built in to make reasonable contacts without distortion.

A search was then made for chemically reactive groupings in the active site that could promote catalysis. Lysozyme was known to

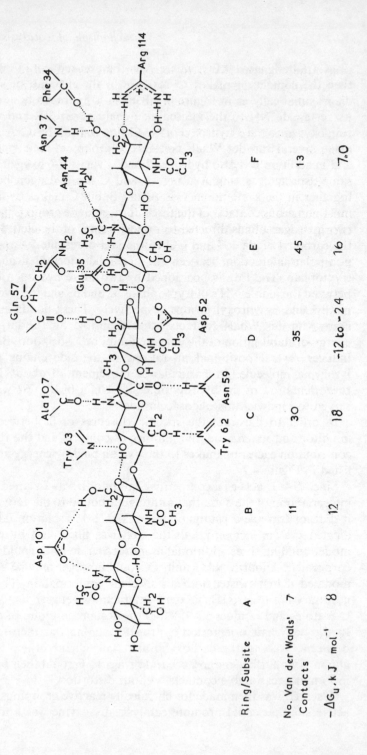

Figure 4.7: Lysozyme-Oligosaccharide Interactions

Ring/Subsite	A	B	C	D	E	F
No. Van der Waals' Contacts	7	11	30	35	45	13
$-\Delta G_u$, kJ mol^{-1}	8	12	18	-12 to -24	16	7.0

cleave the hexamer NAG_6 to tetramer plus dimer and to catalyse the hydrolysis of the glycosidic bond in its natural substrates adjacent to the NAM units. Consequently, the search was focussed on the region between sub-sites D and E. The most likely candidates were the asp 52 and glu 35 side chains previously implicated by chemical modification. The carboxylic acid side chain of glu 35 is 3 Å from the glycosidic oxygen of NAG-D and the acidic arm of asp 52 is slightly less than 3 Å from both C-1 and O-5 of the same monosaccharide. The environments around the two residues are different: the former is in a hydrophobic box which would have the effect of raising its pK_a, whereas asp 52 is in a polar environment enclosed within a hydrogen bonded network which would tend to depress its ionisation. Thus at pH 5, the optimum for lysozyme, asp 52 would be ionised and glu 35 most probably protonated.

'How do these carboxylic acids, singularly or conjunctively, promote bacterial cell wall lysis?' and 'What are the reaction energetics?' Lysozyme catalysed hydrolysis follows Michaelis-Menten kinetics. But because the calculated Michaelis parameters also contain contributions from non-productive binding constants, their mechanistic interpretation is difficult. Non-productive binding does however stimulate transglycosylation (eqs. 4.31, 4.32) and studies of this property have been of use in the analysis of its mechanism of action.

$$NAG_i \cdot NAG_j \; \rightleftharpoons \; NAG_i + NAG_j \qquad (4.31)$$

$$NAG_i \cdot NAG_j + NAG_i \; \rightleftharpoons \; NAG_i \cdot NAG_j \cdot NAG_i \qquad (4.32)$$
$$\text{etc.}$$

During transglycosylation oligosaccharides longer than the starting material are formed, thus the reaction is not a straightforward reversal of hydrolysis. This conclusion is supported by the observed transfer of NAG units to a wide variety of carbohydrate and alcohol acceptors by lysozyme. For example, the enzyme catalyses NAG transfer from NAG_2 to methanol to yield methyl-2-acetamido, 2-deoxy-β-D-glucopyranoside (β-methyl NAG). The rates of transglycosylation to the acceptors are independent of their pK_as, steric requirements or nucleophilicities, and the reactions always proceed with retention of the β-configuration at the anomeric carbon atom.

Consistent with these observations three possible mechanisms can be considered for the action of lysozyme[16] (Figure 4.8):

Figure 4.8: Postulated Mechanisms for Lysozyme Catalysis

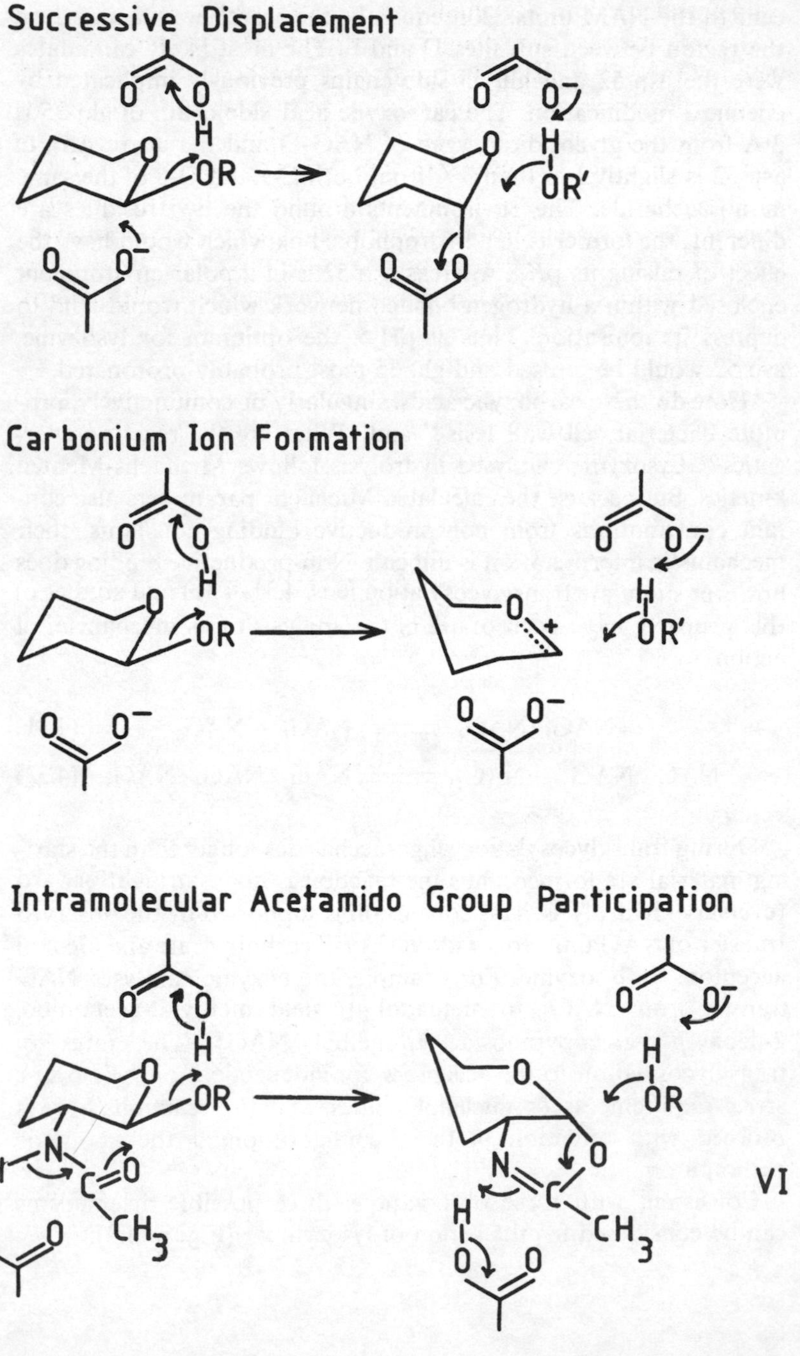

Successive Displacement

Carbonium Ion Formation

Intramolecular Acetamido Group Participation

VI

(a) Successive displacement.
(b) Intermediate carbonium ion formation.
(c) Intramolecular acetamido group participation.

All three consider glutamate 35, which resides in the more apolar region of the active cleft, to be protonated and therefore to be acting as a general acid catalyst. Support for a suitably orientated carboxylic acid assisting glycoside hydrolysis comes from studies on many models such as (III). However the rate accelerations found are still extremely poor compared to lysozyme catalysed hydrolyses. Thus other factors such as substrate distortion, polyfunctional catalysis or transition state stabilisation must also be involved.

Aspartate 52 is situated in the more polar side of the cleft and so would be ionised at the pH of maximum lysozyme activity, and the three mechanisms differ in the role they attribute to this group. Mechanism (a) proposes nucleophilic attack directly at the anomeric carbon atom to form a covalent α-D glycosyl-enzyme. In the second mechanism (b), the anion is envisaged to electrostatically stabilise a positively charged carboxonium ion intermediate, whereas in mechanism (c) the group is proposed to act as a general base, promoting intramolecular nucleophilic attack by the acetamido carbonyl group on the glycosidic carbon atom.

Mechanism (a) accurately predicts (as do the others) that the reaction product would leave with the required stereochemical configuration. However steric constraints in the enzyme sub-site, calculated from the crystal models, limit the movement of the carboxylate side chain to less than the 1.5 Å needed to form a covalent attachment. Mechanism (b) involving carboxonium ion formation is analogous to that accepted for acid catalysed glycoside hydrolysis. The positively charged ion would adopt a half chair conformation, precisely that needed to fit into sub-site D. On these accounts, this mechanism is an attractive possibility particularly as asp 52 is at a suitable distance to electrostatically stabilise the intermediate. However the carboxylate anion is also suitably positioned to remove a proton from the acetamido —NH— and consequently assist cleavage of the glycosidic bond, by enhancing the nucleophilicity of the carbonyl oxygen. The intermediate isoxazole (VI) would also constrain the glycopyranose ring into a flattened half chair conformation. Structures similar to (VI) have been detected in the methanolysis of 2-acetamido,2-deoxy-β-D-glycosides and attributed to breakdown with retention of the β-configuration.[5] Intramolecular catalysis of glycosidic bond cleavage by the 2-acetamido group has also been demonstrated for glycosides containing poor leaving groups, which glycosidic aglycones are.[5] However the assistance is not large in enzymatic terms and formation of an isoxazole ring in the site would, from models, necessitate weakening the association at sub-sites E and F.

Even though lysozyme, in its catalysed hydrolysis of NAG oligo-saccharides, exhibits several characteristics in common with acid catalysed glycoside hydrolyses the energetics of reaction are very different. Its enthalpy of activation, approximately 64 kJ mol^{-1}, is about 40 kJ mol^{-1} smaller. Some idea of the thermodynamic contribution to the overall reaction can be obtained from consideration of the numerical values for the free energy of association to the individual sub-sites A–F (Figure 4.7). The values are in broad agreement with those estimated from the crystal models, except for sub-site D. The binding energy of NAG to this site is unfavourable, but in the cleft as a whole, could be compensated by the favourable free energy of binding to adjacent sites. If in fact steric restrictions at the sub-site are partially relieved by oxidation of the NAG ring to the lactone (which has sp^2 hybridisation at C-1) or by its replacement with N-acetylxylosamine (NAX) (which lacks the hydroxymethyl group of NAG) the corresponding oligosaccharides are calculated to bind around 10^2–10^3 fold, or 10–16 kJ mol^{-1} more effectively than the native oligosaccharides (see Section 3.3.2.3). The total impact at sub-site D is then 20–28 kJ mol^{-1} if the energy needed to convert the chair into the half chair conformation is estimated at 12 kJ mol^{-1}. This amount of energy could be utilised by the enzyme to promote the rate enhancement. In agreement with this estimation, (NAG·NAM)$_2$ which is forced to bind to lysozyme with its reducing terminal NAM in sub-site D, does so with a dissociation constant 130 fold higher than NAG·NAM·NAG which fills the first three sites. And the activation energy for the poor substrate NAG$_2$·NAX·3,5-dinitrophenyl glycoside is approximately 18 kJ mol^{-1} higher than the (NAG)$_3$·3,5-dinitrophenyl glycoside even though it binds with greater affinity.[17] One tentative inference that can be made therefore is that the enzyme translates the association energies between the monosaccharide rings and the binding cleft sub-sites into a lower activation energy for catalysis. By distorting the monosaccharide ring at site D into a conformation closer to that of the transition state of reaction, the expenditure of energy would be reduced and the rate increased.[17]

4.3.2   Chymotrypsin (EC 3.4.4.5)

Pancreatic α-chymotrypsin is one of the most thoroughly investigated enzymes to date. Its amino acid sequence, three-dimensional structure, kinetic properties, chemical modification and mechanism of action have all been subjected to the most intensive scrutiny and the investigations occupy many volumes in the scientific literature.

The enzyme is one of many hydrolases produced by activation of inactive zymogens secreted into the pancreatic juice. Three chymotrypsinogens have been found, all of which give rise to chymotrypsins of similar catalytic specificities. The one investigated in most detail, α-chymotrypsin, is produced from chymotrypsinogen A by tripartite cleavage of its single chain as described in Chapter 3.

The chymotrypsins are proteolytic enzymes catalysing the hydrolysis of peptide bonds at the carbonyl side of the hydrophobic residues tryptophan, phenylalanine, tyrosine and leucine; peptide bonds adjacent to other side chains are attacked on prolonged exposure. The enzymes also possess esterolytic activity, especially towards esters of the above amino acids, studies of which have proved valuable in exploring its mechanism. The pancreatic chymotrypsins are also maximally active at alkaline pH.

Determination of the primary structure of α-chymotrypsin was impeded by its length, almost twice that of previously sequenced chains and by its autolytic behaviour, but its amino acid order was finally reported in 1966[18] (Figure 1.3) and its availability considerably aided interpretation of the 2 Å resolution maps of tosyl-chymotrypsin* (Figure 1.6).[19] The bulky toluene-sulphonyl group was located in a recognisable hydrophobic compartment, the 'tosyl hole', comprised of residues ser 189, ser 190, cys 191, val 213, gly 216, cys 220, gly 226 and tyr 228, with the sulphonyl group attached to ser 195.**

---

* Prepared from α-chymotrypsin and *p*-toluenesulphonyl chloride (tosyl chloride), tosyl-chymotrypsin is inactive and hence non-autolytic.

** From Section 1.1.1 it is clear that chymotrypsin exhibits considerable primary and tertiary homology with the other pancreatic proteases. Homology also extends to the substrate binding pockets of chymotrypsin, trypsin and elastase where almost identical organisations of amino acid side chains are found. The only differences are that in trypsin aspartate replaces ser 189, thus rationalising its specificity for basic side chains in the substrate, and in elastase the bulkier val and thr are at positions 216 and 226 respectively which explains the preference of this enzyme for substrates containing alanyl residues.

Chymotrypsin catalysed peptide hydrolyses follow a Michaelis-Menten pattern, but it was observed that the catalysed hydrolysis of the ester p-nitrophenylacetate did not follow the expected progress curve. An initial 'burst' in product formation was followed by a much slower rate (Figure 2.11). This observation was interpreted as a three step mechanism in which initial Michaelis complex formation was followed by rapid release of one product to give a modified form of the enzyme which then turned over at a slower rate to regenerate the native enzyme and release the second product (see Section 2.3.1.2).

$$E + S \; \rightleftharpoons \; ES \; \xrightarrow{k_2} \; \underset{\substack{+ \\ P_1}}{E'} \; \xrightarrow{k_3} \; E + P_2 \qquad (4.33)$$

In acidic media the reaction steps, and particularly that characterised by rate constant k_3, are slow. Oosterbaan and Hartley *et al.*[20] were able to show, by pepsin* degradation, that the substrate p-nitrophenylacetate acetylated serine 195. The reaction catalysed by chymotrypsin can then be written schematically,

$$E{-}OH + R{\cdot}CO{\cdot}X \; \rightleftharpoons \; E{\overset{\displaystyle ,R{\cdot}CO{\cdot}X}{\underset{\displaystyle OH}{\big<}}} \; \xrightarrow{} \; \begin{array}{l} E{-}O{\cdot}CO{\cdot}R \\ + X^- \end{array} \qquad (4.34)$$

$$\xrightarrow{} E{-}OH + R{\cdot}CO_2H$$

Steady-state analysis yields

$$k_{cat} = k_2 k_3 / (k_2 + k_3) \qquad (4.35)$$

$$K_m = K_s \cdot k_3 / (k_2 + k_3) \qquad (4.36)$$

thus if $\qquad k_2 > k_3$, then $k_{cat} \sim k_3$ and $K_s > K_m \qquad (4.37)$

and if $\qquad k_3 > k_2$, then $k_{cat} \sim k_2$ and $K_m \sim K_s \qquad (4.38)$

Equation 4.37 has been found to describe the behaviour of ester substrates whereas eq. 4.38 is more applicable to the chymotrypsin mediated hydrolysis of amides. Between pH 4 and 8, the pH depen-

---

* A proteolytic enzyme maximally active at acid pH.

dencies of k_2 (acylation) and k_3 (deacylation) are similar. As described in Chapter 3, from k_{cat} may be determined the pK_a values of groups perturbed by substrate binding, and from k_{cat}/K_m those in the free enzyme. With N-acetyl-L-tryptophan amide as substrate, the enzyme exhibits a bell shaped dependence of k_{cat}/K_m on pH with inflexions at pH 7 and 9. Over the same pH range k_{cat} gradually increases, with an inflection at pK_a 6.5, to a constant value independent of pH above 8.0. For the same amide substrate K_m shows the opposite behaviour, remaining constant between pH 6 and 8 but increasing above pH 8 with a mid-point at pK_a 9.5. Acylation and deacylation are therefore dependent on a group with pK_a between 6 and 7, in the range expected for imidazole, and binding of the substrate is a function of the state of dissociation of a group with approximate pK_a 9.0.

From studies on δ-, acetyl- and α-chymotrypsins, Hess *et al.*[21] concluded that the observed pH dependence of K_m for amide substrates was due to an equilibrium between two conformers of the enzyme, controlled by the α-amino group of isoleucine 16, with pK_a 8.5. On protonation the N-terminus was considered to fold into the protein interior and be electrostatically held by the ionised β-carboxylate anion of asp 194. Such an interaction would redirect the backbone in its vicinity and reorientate the next residue, ser 195, into a more reactive conformation.

Crystallographic studies have shown that in the active conformation the hydroxyl oxygen of ser 195 forms a hydrogen bond with a nitrogen atom of his 57. This hydrogen bond is half of a 'charge-relay' system the other hydrogen bond of which joins asp 102, buried in the molecule, to his 57. Asp 102 lies in a hydrophobic locality donated by ala 55, ala 56, cys 58, tyr 94, ile 99 and ser 214. Thus solvent molecules can only approach the conformationally active enzyme from the opposite side of the imidazole. In the active form, in which imidazole is unprotonated, electrons will then be conducted along hydrogen bonds from the interior to the active site. This site is exposed to solvent and so normally would be depressed in reactivity. It has been suggested therefore that the 'charge-relay' system does in fact promote active site reactivity by enhancing the nucleophilicity of the seryl oxygen[22] (Figure 4.9).

A mechanism for chymotrypsin catalysed hydrolyses consistent with most of the kinetic, modification and X-ray crystallographic data,[23] is presented in Figure 4.10. Acylation (VII–IX, Figure 4.10) is shown to occur on ser 195 via a tetrahedral intermediate (VIII,

Figure 4.9: Conformation of a few Amino Acid Side Chains in the Active Site of α-Chymotrypsin (Hydrogen bonds are dashed). The viewpoint is outside the molecule, looking towards the interior

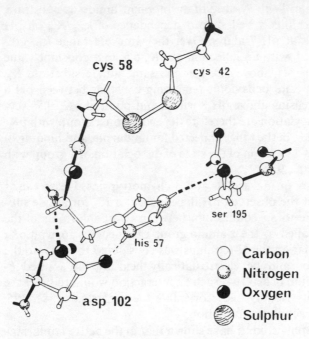

cys 58

cys 42

ser 195

his 57

○ Carbon
◎ Nitrogen
● Oxygen
◉ Sulphur

asp 102

Source: Reproduced from Reference 22 with permission.

Figure 4.10) and deacylation (X–XII, Figure 4.10) to proceed along a pathway the reverse of acylation. The rate limiting process in the hydrolysis of amide substrates is acylation and of the two steps, formation and breakdown of the tetrahedral intermediate, the former is considered to be the slower. Deacylation of the acyl-enzyme by general base catalysis by imidazole is preferred to a nucleophilic mechanism which would give an acyl-imidazole-enzyme intermediate, since the rates of disappearance of acyl-enzymes are independent of the basicity of acyl acceptors added to the medium.

Judged crystallographically the tertiary structures of α-chymotrypsin and its precursor are very similar, and interestingly the 'charge-relay' is formed in both. One major dissimilarity is the orientation of the polypeptide containing the N-terminal ile 16. Its

Figure 4.10: A Possible Mechanism for Chymotrypsin Catalysed Hydrolysis

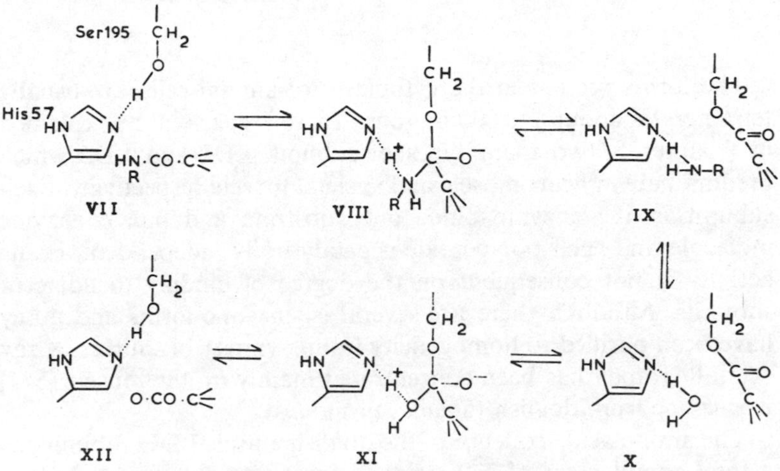

withdrawal into the molecule on activation induces a conformational movement in met 192, amongst others. The thioester side chain then acts as a kind of flexible lid for the specificity pocket,[24] and the most important difference between chymotrypsin and its zymogen appears to be the ability of the former to specifically bind its substrate as a result of this movement. These observations and the fact that the hydrolyses of specific substrates are catalysed with rates at least 10^2 fold faster than non-specific substrates support the theory of Jencks[8] described in Section 4.2 that the binding energy is used to reduce the activation energy of reaction.

4.3.3 L-Lactate Dehydrogenase (EC 1.1.1.27)

Lactate dehydrogenases (LD) catalyse the reaction:

$$CH_3 \cdot CH(OH) \cdot CO_2H + NAD$$

$$= CH_3 \cdot CO \cdot CO_2H + \overset{+}{H} + NADH \quad (4.39)$$

via an ordered mechanism (eq. 4.40), in which the coenzyme first binds to the protein to form a reactive complex for conversion of the substrate,

$$\text{LD} \xrightleftharpoons{\text{NAD}^+} \text{LD}^{\text{NAD}^+} \xrightleftharpoons{\text{lactate}} \text{LD}^{\text{NAD}^+}_{\text{lactate}} \rightleftharpoons \text{LD}^{\text{NADH}}_{\text{pyruvate}} \rightleftharpoons \text{LD}^{\text{NADH}}_{\text{pyruvate}} \rightleftharpoons \text{LD}_{\text{NADH}} \qquad (4.40)$$

The forms present in the cytoplasm of animal cells, are usually tetramers of about 140,000 daltons. There are, as will be explained in Chapter 5, two main types of subunit, LD1 and LD5 which predominate in heart muscle and skeletal muscle respectively. Each subunit in the enzyme binds one substrate and one coenzyme molecule and each polypeptide is catalytically independent, i.e. its activity is not consequent on the degree of binding to adjacent subunits. Although there are several isoenzymic forms and many have been purified to homogeneity from a variety of sources, X-ray crystallography has been concentrated mainly on the muscle (M4) isoenzyme from dogfish (*Squalus acanthius*).[25]

The amino acid sequence of the dogfish muscle LDH subunit has a total complement of 329 residues containing 7 cysteines but no disulphide bridges. Modification of cys 165 by mercurials results in loss of activity, but not coenzyme binding. Other residues found 'essential' to activity are three arginines, modifiable by phenyl-glyoxal and butadione, and a histidine identified at position 195 by affinity labelling with the substrate analogue bromopyruvate and the partial coenzyme analogue 3-bromoacetylpyridine.

The relative positions of these residues in the tertiary conformation were established when the complete three-dimensional structure was reported.[27] A schematic drawing of the dogfish muscle subunit is shown in Figure 4.11. Extensive secondary structure is visible with approximately 40 percent of the residues involved in α-helix and about 23 percent in β-pleated sheet formation. The polypeptide chain can be divided at the 'essential' cys 165 into C-terminal and N-terminal halves. The first, the catalytic domain, is comprised of three pairs of antiparallel sheets. The second, located more in the interior of the molecule is the coenzyme binding domain. This has been discussed in Section 1.1.2. The interactions NAD makes with groups in the enzyme are shown schematically in Figure 4.12. The adenine fits into a fairly non-specific hydrophobic pocket with its N-1 atom hydrogen bonded to a tyrosine phenol. Kinetic evidence indicates that this nucleotide binds first to the subunit and induces a change in its conformation which enables the

Figure 4.11: The Tertiary Structure of the Lactate Dehydrogenase Subunit

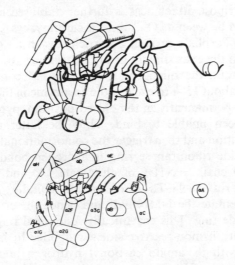

Source: Reproduced from Reference 25b with permission.

Figure 4.12: Diagrammatic Representation of Coenzyme Binding as Appears in the LDH : NAD-Pyruvate Complex

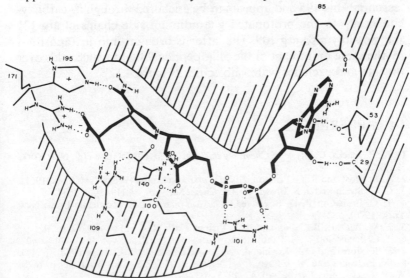

Source: Reproduced from Reference 25a with permission.

rest of the coenzyme molecule to bind. Aspartate-53 moves approximately 2 Å to form a hydrogen bond to O'-2 of the ribofuranose ring of AMP; ribose attachment is further stabilised by hydrogen bond formation between its O'-3 and the carbonyl cys 29. The nicotinamide pyrophosphate grouping participates in an electrostatic interaction with the guanidinium group of arginine 101. Compared to the aposubunit, the oligopeptide loop 98-114 containing this residue moves about 13 Å to enclose the coenzyme in the active site. If arg 101 was permanently in the closed position, pyrophosphate would have been unable to bind. Thus the latter must attain the correct position and then trigger the conformational movement. The nicotinamide ribofuranose ring is hydrogen bonded between O'-2 and main chain $-NH-$ of glutamate 140 and via O'-3 to the carbonyl at residue 98. The whole gamut of binding interactions appears to orientate the NMN half and hence the critical C-4 of the nicotinamide ring. This enzymically functional ring fits into a deep pocket with its non-reactive side surrounded by hydrophobic residues and with its amido carbonyl hydrogen bonded to the ε-amino group of lys 250. The positive charge on the nicotinamide ring nitrogen forms a polar bond with the anionic side chain of glu 140. In contrast its reactive face is exposed to a hydrophilic environment. The substrate, pyruvate, lies between this face and the 'essential' his 195 and appears to be anchored through its carboxylate group to the protonated guanidinium side chains of arg 171 and possibly also arg 109. The latter is brought into juxtaposition by the same movement of the oligopeptide 98-114 which folds over the site and excludes water molecules (Figure 4.12).

4.4   REFERENCES

1. W. P. Jencks, *Catalysis in Chemistry and Enzymology* (McGraw-Hill, New York, 1969).
2. A. A. Frost and R. G. Pearson, *Kinetics and Mechanism* (Wiley, New York, 1961).
3. C. G. Swain and J. F. Brown, *J. Am. Chem. Soc.*, vol. 74 (1952), p. 2534.
4. T. C. Bruice, in P. D. Boyer (ed.) *The Enzymes*, vol. II (Academic Press, New York, 1970), p. 217.
5. F. W. Barchie, B. Capon and R. L. Foster, *Carb. Res.*, vol. 49 (1976), p. 79.
6. T. C. Bruice and U. K. Pandit, *Proc. Natl Acad. Sci. USA*, vol. 46 (1960), p. 402.
7. D. R. Storm and D. E. Koshland, *Proc. Natl Acad. Sci. USA*, vol. 66 (1970), p. 445.
8. W. P. Jencks and M. I. Page, *Proc. Natl Acad. Sci. USA*, vol. 68 (1971), p. 1678; W. P. Jencks, *Adv. Enzym.*, vol. 43 (1975), p. 219.
9. A. R. Fersht, *Proc. Roy. Soc. Lond.*, vol. B187 (1974), p. 397; *Enzyme Structure and Mechanism* (Freeman, Reading, 1977).

10. K. Inouye and J. S. Fruton, *Biochem.*, vol. 6 (1967), p. 1765.
11. M. Levitt and A. Warshal, *J. Mol. Biol.*, vol. 103 (1976), p. 227.
12. F. H. Westheimer, *J. Am. Chem. Soc.*, vol. 83 (1961), p. 1102.
13. D. E. Koshland, *Adv. Enzym.*, vol. 22 (1960), p. 45; *Ann. Rev. Biochem.*, vol. 37 (1968), p. 359.
14. T. Imoto, L. N. Johnson, A. C. T. North, D. C. Phillips and J. Rupley, in P. D. Boyer (ed.) *The Enzymes*, vol. 7 (Academic Press, New York, 1972), p. 712.
15. D. Hoare and D. E. Koshland, *J. Biol. Chem.*, vol. 242 (1967), p. 2447.
16. G. Lowe, *Proc. Roy. Soc. Lond.*, vol. B167 (1967), p. 43.
17. B. Capon and W. M. Dearie, *J. Chem. Soc. D* (1974), p. 370.
18. D. M. Blow in P. D. Boyer (ed.), *The Enzymes*, vol. 3 (Academic Press, New York, 1971), p. 185.
19. D. M. Blow, *Biochem J.*, vol. 112 (1969), p. 261.
20. B. S. Hartley and B. A. Kilbey, *Biochem. J.*, vol. 56 (1954), p. 288; R. A. Oosterbaan and M. E. van Adricken, *Biochem. Biophys. Acta*, vol. 27 (1958), p. 423.
21. A. Himoe, P. C. Parks and G. P. Hess, *J. Biol. Chem.*, vol. 242 (1967), pp. 919, 3963, 3973; *J. Biol. Chem.*, vol. 246 (1971), p. 2918.
22. D. M. Blow, J. J. Birktoft and B. S. Hartley, *Nature*, vol. 221 (1969), p. 337.
23. H. T. Wright, *J. Mol. Biol.*, vol. 79 (1973), p. 13.
24. T. A. Steitz, R. Henderson and D. M. Blow, *J. Mol. Biol.*, vol. 40 (1969), p. 337.
25a. M. J. Adams *et al.*, *Proc. Natl Acad. Sci. USA*, vol. 70 (1973), p. 1793.
25b. J. J. Holbrook, A. Liljas, S. J. Steindel and M. G. Rossman, *The Enzymes*, vol. 11 (1975), p. 191.

5 Enzyme Physiology

5.1 ENZYME ORGANISATION

All enzyme molecules are synthesised within living cells, the range of activities depending on the specialisation of the cell and its stage of development. Some of the enzymes act intracellularly, that is, within the cell in which they are produced, the urea cycle enzymes of hepatocytes for example, while others are secreted by the cell to perform extracellular functions. Cells in the pancreas, for instance, secrete hydrolytic enzymes as inactive precursors which are later activated in the extracellular, digestive juices.

The enzyme complement of a cell is, of course, defined initially by its DNA content and by the transcriptional and translational regulatory factors present. But the activities of these enzymes are subject to further controlling processes. Some of these have been described in previous chapters, for example physical organisation into multienzyme complexes and allosteric modulations were discussed in Chapters 1 and 4 respectively. In this chapter the compartmentation and organisation of enzymes into metabolic pathways and their roles underlying cellular function are discussed. But because the various controlling devices cannot be considered in isolation from each other, specific examples illustrating the integrated nature of metabolic regulation will also be considered.

Two experimental techniques are at hand for the quantitative investigation of the enzyme activities of tissues and cells—histochemical methods and differential fractionation.

5.1.1 Enzyme Histochemistry

A small piece of the tissue of interest is rapidly frozen at about $-100°C$, and from it, small slices are cut by means of a 'cryostat'. These slivers are eventually microscopically analysed for enzyme activity. Before this examination is possible however the cellular and intracellular morphology of the sections must be stabilised by chemical fixation. Two fixatives are in common use, gluteraldehyde $(CHO \cdot (CH_2)_3 \cdot CHO)$ and formalin (formaldehyde solution). Both react predominantly with amino groups and form crosslinked networks which permeate the labile tissue structure. Other advantages are that neither causes extensive enzyme inactivation nor molecular translocation.

The fixed section is then mounted on a microscope coverslip and treated with a buffered solution containing the substrate of the enzyme to be examined. Substrates are chosen that yield readily visualisable products. One potential problem is diffusion of the product away from its site of formation. To avoid this, the 'capture' technique is employed, whereby the product is insolubilised immediately after production. The Gomori procedure for alkaline phosphatase location, for example, traps the phosphate released from β-glycerophosphate as insoluble calcium phosphate by precipitation with calcium ions. Cation exchange with cobalt nitrate followed by anion exchange with ammonium sulphide, then reveals the alkaline phosphatase activity as black spots.

The main advantage of enzyme histochemistry is the facility to examine one cell type at a time, and at their best histological techniques can enable precise location of these activities. For example the alkaline phosphatase activity of the kidney is revealed along the brush border of the proximal convoluted tubules.

Currently the many histochemical reactions for enzyme localisation are being adapted for use with the electron microscope.[1] Electron microscopy has revealed a wealth of structural detail invisible to the light or phase contrast microscope and this technique is becoming of foremost importance for unravelling the complexities of the internal membrane system. This technique is also beginning to yield valuable information on the gross morphology of the larger enzyme complexes such as pyruvate carboxylase.

5.1.2   Subcellular Fractionation[2]

Microscopic techniques permit direct observations to be made of intracellular structures but for quantitative biochemical investigations the various organelles must be isolated and obtained in relatively large, pure quantities. To make this possible, the cell wall must first be broken and the intracellular contents released, a process termed *homogenisation*. The homogenate must then be fractionated into its organellular components. The most convenient and reproduceable technique by which this may be achieved is differential centrifugation.

5.1.2.1 Homogenisation

The objective of the homogenisation process is to rupture the cell membrane and release the organelles, but without destroying their integrity. Several methods have been devised to achieve this difficult balance. Sonification, osmotic shock, high pressure and mechanical blenders are amongst the most common.

Sonicators employing high frequency sound waves have been widely applied in the disruption of bacterial cells, but the ultrasonic vibrations cause extensive damage to the membranes. From electron micrographs they appear to have been punched. This method is therefore of little value when preparations containing membrane-attached ribosomes are required. By comparison, osmotic shock, often applied to rupture reticulocyte and bacterial membranes is relatively mild. After equilibration in isotonic solution, the cells are rapidly removed into a hypotonic medium. This causes the cell walls to burst and the formation of spheroblasts. Any detrimental effect on the organelles is minimised by immediately reinstating isotonic conditions. This procedure is particularly useful for releasing periplasmic enzymes such as the anti-leukemic form of L-asparaginase from *E. coli*.

A method reported to protect proteins during cell rupture is high pressure homogenisation. When high pressures, around 70 bar, of an inert gas are applied to a cell suspension, on attainment of equilibrium, large amounts of the gas will dissolve in the medium and the cells, and the internal pressure of the gas forced into the cells will equal that externally applied. On sudden release of the external pressure, the internal pressure will momentarily remain high but cavitation by explosion will quickly follow causing rupture of the cell membranes.

Two methods of wider applicability are the Waring blender and the Potter-Elvehjam homogeniser. The first is more suitable for the homogenisation of hard, fibrous tissues such as heart whereas the second is usually preferred for the softer tissue such as liver. The Waring blender is the more forceful, akin to the domestic kitchen blender, in which the tissue is subjected to a set of sharp, rapidly rotating blades. Consequently a higher proportion of the intracellular organelles are disrupted.

The Potter homogeniser consists of an outer glass tube into which is fitted a ground glass or teflon plunger, with a gap between the two surfaces of between 0.25 and 0.33 mm. The vertical, piston-like action of the plunger subjects the cell suspension between the surfaces to shearing forces, breaking open the cell walls.

The use of these various methods involves a compromise between the yield of intact organelles and disrupted organelles and cells. Therefore before experiments are carried out using the final preparations, their morphological integrities are microscopically assessed.

5.1.2.2   Differential Centrifugation

After release from their cells, the organelles are separated by programmes of centrifugation in which each organelle is sedimented by a combination of time and centrifugal force in a preparative centrifuge. Centrifugation is a convenient means of applying an increased gravitational field, G, to a particle. The size of this force is expressed in terms of 'g' the earth's gravitational constant and that experienced by a revolving particle is dependent on its distance, r cm., from the axis of rotation and its angular velocity, ω radians per second:

$$G = \omega^2 \cdot r = (\pi \cdot \text{RPM})^2 \cdot r/900 \qquad (5.1)$$
$$\text{(where RPM is revolutions per minute)}$$

The rate of sedimentation of a particle in this applied gravitational field depends on its size, shape and density and since each cellular organelle is unique in at least one of these properties it can in principle be fractionated from the homogenate. The rate at which it sediments depends also on the viscosity and density of the supporting medium. For spherical particles, the relationship between these variables is:

$$t \propto \frac{\eta \cdot \ln\,(r_t/r_b)}{\omega^2 \cdot r_p^2 (p_p - p)} \qquad (5.2)$$

where t is the time taken for a particle, radius r_p and density p_p, to sediment from the top to the bottom of a centrifuge tube in a medium of viscosity η poise and density p. The radial distances from the rotor axis to the top and bottom of the centrifuge tube are respectively r_t and r_b.

The constant quantities can be combined into a single expression and given one symbol, s, thus

$$s \propto r^2 \cdot (p_p - p)/\eta \qquad (5.3)$$

therefore $\qquad\qquad\qquad t \propto s^{-1} \qquad\qquad\qquad\qquad (5.4)$

Table 5.1: Typical Sedimentation Coefficients of Cellular Constituents

Particle	Sedimentation coefficient, Svedbergs
Proteins	2–20
Nucleic Acids	2–100
Ribosomes	30–80
Microsomes	100–15,000
Lysosomes	$1–2 \times 10^4$
Mitochondria	$2–10 \times 10^4$
Nuclei	$5–10 \times 10^6$

s is termed the sedimentation coefficient, with units of s^{-1}. Since its values are therefore rather large a more practical unit, of numerical value 10^{-13} s, is used. This is termed the Svedberg, S, after the Swedish pioneer of ultracentrifugation.

Some typical values of this constant for the cellular organelles are given in Table 5.1, where it is seen that the experimentally determined values correspond closely to their relative particle sizes.*

By centrifugation for different time spans in suitably applied gravitational fields, it should be possible to fractionate the homogenate. The protocol generally adopted for this differential centrifugation is illustrated in Figure 5.1. Fresh rat livers are often the organ of choice, being of a suitable size for easy manipulation but large enough to provide sufficient material. On centrifugation, the heavy nuclei are the first to sediment in a pellet containing the cell wall debris as well. This is followed by first the mitochondrial and then the microsomal fractions, leaving a supernatant containing the cytoplasmic enzymes and free ribosomes. After each sedimentation the pellet is washed by resuspension and recentrifugation, and the supernatants are carried forward.

Even after washing, the fractions obtained in this way are not pure and contain more than one organelle. The mitochondrial fraction, for example, is comprised of mitochondria, lysosomes and

* For proteins the molecular weight can be estimated approximately as $MW = 6500 \, (S_{20,w})^{1.5}$ where $S_{20,w}$ is the sedimentation coefficient in water at 20°C.

Figure 5.1: Differential Fractionation of Rat Liver Cells

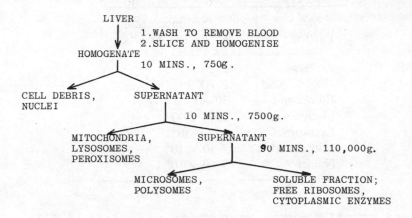

peroxisomes. For this reason, centrifugation procedures affording more precise resolution of the components have been developed. One of these is density gradient centrifugation. Instead of carrying out the sedimentation in a uniformly dense solution, the homogenate is layered on a concentration gradient and as the particles travel down the tube they experience gradually increasing densities. The increase may be step-wise, also called *discontinuous*, arranged by carefully layering solutions of decreasing density one-on-another in the centrifuge tube, or *continuous*, made by a gradient-maker.

Two types of density gradient centrifugation are then possible, zone centrifugation and equilibrium density (or isopycnic) centrifugation. Zone centrifugation employs a shallow gradient of densities smaller than those of the particles to be separated. The purpose of this gradient is to reduce convective mixing of the bands or zones. Alternatively, the gradient may encompass all the individual densities in the homogenate so that on centrifugation each component will come to rest at its buoyant equilibrium density. For both types sucrose solutions between 0 and 2.7 molar have been used, but because of the changes in osmolality involved, ficoll, a sucrose polymer of negligible tonicity is now preferred.

After isolation, the organelles must be assessed for purity and the degree of cross contamination. Effort has therefore been directed towards the identification of 'marker' substances, molecules which can be regarded as localised exclusively in one intracellular compartment. In the main, the histochemical and biochemical search has centred on enzymes because of their ready and sensitive measurement. The establishment of such indicators is far from easy however. Is the measured enzymic activity an impurity or a minor but essential component of the organelle? Nevertheless some enzymes judged from extensive experimentation to be indicators of certain metabolic pathways and organelles are listed in Table 5.2.

Table 5.2: Marker Enzymes and Biochemical Functions of some Cellular Organelles

Fraction	Marker	Biochemical function
Nucleus	NMN adenylyl transferase	DNA, RNA, NAD biosynthesis
Mitochondria	Succinate DH	Electron transfer, oxidative phosphorylation, urea cycle, citric acid cycle, fatty acid oxidation, haem biosynthesis
Lysosomes	Acid phosphatase	Hydrolysis of cellular constituents
'Microsomes' (ribosomes, polysomes, endoplasmic reticulum)	Glucose-6-phosphatase, NADPH-cytochrome reductase	Protein synthesis, drug detoxification, mucopolysaccharide, glucuronide, cholesterol, phospholipid biosynthesis
Soluble fraction	Lactate DH	Amino acid activation, glycolysis, gluconeogenesis, pentose phosphate pathway, fatty acid biosynthesis

A problem encountered on interpretation of the observed enzyme activities of the fractions is their possible failure to reflect the situation in the original cell. From the presence of a specific activity in a fraction it could be inferred that that enzyme is an integral part of its function, but equally it may have become detached during homogenisation and centrifugation and relocated in that fraction. Alternatively the enzyme may be present but not measurable due to the inability of exogeneously provided substrates to permeate the organellular membrane. Such 'latent' or 'cryptic' enzymes can only be assayed after the membrane is disrupted. The mitochondrial envelope is impermeable to NADPH, thus malate dehydrogenase activity is latent, and similarly most of the lysosomal enzymes can only be assayed after lysis of the organelle. Exposure of the enzymes may however considerably change their exhibited kinetic properties.

Nonetheless much understanding of the mechanisms underlying the functions of eukaryotic organelles has come from studies of the enzyme activities of the isolated particles (Table 5.2).

5.2 ENZYME LOCALISATION

Analyses of the fractions prepared by differential centrifugation indicate that some enzymes appear to be associated mainly with one fraction while others are recovered solely in the soluble fraction. This localisation of enzyme activities argues for a relationship between cellular structure and biochemical function and for control of these functions by spatial organisations of the enzyme assemblies in addition to the regulatory mechanisms outlined in Chapter 3.

Several interesting aspects emerge from a study of these enzyme localisations. Where a degradative pathway and reciprocal biosynthetic pathway involve different sets of enzymes, the sequences are usually found in different organelles. For example, fatty acid oxidation, an energy yielding process occurs in the mitochondrion but the enzymes catalysing the reverse reaction, fatty acid biosynthesis remain in the soluble fraction. Obviously, directionally opposed pathways catalysed by the same enzymes (such as glycolysis and gluconeogenesis) can only take place in the same organelle, but in these cases the rate limiting steps and thus the directional fluxes are controlled by different enzymes, separately controlled and located. The immediate breakdown and synthesis of glycogen are catalysed by two different enzymes, glycogen phosphorylase and glycogen synthetase respectively (see Section 5.2.6.2). In the glycolytic direction the rate limiting steps are those catalysed by hexokinase, phosphofructokinase and pyruvate kinase (Table 5.10), whereas for gluconeogenesis, glucose-6-phosphatase, fructose-1,6-bisphosphatase and the enzymes catalysing the phosphorylation of pyruvate are rate controlling (Table 5.10).

Enzymes catalysing the same reactions in different organelles are often non-identical. For example, aspartate aminotransferase activity occurs in both the cytoplasm and the mitochondria, but the two enzymes are different, being coded for by two different genes. Similarly cytoplasmic isocitrate dehydrogenase has a specific requirement for NADP, whereas in the mitochondria, two enzymes are found, one binding NAD, the other NADP.

5.2.1 Membrane Enzymes[3]

The plasma and intracellular membranes are extremely important to the organisation of living cells. They act as compartmental boundaries, allowing restricted access of ions and molecules to particular localities and they are important levels of protein organisation.

The plasma membrane is the semi-permeable matrix separating the internal environment of the cell from external influences. Organised within this limiting membrane are the cell components, the nucleus, mitochondria, lysosomes, golgi apparati etc., all of which are also enclosed within membraneous structures. The endoplasmic reticulum is a membrane system extending throughout the cytoplasmic space. Membranes serve to maintain the dynamic balance of the cell or organelle against fluctuations occurring in its outer environment but far from acting as inert barriers, their function is to allow the interior to respond to changing conditions in its vicinity. The membranes must therefore also be functional and dynamic systems.

Common to all proposed models of membrane structure is a phospholipid bilayer in which proteins are dispersed. Phospholipids (eq. 5.16) are comprised of polar 'heads' plus hydrophobic 'tails', and in the membrane they are thought to aggregate into structures with 'oil-like' interiors and polar surfaces (I). Protein accounts for between 50 to 80 percent by weight of the membrane mass. Those proteins near its surface are classed as 'peripheral' (P) or 'extrinsic' and are similar in properties to the cytoplasmic proteins, which means they can be separated fairly readily; on the other hand, those embedded in its interior, the 'integral' or 'intrinsic' (I) proteins, are less easily isolated and need more drastic methods such as detergent or solvent extraction. The membrane proteins undoubtedly have extensive hydrophobic domains in order to correspond with the membrane, but the degree of their translational mobilities in the membrane and the extent to which they direct its architectural form have yet to be ascertained.

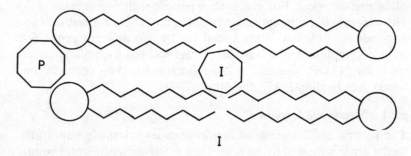

The proteins perform a variety of different functions crucial to intercellular and intercompartmental integration. Some are responsible for the transport of ions such as Na^+ and K^+, while others have been identified as hormone receptors. Many enzymes of metabolism have been located bound to membranes and the mitochondrial enzyme-membrane complexes are crucial to energy conservation. The enzyme topology of many of these membranes will be discussed in the sections dealing with the individual organelles. Here two plasma membrane enzyme systems will be discussed, the sodium ion, potassium ion dependent ATPase system (Na^+, K^+-ATPase) and the adenylate cyclase system.

Na^+, K^+-ATPase maintains the low sodium ion and high potassium ion concentrations inside cells. Sodium ions are cotransported into cells along with glucose, amino acids and dibasic acids, by facilitated diffusion. Since the continual diffusion of these metabolites into the cell is essential to its survival and because protein synthesis (for growth and replication) depends on high potassium ion levels, these ions must be pumped in and the sodium ions pumped out.

A Na^+, K^+-ATPase particle of MW 2×10^5 daltons has been isolated by rigorous extraction from the plasma membrane and found to consist of a polypeptide of MW 9×10^4 and a smaller glycoprotein. The stoichiometry of the particle is probably A_2B_2 although a role for the smaller subunit has yet to be found. The driving force for the ion translocation is provided by ATP. An aspartyl side chain in the larger subunit is phosphorylated by ATP

Figure 5.2: Membrane Bound Na$^+$, K$^+$-*ATPase*

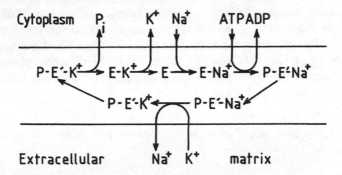

in a Na$^+$ ion dependent reaction and is dephosphorylated in a K$^+$ ion dependent reaction. The presently accepted sequence of steps by which the enzyme pump regulates the ion balance is shown in Figure 5.2. The enzyme is thought to oscillate between two conformers, one which is readily phosphorylated and has a higher affinity for sodium than for potassium ions and a second with a higher affinity for K$^+$ in which dephosphorylation is facilitated.

Cyclic 3',5'-adenosine monophosphate (cAMP) was identified as the heat stable factor which transmitted the extracellular signal brought by the hormones adrenaline (epinephrine) and glucagon to the enzymes promoting glycogen catabolism (Section 5.2.6.2). This 'second messenger' (the hormone is the first messenger) is produced from ATP by the plasma membrane bound adenylate cyclase (eq. 5.5).

$$\text{ATP} \xrightarrow{\text{Adenyl Cyclase}} \text{PP}_i + \text{cAMP} \qquad (5.5)$$

Although much has been revealed concerning the physiological roles of cAMP very little is known of the structure and properties of the mammalian adenylate cyclases producing it. Principally this is because they are at very low concentrations in cells and are very tightly held integral protein complexes. From the available information, there seem to be at least two interacting subunits, a receptor protein on the extracellular side and a catalytic unit on the cellular side of the membrane. The receptor is thought to undergo a conformational change on hormone binding which is transmitted, perhaps allosterically, to the catalytic unit triggering the dephosphorylation of an inactive phosphorylated form of adenylate cyclase; this is shown graphically in Figure 5.3.

Figure 5.3: Schematic Picture of the Membrane Bound Adenylate Cyclase (AC) *Complex*

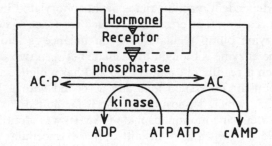

5.2.2 Nuclear Enzymes

The enzymes present in the nucleus and nucleolus are concerned predominantly with the biosynthesis of DNA and its transcription to RNA. These are anabolic reactions, whose energy requirements appear to be fulfilled by the citric acid cycle and glycolytic enzymes, also present in the the nucleus (Table 5.3).

The enzymes of DNA synthesis and replication, DNA polymerases and RNA polymerases respectively, are widely distributed in both eukaryotic and prokaryotic cells. In animal tissues, the DNA polymerases are localised mainly in the nucleus but significant activities are also found in the mitochondria and cytoplasm, although their functions are as yet unclear. No DNA polymerase activity has been observed in the nucleolus, which appears to be the main site of ribosomal but not of messenger or transfer RNA synthesis. These are produced in the nucleus. The precise mode of action of the various polymerases in mammalian cells has yet to be fully elu-

Table 5.3: Nucleoplasm Enzymes

Nucleus:	DNA polymerases, RNA polymerases for mRNA and tRNA, glycolytic pathway enzymes, citric acid cycle enzymes, pentose phosphate pathway enzymes
Nucleolus:	RNA polymerases for ribosomal RNA, NADP pyrophosphorylase (NMN adenylyl transferase), ribonucleases, ATPases, S-adenosylmethionine-RNA transferases

cidated and much more is known at present concerning their roles in *E. coli*.

The DNA polymerase I molecule, discovered by A. Kornberg, exhibits both exonuclease and polymerase activities, which contribute in that order to its observed catalytic activities. The protein molecule is a monomer of molecular weight about 109,000 daltons containing several binding domains (DNA polymerase is therefore a multifunctional enzyme). There is a specific site for deoxyribonucleotide association, one each for the DNA template and the elongating primer and separate sites for the 5'-3' and 3'-5' exonuclease activities. Proteolysis of the enzyme yields a large segment of MW 75,000 daltons which exhibits both 3'-5' exonuclease and polymerase activities and a smaller fragment of MW 35,000 daltons with 5'-3' exonucleolytic activity. In its important function of DNA synthesis the enzyme induces the removal of an RNA fragment from an RNA-DNA primer-template molecule by cleaving the 5'-3' bond, and then replicates the available polymeric DNA strand (Figure 5.4a).

The enzyme also catalyses DNA repair and 'nick' translation. Unpaired DNA is repaired by enzymic digestion of one strand back to the first paired base followed by reconstruction of the correct segment to reform the duplex. This action is shown schematically in Figure 5.4b. The third property of this versatile enzyme is the replacement of missing sections in either strand. This it manages by simultaneously degrading and replacing the strand from the 'nick' (Figure 5.4c).

DNA-dependent-RNA polymerases are localised in the nuclear region in eukaryotes and in the cytoplasm of prokaryotes. They are responsible for the synthesis of RNA and are therefore central to the

Figure 5.4: DNA *Polymerase Activities*

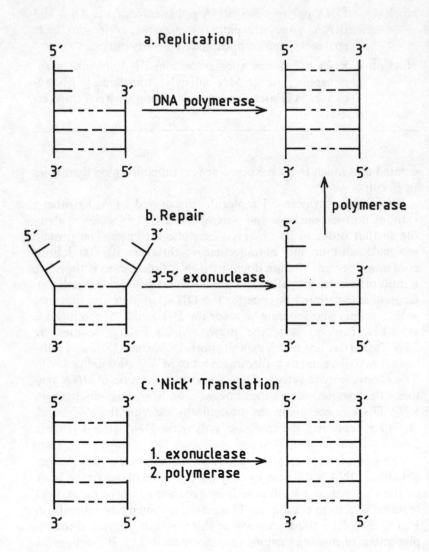

expression of the information stored in DNA. This they achieve by catalysing the synthesis of RNA molecules with primary structures complementary to those of the DNA template.

As purified the *E. coli* enzyme is a polymer of configuration $\alpha_2\beta\beta'\sigma$ but this is probably an oversimplification since for complete metabolic activity the enzyme requires at least five other factors, labelled M, CAP, ρ, ω and ψ. Several of the subunits have been isolated and purified with molecular weights α, 40,000; β, 150,000; β', 160,000; σ, 90,000; ρ, 200,000 and ω, 10,000 daltons, giving a total molecular weight for the enzyme of about 0.5 million daltons.

The sigma subunit is easily removed to give a 'core' enzyme, $\alpha_2\beta\beta'$, which retains polymerase activity although it is unable to initiate RNA formation at specific sites. The subunits β and β' have been implicated in polymerase function; β', unlike β, σ or α can combine with DNA and since β-subunit variants have been isolated from mutants resistant to the antibiotic rifampicin (which inhibits wild type RNA polymerase) the catalytic activity has been ascribed to the β-subunits. Little is known yet concerning the functions of the α or ω units.

Unlike DNA polymerase, RNA polymerase requires double stranded DNA. Only one strand of this is copied, otherwise, of course, two complementary RNAs and hence possibly two proteins would be produced. Deprived of the sigma subunit, the core enzyme is unable to initiate this process. Thus the role of this subunit may be to recognise the base sequence at specific promoter sites on the DNA template and then open the duplex at those points. This action would allow the core to enter and commence asymmetric transcription. Also unlike the DNA polymerases, RNA polymerases transcribe only defined lengths. Thus termination sites, also comprised of unique sequences of bases are presumably recognised by another of the transcription factors, and ρ has been experimentally implicated in this function.

Enzymes catalysing post-transcription modification of single stranded RNA molecules are also present in the nuclear fraction. Ribonucleases split heavy nuclear RNA molecules into the individual ribosomal, messenger and transfer ribonucleic acids. Before translocation to the ribosomes, several of the last are methylated by the S-adenosyl-L-methionine : tRNA methyltransferases. These enzymes catalyse the transfer of methyl groups to specific purine or pyrimidine carbon atoms.

For example tRNA (adenine-1-)methyl transferase catalyses the reaction:

S-adenosyl-L-methionine + tRNA = S-adenosyl-L-homocysteine

+ tRNA containing 1-methyladenine   (5.6)

The enzymes of glycolysis and the citric acid cycle provide energy for the above biosynthetic reactions. These will be discussed in later sections, but a central intermediate in energy metabolism is nicotinamide adenine dinucleotide. Its synthesis is carried out in the nucleus (eq. 5.7), and thus it is probable that by regulating its concentration as well as those of proteins, the nucleus can exercise additional control over cellular metabolism.

$$\text{ATP} + \text{Nicotinamide ribonucleotide} \xrightarrow[\text{transferase}]{\text{NMN adenylyl}}$$

(NMN)

$$\text{Pyrophosphate} + \text{NAD}^+ \quad (5.7)$$

5.2.3 Mitochondrial Enzymes

Mitochondria are present in the majority of mammalian cells, visible under the microscope as rod shaped organelles between 1-3 μm in length, about the size of bacterial cells. These organelles are often described as the 'powerplants of the cell'. Certainly the number present reflect its respiratory activity; rat liver cells with a low respiratory rate contain about 800, accounting for about one fifth of the cytoplasmic volume while mitochondria comprise about half the volume of rat heart cells. Within these organelles were discovered the enzymes and proteins of energy conservation and thus it is not surprising that much effort has been directed to their molecular, structural and functional characterisation.

The double membrane model depicted in Figure 5.5 incorporates most of the current deductions from electron microscopy and fractionation procedures. The surface area of the inner membrane is increased by 'cristae' protruding into the matrix, and the inner surface of this membrane is covered with bush-like protuberances— the inner membrane particles. Table 5.4 gives a summary of the main enzyme systems that have been found to date in the mitochondrial compartments.

The inner membrane accounts for greater than 75 percent of the total mass of the mitochondrial membranes and is approximately

Figure 5.5: Internal Structure of a Mitochondrion

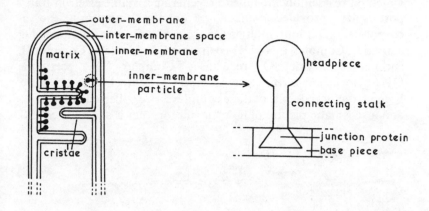

25 percent lipid and 75 percent protein. One of the most fascinating and certainly one of the most complex pathways sited in this membrane is the electron transport system, the sequence of steps whereby reducing equivalents provided by succinate and NADH are transferred to molecular oxygen.

Table 5.4: Compartmentation of Mitochondrial Enzyme Systems

Outer Membrane: Monoamine oxidase, NADH-cytochrome reductase, fatty acyl-CoA synthetase, NADH dehydrogenase, acyl transferases, phospholipase, nucleoside diphosphate kinase

Intermembrane Space: Adenylate kinase

Inner Membrane: Succinate dehydrogenase, cytochrome oxidase, 'ATP synthetase', NADH dehydrogenase, carnitine acyl-transferase, DNA polymerase, cytochromes b, c, c_1, a, a_3, δ-aminolevulinate synthetase ion translocation, nicotinamide nucleotide transhydrogenase

Matrix: Citric acid cycle enzymes, transaminases, pyruvate carboxylase, protein synthesis enzymes, RNA polymerases, fatty acyl-CoA oxidation enzymes

Four high molecular weight lipoprotein complexes have been isolated from the inner membranes of bovine heart mitochondria, which on reassembly are able to regenerate a viable electron transport system, provided a suitable phospholipid mixture is also incorporated. The four multienzyme complexes have the following activities: Complex I, NADH-coenzyme Q* reductase; Complex II, succinate-coenzyme Q reductase; Complex III, coenzyme QH_2-cytochrome c reductase; Complex IV, cytochrome c oxidase. Some structural details of these complexes are collected in Table 5.5 and a schematic picture of how they fit together is shown in eq. 5.8.

$$
\begin{array}{c}
\text{NADH} + \overset{+}{H} \\
\text{NAD}^+ \\
\text{Succinate} \\
\text{Fumarate}
\end{array}
\xrightarrow{2H}
\boxed{\begin{array}{c}\text{Complex I}\\ \text{FMN}\to\text{FeS}\end{array}}
\xrightarrow{2H}
\boxed{\begin{array}{c}\text{Complex II}\\ \text{FAD}\to\text{FeS}\end{array}}
\xrightarrow{2H}
\text{Coenzyme Q}\xrightarrow{2e}
\boxed{\begin{array}{c}\text{Complex III}\\ \text{cyt. b}\to\text{cyt.c}_1\end{array}}
\to\text{cyt.c}\to
\boxed{\begin{array}{c}\text{Complex IV}\\ \text{cyt. a}\to\text{cyt.a}_3\end{array}}
\xrightarrow{2e}
\begin{array}{c}\tfrac{1}{2}O_2\\ H_2O\end{array}
\tag{5.8}
$$

Table 5.5: Components of the Mitochondrial Respiratory Chain

Complex	MW ($\times 10^{-3}$)	Number of subunits
I	850	16
II	97	2
III	240	6–8
IV	200	6

NADH-coenzyme Q reductase is a multienzyme complex of sixteen polypeptides. Two of these bind FMN molecules and at least three form iron-sulphur clusters (FeS)** which span the membrane. Succinate-coenzyme Q reductase is a heterologous dimer of subunits MW 70,000 and 27,000. The larger subunit binds FAD as

* Ubiquinone.

**
$$
\begin{array}{c}
\text{cys} \\
\qquad\diagdown \\
\text{cys}
\end{array}
\begin{array}{c}
\ \ \ \ \ \overset{\displaystyle S}{\diagup\ \ \diagdown} \\
\text{Fe}^{3+}\qquad\qquad\text{Fe}^{3+} \\
\ \ \ \ \ \underset{\displaystyle S}{\diagdown\ \ \diagup}
\end{array}
\begin{array}{c}
\text{cys} \\
\diagup\qquad \\
\text{cys}
\end{array}
$$

coenzyme and both contain FeS clusters. Complex II, which is integrated with the matrix (M) side of the inner membrane, transfers its hydrogens to ubiquinone. Coenzyme Q appears to act as a regulator of their further transport to the cytochromes. While the complexes I–IV are thought to be relatively fixed *in situ*, coenzyme Q and further along the chain cytochrome c, appear to possess some mobility. Coenzyme QH_2-cytochrome c reductase, alternatively called the cytochrome b-c_1 complex contains cytochrome b : cytochrome c_1 : FeS protein : 'antimycin binding' protein : 'core' protein in the molar ratios 2 : 1 : 2 : 1 : 2. Two cytochromes b have been isolated, MW 17,000 and 37,000, and found to be extremely hydrophobic. Thus they are probably integrally localised in the middle of the membrane. Cytochrome c_1, MW 31,000, is also an integral protein, but its haem group is orientated towards the intermembrane (C) side in order to effect electron transfer to cytochrome c. Cytochrome c unlike the others is a peripheral protein and has been observed to undergo a very rapid $(t_{1/2} < 2ms)$ lateral motion to associate with and reduce cytochrome c oxidase. This last complex is an integral multienzyme complex containing cytochrome a, cytochrome a_3 and two copper atoms. Cytochrome c oxidase from beef heart has been differentiated into six subunits of MW 15,000; 11,200; 40,000; 22,500; 9,800 and 7,300 which form, in that order from the M to the C side, a cylindrical quaternary structure spanning and extending beyond each surface of the membrane. Cytochrome a is at the C side and interacts with cytochrome c_1. Cytochrome a_3 is situated on the M side.

The Gibbs free energy of each of the exergonic reactions catalysed by complexes I, III and IV is converted into the chemical energy of ATP, this process being oxidative phosphorylation. Intact mitochondria and a supply of oxidisable substrates, ADP and inorganic phosphate are needed for oxidative phosphorylation but if 2,4-dinitrophenol is added, electron transport and oxidative phosphorylation are uncoupled and ATPase activity is stimulated instead. ATPase is considered to be the 'ATP synthetase' system of intact coupled mitochondrial oxidative phosphorylation.

Racker and coworkers (see reference 4) sonicated mitochondria and isolated vesicles comprised of inner membrane fragments to which the inner membrane particles were still attached. Intact, these submitochondrial vesicles catalysed both electron transport and oxidative phosphorylation, but removal of the membrane particles with trypsin rendered the resulting smooth vesicles incapable of coupling the transport of electrons (which was unaffected) to the phosphorylation of ADP. The unattached membrane particles possessed both ATPase activity and the capacity to restore energy coupling to the vesicles on recombination. They termed these the F_1 coupling factors. The ATP synthesising machinery is thus the mushroom-like protuberances along the inner membrane and cristae. They are composed of three parts (Figure 5.5): (1) a *headpiece* or F_1 unit which is the ATPase proper, containing duplicates of five different subunits with MW 7,500; 17,000; 33,000; 50,000 and 53,000, to give a total of 360,000, (2) a *stalk* or connecting protein, plus (3) a hydrophobic *base piece* integral with the membrane and composed of four polypeptides MW 29,000; 22,000; 12,000 and 7,800.

A question which has eluded an unequivocal answer over the last three to four decades is how the energy of electron transport is coupled to ATP synthesis. The theoretical proposal looked on with most favour at present is Mitchell's chemiosmotic hypothesis.* The main points of the thesis are (1) when two electrons are transported down the chains, six protons move across the membrane and accumulate in the intermembrane space, (2) the inner membrane is impermeable to protons, thus (3) a proton and electrical gradient is set up across the membrane and (4) the energy of electron transport is conserved in this gradient and coupled to ATP synthesis.

Mitchell argues that in the hydrophobic interior of the membrane, the presence of water molecules is energetically unfavourable so that the equilibrium for the reaction $ADP + P_i \rightleftharpoons ATP + H_2O$ will be to the right. The H_2O formed is considered to be removed by dissociation to H^+ plus OH^- which combine with OH^- and H^+ respectively present in excess on the opposite sides of the membrane.

Two other theories have also been put forward. The oldest, the 'chemical coupling' hypothesis proposes intermediate formation of a high energy phosphorylated species which transfers its phosphate to ADP. No such intermediate has yet been found however. According to the third hypothesis, proposing 'conformational coupling' the energy of electron transport is translated into a conformational change in the ATPase particle which induces its association with ADP and P_i to release bound ATP. Research concerning these mechanisms has been reviewed by leaders in the field.[4]

* Mitchell was awarded the 1978 Nobel Prize for this work.

The inner and outer membranes exhibit quite different permeabilities. The external envelope is permeable to most molecules of molecular weight below 10^4 daltons, but the internal membrane is less freely permeable and shows greater selectivity. Thus whereas ATP, ADP, inorganic phosphate, hydroxide ions, citrate, malate, glutamate, aspartate and pyruvate are all able to pass through the membrane, no translocation of NADPH, NAD^+, NADH, $NADP^+$, AMP, GTP, CTP or acetyl-CoA has been observed. This differentiation is attributed to selective transport systems for the former in the membrane.* These carriers appear to allow both the passive transport of ions in response to their concentration gradients on either side of the membrane and also their translocation against these concentration gradients. Certainly, phosphate ions are accumulated by respiring mitochondria—a capacity possibly linked to proton translocation and chemiosmosis.

The segregation of some cytoplasmic and mitochondrial metabolite pools and the lack of specific carrier systems raises the problem of how those pathways involving common intermediates such as reducing equivalents or acetyl-CoA molecules are linked. Such considerations have led to the postulation of indirect shuttle systems, series of enzyme catalysed reactions which link these intra- and extra-mitochondrial metabolite pools through their conversion to more readily permeable ions. Possibly the earliest to be described for reducing equivalent translocation in muscle and nerve cells was the 'glycerol phosphate shuttle'. In the cytoplasm, glycerol phosphate dehydrogenase catalyses the addition of hydrogen from NADH to dihydroxyacetone phosphate, giving glycerol phosphate. This diffuses through the outer mitochondrial membrane to the outer surface of the inner membrane but no further. Here a second, inner-membrane bound, glycerol phosphate dehydrogenase activity reforms dihydroxyacetone phosphate, simultaneously passing the hydrogens to its prosthetic flavin group. The reducing equivalents are then transmitted along the cytochrome chain and yield two ATP molecules. Dihydroxyacetone phosphate returns to the cytoplasm and is thus needed in only co-catalytic quantities.

In heart and liver cells, the aspartate-malate shuttle serves a similar functional role, but unlike that described above, this is reversible, transporting electrons both in and out of mitochondria,

* The absence of a nicotinamide dinucleotide carrier explains in part the latency of the mitochondrial dehydrogenases.

Figure 5.6: The Aspartate-Malate Shuttle System for Reducing Equivalents

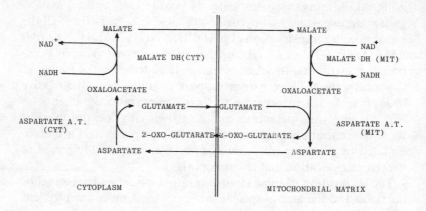

Figure 5.7: Pyruvate-Citrate Shunt

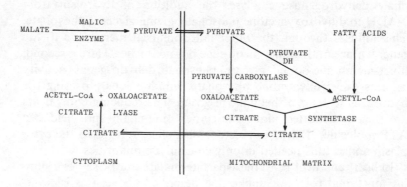

through the action of organelle specific enzymes catalysing the same, reversible reactions in opposite directions. The system is described in Figure 5.6 and the individual enzymes involved are discussed in detail in Section 5.3.3. The proposed system linking acetyl-CoA availability and hence fatty acid metabolism in the two compartments is described in Figure 5.7.

Table 5.6: Citric Acid Cycle Enzymes

Reaction	Enzyme, EC number	Source	Cofactors	Molecular weight; Number of subunits
Oxaloacetate + Acetyl-CoA ↓ Citrate	Citrate(si) synthase, 4.1.3.7	Rat E. coli	—	100,000; 2 280,000
↓ Isocitrate	Aconitate hydratase, 4.2.1.3	Porcine heart	Fe(II)	89,000; 2
↓ 2-Oxoglutarate	Isocitrate DH, 1.1.1.41	Yeast Bovine	NAD, Mg(II)	300,000; 8 670,000; 16
↓ Succinyl-CoA	Oxoglutarate DH, 1.2.4.2	E. coli	CoA, NAD, TPP Lipoate	2×10^6
↓ Succinate	Succinyl-CoA synthetase, 6.2.1.4	E. coli	GDP, P_i	140,000; 4
↓ Fumarate	Succinate DH, 1.3.99.1	Bovine	FAD, NHI	97,000; 2
↓ L-Malate	Fumarate hydratase, 4.2.1.2	Porcine	H_2O	194,000; 4
↓ Oxaloacetate	Malate DH, 1.1.1.37	*Neurospora* Bacilli Heart	NAD	54,000; 4 148,000; 4 194,000; 4

The mitochondrially based intermediates in these systems are those of the citric acid cycle.* The functions of this important amphibolic cycle are the catabolism of acetyl-CoA (produced from the breakdown of carbohydrates and fatty acids) to carbon dioxide plus water and the anabolic provision of carbon skeletons for porphyrin, amino acid and carbohydrate synthesis. The molecular weights and subunit structures of the enzymes catalysing the reactions that constitute the citric acid cycle are shown in Table 5.6.

As part of its central role in metabolism, many enzyme mediated interconnections are made between the Krebs cycle intermediates and the major metabolite classes. Thus for example, fatty acids are catabolised to acetyl-CoA and the branched chain amino acids valine and isoleucine are converted to succinyl-CoA. A major group of enzymes connecting the cycle and amino acid metabolism are the aminotransferases. Fifty-three of these have been documented. Not all occur in the mitochondria. The majority employ the α-keto acids 2-oxoglutarate or pyruvate as amino group acceptors, eq. 5.9. Nearly all also require the cofactor pyridoxal phosphate.

$$
\begin{array}{ccccc}
R & R & R & & R' \\
| & | & | & & | \\
CH \cdot NH_2 & + \ CO & = CO & + & CH \cdot NH_2 \\
| & | & | & & | \\
CO_2^- & CO_2^- & CO_2^- & & CO_2^-
\end{array} \qquad (5.9)
$$

$R' = -(CH_2)_2CO_2H$ 2-Oxoglutarate   L-Glutamate

$R' = -CH_3$ Pyruvate   L-Alanine

In mammalian liver cells transamination reactions probably take place serially, integrating several pathways and cycles (eq. 5.10):

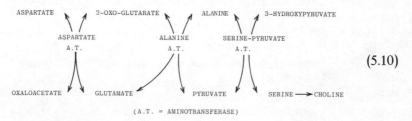

(5.10)

(A.T. = AMINOTRANSFERASE)

---

* Also called the tricarboxylic acid (TCA) cycle and the Krebs cycle.

The multitude of pathways emanating from the Krebs cycle would cause a considerable depletion of its intermediates if pathways to replenish these had not evolved. H. Kornberg has described these as *anaplerotic* pathways.[5] Two anaplerotic enzymes found in mitochondria which augment the production of Krebs cycle intermediates are pyruvate carboxylase which catalyses eq. 5.11, and malic enzyme, malate DH(carboxylating, NADP) catalysing eq. 5.12:

$$\text{Pyruvate} + CO_2 + ATP = \text{Oxaloacetate} + ADP + P_i \quad (5.11)$$

$$\text{Pyruvate} + CO_2 + NADPH = \text{Malate} + NADP \quad (5.12)$$

A large proportion of the TCA cycle intermediates are derived from acetyl-CoA, its two main sources being pyruvate and the almost universal β-oxidation pathway of fatty acid degradation. The enzymes of β-oxidation occur in most cells. In the liver they are localised in the inner membrane and matrix of the mitochondria. Free fatty acids enter the liver complexed to albumin, but the complexes rapidly dissociate and free fatty acids are actively transported across the two mitochondrial membranes. An acyl-CoA synthetase, present in the outer membrane first converts the fatty acids to CoA derivatives, which then either diffuse or are transported across the inter-membrane space to the inner membrane. Here two transacylation reactions catalysed by two distinct inner membrane bound carnitine : acyl-CoA transferases* are responsible for their entry into the matrix. The first synthesises O-acyl carnitine, releasing CoA into the intermembrane space for re-use by acyl-CoA synthetase, and the second, with access to the matrix, transfers the fatty acyl group to acetyl-CoA, simultaneously releasing carnitine which is then available for the first transferase. The resynthesised fatty acyl-CoA molecules then undergo β-oxidation in the mitochondrial matrix, Table 5.7.

Three fatty acid activating enzymes, fatty acid-CoA synthases, are present in the cell. One activates acetate and propionate, a second acts on C_4 to C_{10} fatty acids while a third, present in the endoplasmic reticulum activates those of between ten and twenty carbon atoms. Three acyl-CoA dehydrogenases are also found in the mitochondrial matrix and have C_4–C_6, C_6–C_{12} and C_6–C_8

---

* Carnitine is the common name for 3-hydroxy-4-trimethylammonium butyrate.

Table 5.7: Fatty Acid β-Oxidation Enzymes

Reaction	Enzyme
$R \cdot CH_2 \cdot CH_2 \cdot CO \cdot S \cdot CoA$	
↓	Acyl-CoA dehydrogenase
$R \cdot CH = CH \cdot CO \cdot S \cdot CoA$	
↓	Enoyl-CoA hydratase
$R \cdot CH(OH) \cdot CH_2 \cdot CO \cdot S \cdot CoA$	
↓ NAD	3-Hydroxyacyl-CoA DH
$R \cdot CO \cdot CH_2 \cdot CO \cdot S \cdot CoA$	
↓	Acetyl·CoA acyltransferase
$R \cdot CO \cdot S \cdot CoA + CH_3 \cdot CO \cdot CoA$	

specificities. All utilise FAD as a prosthetic group which after reduction transfers the electrons to cytochrome b. The next enzyme, enoyl-CoA hydrolase hydrates a wide range of trans-α,β-unsaturated acyl-CoAs but it always produces the corresponding L-β-hydroxyacyl-CoA derivatives. With the *cis*-geometric isomers the other optical isomer is produced. Stereospecificity is retained by the pathway as a whole however because the next enzyme in sequence, 3-hydroxyacyl-CoA dehydrogenase, is specific for the L-isomer. Finally, thiolase catalysed cleavage of the 2-oxo-acyl-CoAs, leaves the enzyme acylated at an essential cysteine group which must undergo thiol interchange with a coenzyme A molecule prior to the next circuit of the pathway.

Localised in the outer membrane is the flavoprotein monoamine oxidase (MAO, monoamine : O_2 oxidoreductase (deaminating), EC 1.4.3.4). This extensively investigated enzyme[6] catalyses the oxidative deamination of biogenic amines, eq. 5.13.

$$R \cdot CH \cdot (NH_2) \cdot CO_2H \xrightarrow{MAO}$$
$$R \cdot C \cdot (=NH) \cdot CO_2H \longrightarrow R \cdot CO \cdot CO_2H \quad (5.13)$$

Its main biological role is the determination and regulation of serotonin : catecholamine ratios in the brain although it also contributes to the elimination of unwanted amines. The levels of serotonin and catecholamine are found to contribute to mammalian behaviour patterns, influencing sleep, depressive states, aggressiveness etc.

Indeed the study of natural and artificial regulation of MAO activity by the thyroid and sex hormones and by synthetic inhibitors such as N-benzyl-N-methyl-propynylamine, phenalzine or tranylcypromine is at present an interesting area of psychopharmacology.[6]

5.2.4 Lysosomal Enzymes[7]

Lysosomes are present in all eukaryotic cells, except red blood cells, sedimenting in the mitochondrial fraction. They are essentially intracellular vesicles, comprising a single 'thick' membrane surrounding a wide variety of latent hydrolytic enzymes. At least fifty different enzymes have been described (Table 5.8), the majority of

Table 5.8: Some Lysosomal Enzymes

Enzyme	Hydrolytic reaction
Triacylglycerol lipase	Triacylglycerol = Diacylglycerol + Fatty acid
Acid phosphatase	Mono-organophosphate = Orthophosphate + Alcohol
Deoxyribonuclease	DNA = Oligodeoxyribonucleotides
Sphingomyelin phosphodiesterase	Sphingomyelin = Choline phosphate + N-Acylsphingosine
Sulphatases	Organosulphate = Sulphate + Organic group
Sialidase	Sialic acid hydrolysis
Glycosidases	Release of monosaccharides*
β-Glucuronidase	β-D-Glucuronide = Alcohol + D-Glucuronate
β-N-Acetylhexosaminidase	β-N-Acetylglucosaminide = Alcohol + NAG**
Cathepsins B, C and D	Protein hydrolysis
Pyrophosphatase	Pyrophosphate = 2,Orthophosphates

* From the non-reducing terminus of the polysaccharide.
** N-acetyl-glucosamine.

which exhibit acidic pH-activity optima, although a few, notably elastase, cathepsin and some lipolytic enzymes are maximally active around neutrality. The first two mentioned are found attached to the external surface of the membrane.

The lysosomal membrane is similar with respect to many of its properties to the plasma membrane. For example Na^+, K^+-dependent ATP-ases are present, but whether these pump to maintain an acidic internal environment in the lysosome is unclear.

232   *Enzyme Physiology*

Unlike the plasma membrane the lysosomal membrane contains a higher proportion of sialic acid. Many cationic proteins e.g. NADH dehydrogenase, are located in this membrane, as distinct from the interior matrix which contains a higher proportion of anionic enzymes. Most of the enzymes are glycoproteins and comparisons of the rates of protein and carbohydrate turnover indicate that the amount of bound oligosaccharide varies during the lifetime of an enzyme molecule. The glycosyl groups may be responsible for controlling the enzyme distribution in the organelle and for stabilising the molecule to proteolytic degradation, although the lysosomal enzymes will undoubtedly have evolved some resistance to protease and glycosidase attack. After isolation they show considerable resistance to autolysis.

The lysosome has been proposed as the main site of intracellular protein digestion, a proposal that has been extended to include a role in the autophagic digestion of the other components of expired cellular organelles. Lysosomes have been observed to digest endocytosed particulate matter, bacteria, viruses and soluble polymers, reducing these with enzymes such as bacteriocidal lysozyme to lower molecular weight components which are then released into the cytoplasm for reutilisation. These degradative enzyme mixtures are potentially lethal to the cell, and their release, induced by lipid soluble drugs, bacterial toxins or physical trauma, all of which can disrupt membrane equilibria, can cause secondary damage, in many cases as severe as the primarily acknowledged cause. Commensurate with this, a role for lysosomal enzymes in inflammatory diseases has been suggested, in which the lysosomal proteases act as mediators of the inflammation after their release, stimulated by immune complexes.[7]

5.2.5 Microsomal Enzymes

Microsomes are not microscopically visible organelles but are artifacts of the homogenisation process, balloon shaped segments and vesicles of the intracellular endoplasmic reticulum, ER. This bifurcate membrane system, present in all eukaryotic cells except mature erythrocytes, is an extensive network continuous with the cell and nuclear membranes and the Golgi apparatus. It appears in two forms, rough and smooth, depending on the presence or absence of ribosomes. Since the ribosomes, protein plus ribonucleic acid, are the seats of protein synthesis, a role in the organisation of this crucial metabolic process is also indicated for the ER. The endoplasmic reticulum enzymes are also responsible for the biosynthesis of mucopolysaccharides and lipids, and for the disposal of unwanted molecules such as bilirubin and synthetic drugs.

Dietary lipids are broken down in the lumen of the small intestine to free fatty acids and acylglycerols by pancreatic lipase. In the ER of the intestinal epithelial cells, these fatty acids are activated by acyl-CoA synthetase (eq. 5.14) and the CoA derivatives are then combined with the acyl-glycerols to reform the triacylglycerols, eq. 5.15.

$$R \cdot CH_2 \cdot CO_2H + CoA + ATP$$
$$= R \cdot CH_2 \cdot CO \cdot S \cdot CoA + PP_i + AMP \quad (5.14)$$

$$\begin{array}{l} CH_2 \cdot OH \\ | \\ CH \cdot OCO \cdot R_1 \\ | \\ CH_2 \cdot OH \end{array} \xrightarrow[\substack{EC\ 2.3.1.22 \\ +R_2 \cdot CO \cdot SCoA}]{\text{Monoglyceride acyltransferase}}$$

$$\begin{array}{l} CH_2 \cdot OH \\ | \\ CH \cdot O \cdot COR_1 \\ | \\ CH_2 \cdot O \cdot CO \cdot R_2 \end{array} \xrightarrow[\substack{EC\ 2.3.1.20 \\ +R_3 \cdot CO \cdot SCoA}]{\text{Diglyceride acyltransferase}} \begin{array}{l} CH_2 \cdot O \cdot CO \cdot R_3 \\ | \\ CH \cdot O \cdot CO \cdot R_2 \\ | \\ CH_2 \cdot O \cdot CO \cdot R_1 \end{array} \quad (5.15)$$

Phospholipids, the building blocks of all membranes are synthesised in the ER of eukaryotic cells (in prokaryotes the corresponding enzymes are integrated with the plasma membrane). The major phospholipid, lecithin, 3-sn-phosphatidylcholine, is synthesised from sn-1,2-diacylglycerol and cytidine diphosphate choline (CDP-choline), by the action of cholinephosphotransferase, eq. 5.16.

$$\begin{array}{l} CH_2 \cdot O \cdot CO \cdot R_1 \\ | \\ CH \cdot O \cdot CO \cdot R_2 \\ | \\ CH_2 \cdot OH \end{array} \xrightarrow{\text{CDP-Choline}} \begin{array}{l} CH_2 \cdot O \cdot CO \cdot R_1 \\ | \\ CH \cdot O \cdot CO \cdot R_2 \\ | \\ CH_2 \cdot OPO_2H \cdot O \cdot CH_2 \cdot CH_2 \cdot \overset{+}{N}(CH_3)_3 \end{array}$$
$$(5.16)$$

The triacylglycerols and phospholipids are stored in adipose tissue but in response to starvation or an energy demand by the organism they are rapidly mobilised by conversion to free fatty acids and acylglycerols by lipoprotein lipase. The most important fatty acids for both animal and plant cells are stearic,,oleic and linoleic acids, all of which have C_{18} structures. Enzymes for the *de novo* synthesis of the last two are absent from animal cells; thus they must be provided in the diet. The fatty acid synthesising enzymes present in the cytoplasm produce palmityl-CoA as end product and this is elongated in the ER to stearyl-CoA by malonyl-CoA and an NADPH-dependent elongating system.

The ER also disposes of hydrophobic molecules such as bilirubin and administered drugs. This process is entitled detoxification.

5.2.5.1   Drug Metabolism

Liver microsomes perform major roles in drug detoxification. Here lyophilic and toxic substances are oxidised, reduced or hydrolysed and converted into water-soluble derivatives which can more readily be excreted in the urine.

The Microsomal Oxidation System.   Most exogenous lipid soluble substances are oxidised by the NADPH-linked enzymes in the ER of hepatic parenchyma. These enzymes have yet to be characterised in detail but their functional variety is known, in catalysing aromatic hydroxylation, alkyl oxidation, deamination, epoxidation of double bonds, and dealkylation. A feature of the system is the non-mutual exclusivity of their conversions; thus chlorpromazine(II) a synthetic major tranquillizer is *N*-oxidised, *N*-dealkylated, hydroxylated on the benzene rings and conjugated.

Present evidence indicates that the main transformations are carried out by an electron transport chain constituting approximately one-fifth of the microsomal membrane (Figure 5.8). An essential component of this system is cytochrome P_{450}, named after the wavelength of maximum absorption of its complex with carbon monoxide. The postulated function of this protein-iron-porphyrin complex is to activate molecular oxygen which in its normal state is relatively unreactive. At least two other enzymatic complexes are also involved in this mixed function oxidase system, NADPH-cytochrome c reductase, which transfers electrons from NADPH to non-haem iron, and NADPH-cytochrome P_{450} reductase which maintains an appropriate level of reduced cytochrome P_{450}. A

Figure 5.8: Microsomal Electron Transport

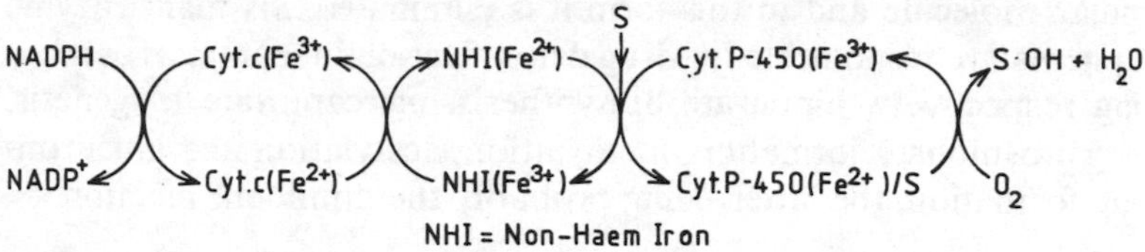

NHI = Non-Haem Iron

cyclic oxidative pathway guided by cytochrome P_{450} in which these enzyme systems are integrated has been suggested by Cohen and Estabrook[8] (Figure 5.9).

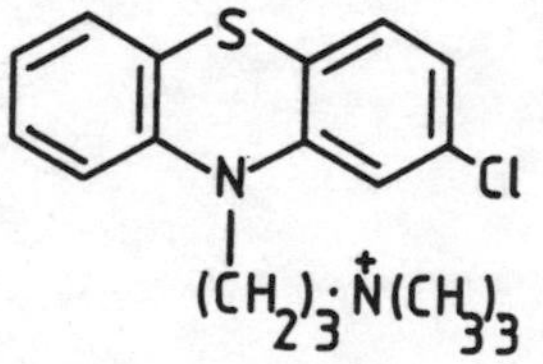

II

Liver Microsome Reducing Enzymes. Aliphatic and aromatic nitro groups are reduced by NADPH dependent reductases to the amines. *In vitro* the reaction has been shown to be dependent on cytochrome P_{450} and NADPH-cytochrome c reductase, but whether these complexes are identical to those of the oxidative system has as yet to be determined. The reducing system possesses wide specificity, reducing ketones, aldehydes, disulphides and alkenes to alcohols, alcohols, mercaptans and alkanes respectively. Chloral hydrate is reduced to trichloroethanol.

Figure 5.9: Cytochrome P_{450} Cycle

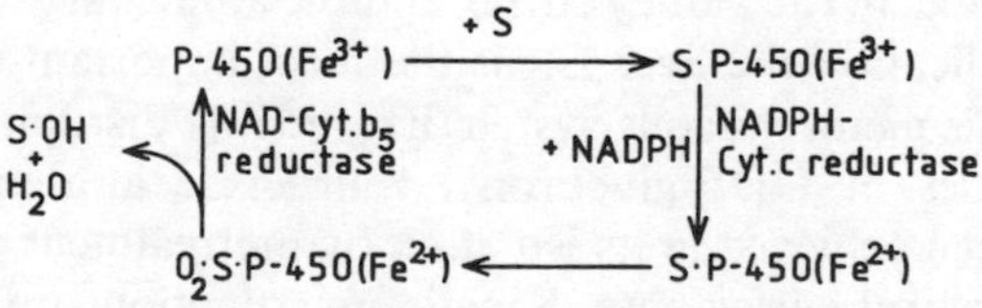

Conjugation. In the second phase of drug metabolism, detoxification, the substance is inactivated by conjugation with a polar molecule and in this form it is eliminated. Six main enzyme systems are responsible for drug detoxification in mammals catalysing respectively, hippurate biosynthesis, mercapturate biogenesis, organosulphate formation, methylation, acetylation and glucuronide formation, the latter being probably the dominant reaction.

Glucuronide Conjugating Enzymes. Conjugation of physiologically and pharmacologically active substances with β-D-glucosiduronic acid is catalysed by several microsomal glucuronyl transferases. The major complex involved exhibits UDP-glucuronyl transferase activity, catalysing the transfer of the glucuronyl group from uridine diphosphate-α-D-glucuronic acid (UDPGA) to a wide variety of acceptors, eq. 5.17.

$$(5.17)$$

In the endoplasmic reticulum there is probably a family of closely related UDP-glucuronyl transferases each one of which differs in its substrate specificity. Their activities have been found in all mammals investigated, in the kidney cortex and the alimentary canal, but for drug modification the liver ER is the most important site. Disturbance of this membraneous system irreversibly disrupts the catalytic properties of UDP-glucuronyl transferase although some success has been achieved in its isolation by pretreatment of microsomes with neutral surfactants. Kinetic investigations on the partially pure fractions are plagued by their instability but the results indicate that the transfer reaction is irreversible and highly efficient. Little specificity for the subsequent aglycone $(R \cdot X-)$ is apparent, the enzymes catalysing glucuronyl transfer to a wide variety of hydroxyl, carboxyl, sulphydryl, amino, imino, aromatic, aliphatic and heterocyclic acceptors.

Mercapturic Acid Biosynthesis. Important pathways for the dispatch of lipophilic compounds are catalysed by a group of related S-glutathione transferases. These transfer glutathione to the substance, forming substituted N-acetyl-cysteines(III), which are then probably metabolised via the same route as glutathione itself, eq. 5.18.

$$\underset{\text{Glutathione}}{\overset{\displaystyle CO \cdot R_2}{\underset{|}{HS \cdot CH_2 \cdot CH \cdot NH \cdot R_3}}} \xrightarrow{R_1 \cdot X} \underset{\text{III}}{\overset{\displaystyle CO \cdot R_2}{\underset{|}{R_1 \cdot S \cdot CH_2 \cdot CH \cdot NH \cdot R_3}}} \rightarrow \rightarrow \rightarrow$$

$$\overset{\displaystyle COOH}{\underset{|}{R_1 \cdot S \cdot CH_2 \cdot CH \cdot NH \cdot CO \cdot CH_3}} \quad (5.18)$$

$$R_2 = -NH \cdot CH_2 \cdot COOH;$$

$$R_3 = -CO \cdot (CH_2)_2 \cdot CH(NH_2)COOH$$

Sulphate Ester Formation. Sulphation confers increased water solubility on a substance so aiding its urinary excretion. The soluble fraction of the liver has been found to prepare sulphate for this by catalysing its activation to the mixed anhydride, adenylyl sulphate, APS(IV), eq. 5.19, followed by phosphorylation to 3'-phosphoadenylyl sulphate, PAPS(V), eq. 5.20; the overall activation requires two ATP molecules per sulphate ion.

$$ATP + SO_4^{2-} \xrightarrow[\substack{\text{adenylyltransferase} \\ \text{E.C. 2.7.7.4}}]{\text{ATP: sulphate}} APS + PP_i \quad (5.19)$$

$$APS + ATP \xrightarrow[\substack{-3'\text{-phosphotransferase} \\ \text{EC 2.7.1.25}}]{\text{ATP: adenylylsulphate}} PAPS + ADP \quad (5.20)$$

$$RXH + PAPS \xrightarrow{\text{sulphotransferases}} PAP + RXSO_3H_2 \quad (5.21)$$

From the microsomes different sulphotransferase preparations have been isolated which catalyse the irreversible transfer of the sulphate group from PAPS to phenols, alcohols, steroids, cerebrosides, choline and mucopolysaccharides, eq. 5.21.

Adenine $CH_2OP(O_2H)\cdot OP(O_2H)\cdot OSO_3H_2$

OR OH

IV   R=H , Adenylyl Sulphate, APS
V   R=PO$_3$H$_2$, Phosphoadenylyl Sulphate, PAPS

Acetylation. Drugs containing aliphatic or aromatic amino groups are usually *N*-acetylated prior to elimination. The acetyl group is invariably derived from acetyl-CoA and enzymes catalysing its transfer have been identified in the epithelial mucosa of the gastrointestinal tract, and in liver parenchymal cells. These acetyl CoA: arylamine *N*-acetyl transferases are probably polymorphic, their molecular forms depending on genetic and developmental factors, but all catalyse the same overall reaction, eq. 5.22.

$$R\cdot NH_2 + CH_3\cdot CO\cdot S\cdot CoA \longrightarrow$$
$$R\cdot NH\cdot CO\cdot CH_3 + CoA\cdot SH \quad (5.22)$$

Methylation. Several methyl transferases (MT) of varying specificities are present in biological tissues, e.g. catechol-*O*-MT, phenol-*O*-MT, hydroxyindole-*O*-MT, phenylethanolamine-*N*-MT and histamine-*N*-MT, all of which in common catalyse transfer of the methyl group from *S*-adenosylmethionine to the substrate.

5.2.5.2   Protein Biosynthesis

One of the most significant advances arising from the application of differential centrifugation has been the identification of the sites of protein biosynthesis in the cell. After incubation of liver cells with radioactively labelled amino acids, the tracer was found localised mainly in the microsomal fraction. On subfractionation of these labelled vesicles, most of the activity was recovered in ribonucleoprotein particles—the ribosomes.

Protein synthesis takes place in four separate stages, each enzyme catalysed and each requiring a series of cofactors. The individual stages are: (1) *activation* of the amino acids to aminoacyl-tRNA derivatives, (2) *initiation*, binding of the first aminoacyl-tRNA and mRNA to the ribosome, (3) *elongation*, lengthening of the polypeptide chain by addition of subsequent aminoacyl-tRNAs, and (4) *termination* and release of the completed polypeptide. Specific protein factors propel each of the four steps. These are respectively aminoacyl-tRNA synthetases, three initiation factors (IF·1, IF·2, IF·3), two elongation factors (ET·T and ET·G) and several releasing factors. Some of these factors may eventually prove to be enzymes but at present the step most thoroughly characterised enzymically is the first, the activation of amino acids by esterification with transfer RNAs.

5.2.6 Cytoplasmic Enzymes

The enzymes of four major pathways are recovered in the supernatant, those of amino acid activation, glycolysis, gluconeogenesis, and fatty acid biosynthesis.

5.2.6.1 Amino Acid Activation

The aminoacyl-tRNA synthetases, (amino acid : tRNA ligases, AMP forming, EC class 6.1.1) catalyse the formation of aminoacyl-tRNAs from amino acids and their cognate transfer ribonucleic acids in preparation for their incorporation into the growing polypeptide chain, eq. 5.23.

$$\text{Amino acid} + \text{tRNA} + \text{ATP} = \text{Amino acyl-tRNA} + \text{AMP} + \text{PP}_i$$

$$(5.23)$$

Ligases acting on all the twenty amino acids have been isolated and those specific for lysine, methionine, leucine, tyrosine and tryptophan have also been crystallised. Summaries of the quarternary structures of those purified to homogeneity are tabulated in Table 5.9. There appears to be no pattern to the subunit structures. As a generality, the number of substrate binding sites equals the number of subunits. Thus one molecule of methionine : tRNA^{met} ligase or lysine : tRNA^{lys} ligase binds two molecules of ATP, two amino acid and two tRNA molecules, and the monomeric valine : tRNA^{val} ligase binds a single molecule of each substrate. A possible exception is phenylalanine : tRNA^{phe} ligase which is a conglomerate of

Table 5.9: Quarternary Structure of some Aminoacyl-tRNA Synthetases

Amino Acid	Source	Quarternary structure type	Subunit molecular weight
Alanine	*E. coli*	A_2	80,000
Arginine	*E. coli*	A_2	40,000
Arginine	*B. stearother-mophilus*	A	78,000
Aspartate	Yeast	A	100,000
Glutamate	*E. coli*	AB	56,000; 46,000
Glutamine	*E. coli*	A	69,000
Glycine	*E. coli*	A_2B_2	33,000; 80,000
Histidine	*S. typhimurium*	A_2	40,000
Isoleucine	*E. coli*	A	110,000
Leucine	*E. coli*	A	105,000
Leucine	Yeast	A_2	60,000
Lysine	*E. coli*	A_2	52,000
Lysine	Yeast	A_2	70,000
Methionine	*E. coli*	A_2	90,000
Phenylalanine	*E. coli*	A_4	43,000
Phenylalanine	Yeast	A_2B_2	50,000; 60,000
Phenylalanine	Rat liver	A_2B_2	69,000; 74,000
Proline	*E. coli*	A_2	47,000
Serine	*E. coli*	A_2	48,000
Serine	Yeast	A_2	60,000
Tryptophan	*E. coli*	A_2	37,000
Tryptophan	Human placenta	A_2	58,000
Tyrosine	*E. coli*	A_2	48,000
Tyrosine	Yeast	A	40,000
Tyrosine	*S. cerevisiae*	A_4	31,500
Valine	*E. coli*	A	110,000

Source: Reference 26.

four identical subunits but its associations with only one phenyl-alanine and one tRNA[phe] molecule have been observed. All require Mg(II) for complete activity and ATP and dATP are the only nucleotides capable of activating the enzyme catalysed reactions.

As eq. 5.23 shows, all the amino acid : tRNA ligases require three substrates, tRNA, amino acid and ATP. It is necessary therefore to decide their orders of binding since it is unlikely that all three bind simultaneously to the enzymes. With ATP plus amino acid, many of the ligases form a bound aminoadenylate complex; in certain cases this complex has been isolated and shown to have the structure in eq. 5.24 (VI).

$$R \cdot CH(\overset{+}{N}H_3) \cdot CO_2^- + ATP \xrightarrow{\quad E \quad}$$

$$[E][Adenine-ribose-CH_2 \cdot O \cdot PO_2 \cdot O \cdot CO \cdot CH(\overset{+}{N}H_3)R] \quad (5.24)$$

$$VI$$

The carboxyl group of this mixed anhydride is a powerful electrophile and accordingly the second step postulated for synthetase action is directed attack by the ribose 3'-hydroxyl nucleophile of the terminal tRNA adenosine at this position:

$$(VI) + tRNA \xrightarrow{\quad\quad} E + AMP + tRNA-O \cdot CO \cdot CH(\overset{+}{N}H_3)R$$

$$(5.25)$$

Part of the evidence cited to support inclusion of this step is the experimental observation that the isolated complexes (VI) give the expected products with rate constants similar to those of the overall reaction. A sequence consistent with most of the available experimental data can then be written

$$E \underset{\quad}{\overset{ATP}{\rightleftharpoons}} E \cdot ATP \underset{PP_i}{\overset{AA}{\rightleftharpoons}} E \cdot AMP-AA \overset{tRNA}{\rightleftharpoons}$$

$$E + AMP + tRNA-AA \quad (5.26)$$

In vitro rapid migration of the tRNA attached amino acyl group gives an equilibrium mixture of the 2' and 3' derivatives (eq. 5.27). For this reason the final structure is very reactive with a free energy of hydrolysis numerically similar to that of ATP; the equilibrium constant for the overall reaction eq. 5.23 is therefore close to unity.

tRNA   Adenine   tRNA   Adenine   tRNA   Adenine

$$(5.27)$$

A remarkable feature of the synthetases is the high degrees of specificity they exhibit for their substrates, which is of course of paramount importance to the fidelity of protein synthesis. The association constants for the tRNA molecules with their synthetases are of the order $10^7 M^{-1}$ i.e. a high affinity interaction is involved. All the tRNA molecules are of approximately the same size, between 75–80 nucleotides and they possess similar secondary and probably also similar tertiary structures. One feature of this type of ribonucleic acid is the multitude of minor bases resulting from post-translational modification (eq. 5.5). These were originally thought to determine synthetase recognition but from present evidence this tenet seems unlikely.

One of the current, most interesting research topics therefore is to determine how this specificity is attained and the features on the enzyme and the tRNA molecules responsible. Kim[9] has suggested that an enzyme and its cognate tRNA bind as a consequence of corresponding symmetry functions in the two molecules. This 'symmetry recognition' hypothesis he bases on (a) the known three-dimensional structure of yeast tRNA[phe] which is L-shaped with the anticodon (mRNA recognition site) situated towards the end of one leg and the enzyme binding regions lying on its inside surface and (b) the known quaternary structures of the synthetases, which are either even numbers of subunits or are monomers containing homologous sequences; trypsin digests of the monomeric enzymes gave only half the expected number of peptides, indicating that these synthetases are probably composed of two large repeating units covalently joined. Kim has proposed therefore that for stability the pseudo two-fold symmetry of a tRNA molecule complements that of a synthetase and recognition is achieved by specific interactions in the vicinities of the contact areas, Figure 5.10.

Synthetase-tRNA recognition is only half the specificity problem

Figure 5.10: Proposed Structure of a tRNA-*Amino Acid* : tRNA *Ligase Complex*

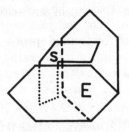

that has to be explained. The fidelity of amino acid incorporation is of the order of 2999 in 3000, and thus the selectivity a synthetase shows for its amino acid must be higher than this, which is much greater than the differences in K_S exhibited by many synthetases for amino acids of similar structure. For example addition or deletion of a methyl group in the series valine, norvaline, isoleucine, increases K_S by only 5×10^2 at the most, relative to the cognate amino acid.*

Based on their experiments with isoleucyl-tRNA^{ile} synthetase and valyl-tRNA^{val} synthetase, Fersht and Kaethner[10] have attributed the high degree of selectivity to the evolution of an 'editing mechanism'. Both enzymes follow the two step mechanism for amino acid-tRNA synthesis described above but in addition both are able to form stable aminoacyl-adenylate complexes with non-cognate amino acids. Isoleucyl-tRNA^{ile} synthetase forms and binds valyl-adenylate, although with a smaller k_{cat}/K_m value; addition of tRNA^{ile} to this complex results in its breakdown to valine plus AMP rather than transfer of the valyl group to the ribonucleic acid, i.e.

$$E + Ile + ATP$$

$$= E[AMP\text{-}Ile] \xrightarrow{\text{tRNA}^{ile}} tRNA^{ile}\text{-}Ile + AMP + E \quad (5.28)$$

and

$$E + Val + ATP$$

$$= E[AMP\text{-}Val] \xrightarrow{\text{tRNA}^{ile}} tRNA^{ile} + Val + AMP + E \quad (5.29)$$

---

* Most of the synthetases will transfer to their tRNAs compounds normally absent from the cell such as amino alcohols derived by reduction of the corresponding amino acids.

A similar 'editing mechanism' by which the enzyme is induced to act as an ATP pyrophosphorylase has also been substantiated for valyl-tRNA synthetase catalysis in the presence of threonine. This enzyme forms a stable AMP-Thr complex, with k_{cat}/K_m some 500 fold smaller than the value for formation of the correct complex; when $tRNA^{val}$ is added the enzyme transfers the threonyl group to a water molecule rather than to the $tRNA^{val}$. Rapid quenching experiments have shown that the hydrolysis occurs via a transient enzyme-[Thr-tRNA^{val}] species, which breaks down approximately 2600 times faster than enzyme-[Val-tRNA^{val}]. A rationalisation consistent with the quenched flow observations envisages the synthetase to possess two hydrophobic sites, one for the correct substrate, valine, and a second, possibly displaced to allow the 2'-3' migration (eq. 5.27), containing an extra hydrogen bonding facility in a suitable position to interact with the threonine hydroxyl group.[10] The binding energy at this adjacent site may then be used to effect hydrolysis of the mischarged $tRNA^{val}$-Thr.

There have evolved therefore two contributory mechanisms for the amino acid specificity of the synthetases, the initial k_{cat}/K_m ratios which discriminate for amino acids in the first step (see Chapter 4) and an 'editing mechanism' which functions in the second step.

5.2.6.2   The Enzymes of Glucose Metabolism

Table 5.10 details the enzymes of glycolysis, the main energy producing catabolic pathway of carbohydrate metabolism, and its chemical reverse, gluconeogenesis. Glucose and its storage form, glycogen, are among the most important energy sources for living systems. The initial enzyme catalysed stages in glucose catabolism occur in the cytoplasmic space of both aerobic and anaerobic cells. In aerobic cells such as those of liver and heart, the product of the glycolytic pathway, pyruvate, is further catabolised to acetyl—CoA which then enters the Krebs cycle. This route is not available in the anaerobic cells of muscle and here the tissue specific lactate dehydrogenase-5 converts the pyruvate to lactate, two molecules being produced for every one of glucose.* The lactate is secreted

---

* Anaerobic microorganisms promoting alcoholic fermentation have yet an alternative fate for pyruvate. Pyruvate decarboxylase, a TPP dependent enzyme, catalyses the conversion of pyruvate to acetaldehyde which is then further enzymically transformed to alcohol, eq. 5.30.

$$CH_3 \cdot CO \cdot CO_2H \xrightarrow[\text{decarboxylase}]{\text{Pyruvate}} CH_3 \cdot CHO \xrightarrow[\text{+ NADH}]{\text{Alcohol DH}} CH_3 \cdot CH_2 \cdot OH \quad (5.30)$$

Table 5.10: Enzymes of the Glycolytic Pathway

Reaction	Enzyme recommended name, EC No.	Cofactors	source*	Molecular weight; Number of subunits
D-Glucose				
↓	Hexokinase, 2.7.1.1	ATP, Mg(II)	Y	102,000; 2
D-Glucose-6-P				
↓	Phosphoglucoiso-merase, 5.3.1.9	—	M	125,000; 2
			Y	145,000
D-Fructose-6-P				
↓	Phosphofructo-kinase, 2.7.1.11	ATP, Mg(II)	M	190,000; 2
			Y	600,000; 6
D-Fructose-1,6-bisP				
↓	Aldolase, 4.1.2.13	—	Y	80,000; 2
			M	160,000; 4
Dihydroxyacetone-P				
+ ↓↓	Triose phosphate isomerase, 5.3.1.1	—	M	53,000; 2
Glyceraldehyde-3-P				
↓	Glyceraldehyde-3-P DH, 1.2.1.12	P_i, NAD	M	145,000; 4
1,3-Biphosphoglycerate				
↓	Phosphoglycero-kinase, 2.7.2.3	ADP, NAD, P_i	M	34,000
3-Phosphoglycerate				
↓	Phosphoglycero-mutase, 5.4.2.1		M	66,000; 2
			Y	110,000; 4
2-Phosphoglycerate				
↓	Enolase, 4.2.1.11	Mg(II)	M	82,000; 2
			Y	88,000; 2
Phosphoenol pyruvate				
↓	Pyruvate kinase, 2.7.1.40	ADP, Mg(II)	Y	162,000; 8
			M	240,000; 4
Pyruvate				
↓	Lactate DH, 1.1.1.27	NADH	M	140,000; 4
Lactate				

* Y = Yeast, M = Rabbit muscle.

into the blood stream and is transported to the liver where it is converted into glucose and glycogen. Although the majority of enzymes are common to both pathways, three important ATP-linked steps require different sets of enzymes in the two directions. In the catabolic direction, glucose is phosphorylated by ATP and hexokinase, followed in the third step by a second endergonic reaction, the phosphofructokinase catalysed phosphorylation of fructose phosphate by ATP. The penultimate step is the exergonic conversion of phosphoenol pyruvate (PEP) to pyruvate, catalysed by pyruvate kinase. In gluconeogenesis, the first two reactions mentioned become exergonic and are reversed by relatively non-specific phosphatases, glucose-6-phosphatase and hexosediphosphatase respectively. The enzymatic reversal of PEP hydrolysis is more complex however since the free energy change is about twice that for ATP hydrolysis; principally because the transient enol form of pyruvate, produced as an intermediate in the hydrolysis of PEP, tautomerises to the keto form with an additional decrease in free energy of about 30 kJ mol^{-1}. This step is bypassed by the concerted action of pyruvate carboxylase and PEP carboxykinase. Mitochondrial pyruvate carboxylase, an allosteric enzyme of MW 5×10^5 requiring biotin and acetyl-CoA as coenzymes, catalyses the fixation of carbon dioxide to pyruvate to yield oxaloacetate (OAA). The dicarboxylic acid is then catalytically decarboxylated and phosphorylated to PEP by PEP carboxykinase, also an allosteric enzyme. This reaction is expensive in energy but is paid for by the nucleotide guanosine triphosphate (GTP):

$$\text{Pyruvate} + CO_2 + \text{ATP} = \text{OAA} + \text{ADP} + P_i \qquad (5.31)$$

$$\text{OAA} + \text{GTP} = \text{PEP} + CO_2 + \text{GDP} \qquad (5.32)$$

The net reaction thus takes place with little total change in free energy ($+0.8$ kJ mol^{-1}).

Phosphorylase 'a' catalyses the progenerative step of glycolysis, the removal and phosphorylation of a glucose residue from glycogen (eq. 5.33). The control of its activity, crucial to the regulation of glucose metabolism, is an important example of energy linked covalent modulation of enzyme activity (Section 3.4.3.1).

$$\text{Glucose}_{n+1} \xrightarrow{\text{Phosphorylase 'a'}} \text{Glucose-1-phosphate} + \text{Glucose}_n$$
$$(5.33)$$

Phosphorylases 'a' are key enzymes in the mobilisation of glucose and accordingly have been extensively studied, originally by C. Cori and G. Cori and later by Sutherland. In its active form, the muscle enzyme is a tetramer of MW 370,000 daltons. Each subunit is identical and contains a phosphate group attached to a serine hydroxyl and a pyridoxal phosphate linked to the ε-amino group of lysine. Active phosphorylase 'a' probably results from the dimerisation of a partially active unphosphorylated dimeric precursor phosphorylase 'b' catalysed by phosphorylase 'b' kinase, eq. 5.34.

$$2 \times \text{HO} - \bigcirc\bigcirc - \text{OH} \xrightarrow[+\ 4\,\text{ATP}]{\text{Phosphorylase 'b' kinase}} \quad \text{(5.34)}$$

Phosphorylase 'b'   Phosphorylase 'a'

Phosphorylase 'b' can also be partially activated by several organophosphates such as 5'-AMP, glucose-6-phosphate and UDP-glucose. Accordingly there are suggestions that phosphorylase 'b' kinase alters the allosteric response of phosphorylase 'b' to these, rather than inducing an inactive-active transformation. (The tetrameric state of phosphorylase 'a' may not in fact be the physiologically active species in muscle since the corresponding enzyme in liver is a dimer exhibiting a higher affinity for glycogen and a lower activation energy of reaction.) The activation of phosphorylase 'b' is reversed by the specific enzyme, phosphorylase phosphatase, which removes the four phosphate groups. *In vitro* this results in dissociation of the tetramer back to the dimer.

Muscle phosphorylase 'b' kinase, MW 1.3×10^6 daltons, is comprised of three different subunits and probably has a structure $A_4B_4C_8$. Like phosphorylase it also exists in interconvertible forms of differing kinetic properties. The more active is produced by protein kinase catalysed phosphorylation of two serine residues in the B-subunits. Protein kinase promotes simultaneous phosphorylation of the A-subunits, but more slowly, to yield an enzyme a hundred-fold more susceptible to phosphorylase kinase phosphatase. It may be that the consequent lag period is part of a control device whereby sufficient glycogen is broken down before the activating enzyme is itself inactivated.

The activity of protein kinase is triggered by increases in cellular levels of the positive modifier cyclic AMP (cAMP, see Section 5.2.1).

Rabbit muscle protein kinase binds cAMP with a K_m value of about 10^{-7}M, i.e. a strong interaction is involved. The activity stimulated in protein kinase by cAMP is the catalysed attachment of a phosphate group on specific serines in proteins (like phosphorylase kinase) which have one or more arginines within 2 to 5 residues on the N-side of the modifiable group. Two types of protein kinase have been found in mammalian tissues, type I, present in heart and brain tissue and type II present in skeletal muscle and adipose tissue. Both types are tetrameric, comprising two different polypeptides, with regulatory (R) and catalytic (C) properties respectively, and both undergo reversible dissociation to free, active catalytic units in the presence of cAMP. However type I protein kinases catalyse their own phosphorylation whereas there is no phosphorylation step with type II. In type I, phosphorylation acts as a further modulation of activity since only the phospho-form can be activated by cAMP:

$$\text{Type I} \quad C_2R_2 \rightarrow C_2(R \cdot P)_2 \underset{}{\overset{2 \cdot cAMP}{\rightleftharpoons}} (R \cdot P)_2(cAMP)_2 + C_2$$
$$\text{Inactive} \qquad\qquad\qquad\qquad\qquad \text{Active}$$

$$\text{(5.35)}$$

$$\text{Type II} \quad C_2R_2 \underset{}{\overset{2 \cdot cAMP}{\rightleftharpoons}} R_2(cAMP)_2 + C_2 \qquad \text{(5.36)}$$
$$\text{Inactive} \qquad\qquad \text{Active}$$

The complete enzyme mediated route from the hormone-induced cAMP to the first step of glycolysis is shown in Figure 5.11.

The metabolic interconversions of glycogen provide excellent illustrations of the intricacy of metabolic control exercised at the level of enzyme activity. The synthesis of glycogen is catalysed by glycogen synthase (UDP-glucose-glycogen glucosyl transferase) which adds a glucose unit to an existing glycogen primer, eq. 5.37.

$$+ \text{ Glucose}_n \longrightarrow \text{Glucose}_{n+1} + \text{UDP} \qquad \text{(5.37)}$$

In common with the other enzymes of glycogen metabolism, the synthase also exists in two interconvertible forms, one of which is

Figure 5.11: Enzyme Mediated Route to Glycolysis

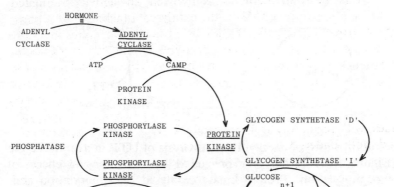

Note: Enzymes underlined are more active

dependent on glucose-6-phosphate for activity (designated D, or 'b' by analogy with phosphorylase) and a glucose-6-phosphate independent form (I or 'a'). Like the phosphorylases, the two forms are reversibly interconverted by enzyme catalysed phosphorylation-dephosphorylation reactions, but in contrast, phosphorylation of active glycogen synthase I produces the partially active D form, which is dependent on allosteric interactions with glucose-6-phosphate for full activity, eqs. 5.38, 5.39.

$$\text{GSD} \underset{\text{Glucose-6-P}}{\overset{}{\rightleftharpoons}} \text{GSD} \underset{\text{ATP}}{\overset{P_i}{\rightleftharpoons}} \text{GSI}$$

GSD		GSD		GSI
Active	Glucose-6-P	Partially active	ATP	Active

(5.38, 5.39)

The phosphorylation of glycogen synthase is catalysed by a kinase whose properties are identical to those of protein kinase. Both glycogen synthesis and its breakdown are thus coupled and triggered by the same hormonal message carried by cyclic AMP. A

Figure 5.12: A Futile Cycle

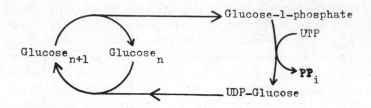

resultant waste of energy, the hydrolysis of UTP in a 'futile' cycle (Figure 5.12), is however precluded by the opposing effects of phosphorylation. Protein kinase catalysed phosphorylation activates phosphorylase 'b' but deactivates glycogen synthase. Thus increased levels of cAMP stimulate glycogen utilisation but depress glycogen synthesis, Figure 5.11.

5.2.6.3 Fatty Acid Synthesis

Long chain fatty acids are needed for adipose storage and are essential components of membranes. Their biosynthesis in mammalian and yeast cells is catalysed by polyfunctional fatty acid synthetase (FAS) complexes, often described as type I, and in bacteria by loosely aggregated series of enzymes termed FAS type II. However, the overall reactions and step-wise chemical interconversions from acetyl-CoA to palmitate are similar in both. Stoichiometrically the reaction is:

$$8 \text{ Acetyl-CoA} + 7 \text{ ATP} + 14 \text{ NADPH} + 14 \text{ H}^+$$

$$= \text{Palmitate} + 7 \text{ ATP} + 7 \text{ P}_i + 14 \text{ NADP}^+ + 8 \text{ CoA} \quad (5.40)$$

A direct link via acetyl-CoA thus exists for the conversion of carbohydrates into fatty acids. Except for the first stage of fatty acid synthesis acetyl-CoA is not incorporated directly into the polymethylene chain but is first carboxylated to malonyl-CoA by acetyl-CoA carboxylase. A regulatory function determining the rate of fatty acid synthesis is generally accepted for this biotin dependent enzyme. That from avian liver is allosteric, its aggregation being positively effected by citrate but negatively effected by palmityl-CoA (the end product). In the presence of citrate both the rat and avian liver enzymes have been isolated as filaments approximately

100×4000 Å and MW $4\text{-}5 \times 10^6$ daltons. Removal of citrate induces their disaggregation into inactive protomers of 4×10^5 daltons containing two biotin molecules. Each protomer is tetrameric. The yeast enzyme is also a tetramer, of subunit size 150,000 daltons and like the liver enzymes its carboxylase, transferase and biotin carrier properties appear to be located on the same polypeptide chains. By comparison, *E. coli* acetyl-CoA carboxylase is comprised of three functionally distinct proteins: a carrier polypeptide of 82 residues containing one biotin molecule, a dimeric carboxy-biotin synthetase of MW 100,000 containing the ATP and bicarbonate binding sites and a transcarboxylase of two subunits each of 90,000 daltons.

The sequence of reactions by which acetyl-CoA and the malonyl-CoA molecules are incorporated into the growing fatty acid chain are shown in Figure 5.13. The central component of *E. coli* FAS is the acyl carrier protein, ACP. Except for the condensation stage this polypeptide transmits the growing acyl function through all the enzymic steps of each cycle of FAS by a series of protein-protein interactions. The substrates are linked as thioesters to 4′-phosphopantotheines bound to ACP, this 'prosthetic group' being attached to a serine residue approximately halfway along the

Figure 5.13: Fatty Acid Biosynthesis (ACP *is acyl carrier protein*)

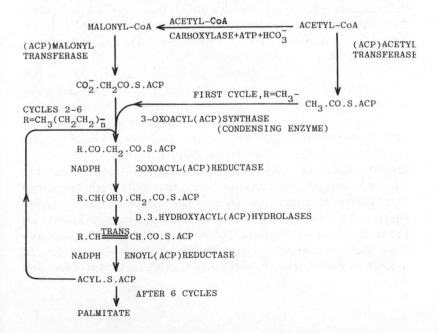

primary sequence of 77 amino acids. The sequence around this residue is -gly-ala-asp-ser36-leu-asp-.

Transfer of acetyl-CoA and malonyl-CoA to ACP is catalysed by two different enzymes, specific for their respective substrates. They both promote transacylation with the intermediate formation of an acyl enzyme. In (ACP) acetyl transferase this is probably via a sulphydryl group, as judged by its sensitivity to periodate and iodoacetate but in (ACP) malonyl transferase the intermediate ester is probably via serine. The molecular weight of the latter enzyme from *E. coli* is 36,500 daltons.

Condensation is a critical step in *E. coli* fatty acid synthesis since it is at this stage that the chain is increased in length. The acetyl group is first transferred from ACP to a sulphydryl group of the enzyme which then catalyses its condensation with malonyl-CoA. The second activated ester is introduced by the ACP freed in the first step. Free enzyme and carbon dioxide are then released. *E. coli* oxoacyl synthetase is a dimer of 66,500 daltons containing identical subunits. It exhibits similar K_m values for saturated fatty acids shorter than palmityl but little reactivity with this group; hence the enzyme also limits the range of the pathway. The NADP dependent reductase and hydrolase are both specific for the D(—) isomer. Three distinct hydrolases have been found in *E. coli* acting on short, intermediate and long chain 3-hydroxy fatty acyl groups. After the second reduction step the fatty acyl group is transferred to the condensing enzyme to which a further malonyl-CoA molecule is then directed by the carrier protein. On completion of the palmityl homologue a specific ACP thioesterase cleaves the thioester bond to yield, in *E. coli*, the free fatty acid.

The type I fatty acid synthetases are multifunctional catalysts (see Chapter 1) and although they catalyse the reactions described above the enzymes are close agglomerates of fewer proteins. The yeast enzyme is ellipsoidal, MW approximately 2.3×10^6 daltons and possesses a subunit structure A_6B_6, where A and B are 185,000 and 175,000 daltons respectively. Support for the existence of several copies of only two multifunctional polypeptides has emerged from genetic analysis of yeast fatty acid synthetase auxotrophs. Two gene loci have been found,[11] fas-1 and fas-2, each of which codes for a series of enzymic activities. Gene fas-1 codes for the series acetyl transacylase, malonyl transacylase, dehydrase and enoyl reductase, and polycistronic gene cluster fas-2 codes for the 4'-phosphopantotheine binding region, 3-oxo-acyl synthetase and

3-oxo-acyl reductase. Labelling studies show fas-1 to code the B-subunit and fas-2 the A-subunit.

The A-polypeptide contains the two classes of sulphydryl groups found by Lynen.[12] A 'peripheral' thiol reacted rapidly with iodo-acetamide but was protected by acetyl CoA and a 'central' thiol showed lower reactivity towards alkylating reagents. The former has since been shown to be a cysteine of the 3-oxoacyl synthetase domain and the latter 4'-phosphopantotheine. As for the type II FAS the prosthetic group is bound to a serine group of the carrier protein and between 4 and 6 have been counted per FAS molecule. The initial charging of yeast fatty acid synthetase with acetyl-CoA and malonyl-CoA follows a course similar to that of the *E. coli* enzyme. However the enzymic activities are all contained within one complex and Lynen[12] has proposed that the growing polymeric chain is attached to the flexible 'central' pantotheine group which travels from one site of enzymic activity to the next with a concentric motion.* A mechanism of this type it is argued would increase catalytic efficiency since the intermediates could not diffuse away from the reaction centre and undergo competing reactions.

5.2.7 Control of Metabolism

The foregoing sections have stressed that compartmentation, allosteric modulation, physical organisation and cascade reactions all contribute to metabolic regulation. Individually, many of these controlling factors are understood in considerable detail but what is needed is a detailed quantitative picture of cellular metabolism into which these factors can be integrated. Such a picture would include the enzymes of the metabolic pathways, their concentrations and rates of reaction, and the concentrations of their substrates, products and modifiers. A mathematical description must be able to predict changes in the rates of the various aspects of cellular metabolism when different limiting conditions are applied.

An approach to a model for the control of metabolic rate has been framed by Kacser and Burns,[13] but before their theory is outlined some of the underlying principles are first discussed.

Consider the addition of the first metabolite (G) to the metabolic pathway

$$G \underset{E_g}{\rightleftharpoons} H \underset{E_h}{\rightleftharpoons} I \underset{E_i}{\rightleftharpoons} J \qquad (5.41)$$

---

* Mechanically similar to that suggested for pyruvate dehydrogenase (Chapter 1).

In a closed system, i.e. one that allows no exchange of material with its environment, equilibrium will be reached after some time with $J/G = K_e$, and in which, although each individual reaction will still be catalysed, each *net* reaction $G \rightleftharpoons H$, $H \rightleftharpoons I$, etc. will be zero (i.e. each will be dynamically balanced). If the system is converted to an open one where J is continually removed and G continually added at constant rates, an equilibrium position will not be reached but instead a steady state will prevail in which the rate of synthesis of each intermediate will equal its rate of utilisation, i.e. there will be a constant flow or state of flux, F.

A net flux from G to H will require non-equilibrium conditions (i.e. $J/G < K_e$) and an overall negative Gibbs free energy change. The extent to which each step in the series is in a state of non-equilibrium will be different but each can be described by a disequilibrium ratio (ρ). For any step $X \underset{v_{-1}}{\overset{v_1}{\rightleftharpoons}} Y$ (where v_1 and v_{-1} are respectively the rates of the forward and back reactions) ρ is defined as $Y/(XK_e)$ and the flux of the reaction is $F = v_1 - v_{-1}$. It can be shown that $\rho = v_{-1}/v_1$* and this ratio is a measure of the closeness with which a step approaches its equilibrium position (when equilibrium is attained $\rho = 1$). Thus it can be used as an aid to the identification of the possible step(s) in a sequence where flux regulation could occur. A near-equilibrium reaction, with a high value of ρ, would not be a good site for regulation, since decreasing the enzyme activity would alter the rates of both forward and back reactions to virtually the same extent. Steps characterised by disequilibrium ratios less than 0.05 are possibilities for regulation since the flux is determined essentially by v_1 and altering the enzyme activity or concentration would lead to large changes in rate. (This would be especially so if the physiological substrate concentration was saturating.) Such steps would be characterised by large negative Gibbs free energies and these types are often found near the termini of metabolic routes (see for example the glycolytic pathway).

In their theory for the control of flux, Kacser and Burns[13] first distinguished between 'parameters', the factors which can be set in an experiment, and 'variables' the factors not under control and which respond to the conditions applied. They then introduced a

* For a unimolecular reaction $X \underset{k_{-1}}{\overset{k_1}{\rightleftharpoons}} Y$, $v_{-1}/v_1 = k_{-1}Y/k_1X = Y/(X \cdot K_e) = \rho$; the algebraic manipulations involving enzyme intermediacy are more complicated but the result is the same.

response coefficient, R, as a measure of the control exercised by a parameter on an enzyme catalysed reaction. R is defined as the small fractional change in flux (F) resulting from a small fractional change in concentration of a parameter (P) (such as an inhibitor).

$$\delta F/F = R \cdot \delta P/P \qquad (5.42)$$

This response they further considered to be the product of an enzyme 'controllability' factor and a system 'sensitivity' factor. The 'controllability' coefficient (K) of an enzyme they defined as the fractional change in rate produced by a fractional change in local parameter concentration.

$$\delta v/v = K \cdot \delta P/P \qquad (5.43)$$

The enzymic system will respond to a change in parameter concentration and the fractional change in flux that results from a change in enzyme activity levels they called the 'sensitivity coefficient' (Z).

$$\delta F/F = Z \cdot \delta E/E \qquad (5.44)$$

The numerical value of Z is taken to indicate the sensitivity of a step to a regulatory influence and thus the extent to which the step is controllable. Summation of the sensitivity coefficients along a pathway yields a numerical value of unity. This means that all the contributing enzymes could have fairly low values. Alternatively one could have a relatively high sensitivity coefficient in which case that is catalysing the controllable step.

Eliminating $\delta P/P$ between eqs. 5.42 and 5.43 and substituting the result into eq. 5.44 gives $R = ZK$. Insertion of this into eq. 5.42 yields,

$$\delta F/F = ZK \cdot \delta P/P \qquad (5.45)$$

which predicts that a change in flux is linked to the sensitivity of the step, its controllability and the relative size of the alteration in parameter concentration.

The extent to which the enzyme rate responds to changes in concentrations of a substrate, product, modulator or other *variable* (M) is defined as the 'elasticity' (E), thus,

$$\delta v/v = E \cdot \delta M/M \qquad (5.46)$$

The elasticity coefficient can be determined from the enzyme kinetic properties. For example, the rate equation for the reaction $E_g + G \rightleftharpoons E_g G \rightleftharpoons E_g H \rightleftharpoons E_g + H$ is:

$$v = \frac{G \cdot V_{max}^G/K_m^G - H \cdot V_{max}^H/K_m^H}{1 + G/K_m^G + H/K_m^H} \quad \text{(see eq. 2.38)} \qquad (5.47)$$

where V_{max}^G and K_m^G are the Michaelis parameters for the forward direction and V_{max}^H and K_m^H are those for the reverse direction. From eq. 5.46 $E^H = (\delta v/v)/(\delta H/H)$ thus the elasticity is obtained by partial differentiation of the rate equation eq. 5.47. If non-saturating conditions are employed the following relationship is obtained:

$$E_g^H = - \frac{H/(K_g \cdot G)}{1 - H/(K_g \cdot G)} \qquad (5.48)$$

where K_g is the equilibrium constant. The quotient $H/(K_g \cdot G)$ is however the disequilibrium ratio, ρ_g, for the step so that

$$E_g^H = -\rho_g/(1 - \rho_g) \qquad (5.49)$$

and by a similar process the elasticity of E_h with respect to H is

$$E_h^H = 1/(1 - \rho_h) \qquad (5.50)$$

Kacser and Burns showed that the elasticities of an enzyme for its substrate regulate its sensitivities. For the reaction sequence in eq. 5.41, if Z_g and Z_h are the sensitivities of enzymes E_g and E_h respectively and if E_g^H and E_h^H are their respective elasticities for the intermediate H, then:

$$Z_g/Z_h = -E_h^H/E_g^H \qquad (5.51)$$

i.e. the sensitivities are inversely proportional to the ratio of their elasticities towards the intermediate. Substituting eqs. 5.49 and 5.50 into this relationship yields

$$Z_g/Z_h = (1 - \rho_g)/\rho_g(1 - \rho_h) \qquad (5.52)$$

Thus the sensitivities of the enzymes in the pathway are related to their disequilibrium ratios. More generally,

$$Z_g : Z_h : Z_i : Z_j : \dots$$
$$= 1 - \rho_g : \rho_g(1 - \rho_h) : \rho_g\rho_h(1 - \rho_i) : \rho_g\rho_h\rho_i(1 - \rho_j) : \dots \quad (5.53)$$

If any step is at equilibrium, its disequilibrium ratio is unity and thus the term corresponding to its sensitivity coefficient becomes zero, i.e. the step is insensitive and non-controlling. Kacser and Burns thus propose that metabolic pathways as integrated wholes should be analysed and terms like 'pacemaker' enzymes, etc. replaced by more quantitative specifications such as sensitivity ratios.

5.3 ENZYME MULTIPLICITY[14,15]

There is a striking similarity in the biochemical sequences and types of reaction occurring in living organisms. Most of the chemical reactions that constitute glycolysis in muscle and liver cells, for example, are identical to those of alcoholic fermentation carried out by yeast and similar reaction sequences controlling the synthesis of purines and pyrimidines from amino acid precursors are found in both bacterial cells and those from higher animals.

However, even though the series of chemical steps may be the same in the species of these widely divergent groups, the individual enzymes catalysing the common steps are dissimilar. A considerable amount of research has shown that more often than not, reactions are catalysed by location specific enzymes, and that enzymes catalysing the same reaction in different tissues are often very different proteins. Such enzymes are entitled *isodynamic*. For example, isodynamic proteins with alcohol dehydrogenase activity are present in both yeast and liver cells, yet the latter with a molecular weight of 70,000 daltons is only half the size of that from the microorganism. Such isodynamic enzymes occurring in different species are frequently termed *heteroenzymes*.

Even when individuals within a given species are considered, different forms of an enzyme are found. Human heart muscle lactate dehydrogenase for example, exhibits a very different electrophoretic mobility and substrate selectivity to that isolated from human skeletal muscle; and isodynamic creatine kinases have been purified from human brain and muscle. The same is true for many other enzymes catalysing identical reactions in different organs within a single species.

Markert has pointed out that the genes coding for the enzyme molecules will be subject to different evolutionary pressures in the specialized locations of the cell, shaping the structures and roles of the molecules.[14] It is unlikely that the influences will be the same in each of these locations and thus it is to be expected that dissimilar forms of an enzyme will have developed. It is also to be expected that more than one mechanism for the formation of these different forms will have evolved. This view is borne out by the differences in physico-chemical character found to underly the observations of enzyme multiplicity. These can be categorised under two broad headings (Table 5.11), multiplicities within a species attributable to genetically determined differences in primary structure (groups 1, 2 and 3) and to post-translational (epigenetic) changes in the proteins (groups 4, 5 and 6).

Originally, isodynamic enzymes occurring within the same species were called *isoenzymes* or *isozymes*.[16] These terms are still used operationally to describe enzyme fractions exhibiting similar

Table 5.11: Multiple Forms of Enzymes

	Group	Example
1.	Genetically independent proteins	Malate dehydrogenase in mitochondria and cytosol
2.	Heteropolymers (hybrids) of two or more polypeptide chains, non-covalently bound	Lactate dehydrogenase
3.	Genetic variants (allelozymes)	Glucose-6-phosphate dehydrogenase in man
4. (a)	Proteins conjugated with other groups	Phosphorylase a and b
(b)	Proteins derived from one polypeptide chain	The family of chymotrypsins arising from chymotrypsinogen
5.	Polymers of a single subunit	Glutamate dehydrogenase of MW 1,000,000 and 250,000
6.	Conformationally different forms	All allosteric modifications of enzymes

Source: See Section 1.7, Reference 21.

catalytic activities where the exact cause of the multiplicity has yet to be determined. But where the cause of the multiplicity is known the definition, isoenzyme, is restricted to those proteins falling into groups 1, 2 and 3 of Table 5.11. Isoenzymes can then be described[15] as 'enzymically active proteins, catalysing the same reaction and occurring in the same species but differing in certain physicochemical properties arising from genetically determined differences in primary structure'.

Enzymes which undergo self polymerisation, e.g. pseudocholinesterase from blood plasma, which can be separated into as many as seven bands on gel electrophoresis, are thus excluded from this definition. Serum cholinesterases do however exhibit genetic polymorphism, inherited variations in individuals. This phenomenon was first detected in the cholinesterases after variable sensitivities were detected in patients treated with the muscle relaxant suxamethonium

$$(CH_3)_3 \cdot N^+ \cdot (CH_2)_2 \cdot O \cdot CO \cdot (CH_2)_2 \cdot CO \cdot O \cdot (CH_2)_2 \cdot N^+ \cdot (CH_3)_3$$

Three phenotypes have been observed in the population: individuals possessing the typical gene and who normally and rapidly hydrolyse suxamethonium, heterozygotes with decreased levels of activity and those who are homozygous for the abnormal allelic gene, who are thus sensitive to the drug. Also excluded from classification as isoenzymes are the family of chymotrypsins (all of which are formed epigenetically by enzymic degradation of the much less active precursor chymotrypsinogen), allosteric modifications of enzymes, and apoenzymes conjugated to different non-protein groups.

5.3.1 Separation, Localisation and Nomenclature

Isoenzymes are species specific, isodynamic proteins which can differ in their primary structures, affinities for substrates, inhibitors and coenzymes and in their stabilities to thermal and chemical denaturation. All these variations have been used for their differentiation and characterisation, but the predominant method employed to indicate multiplicity and to distinguish between isoenzymes is electrophoresis. This technique is insensitive to the degree of purity in the original sample and when coupled to selective histological staining, has the advantage of high resolution and rapid performance.

Figure 5.14: *Schematic Electrophoretogram of Lactate Dehydrogenase* (LD) *Isoenzymes*

For these reasons international bodies have recommended that isoenzyme nomenclature be based on electrophoretic behaviour. Each isoenzyme band is given a number consecutively from the zone of greatest mobility towards the anode. The isoenzyme is then described by the enzyme name followed by its number. For example, lactate dehydrogenase-1 (LD1) has the fastest and LD5 the slowest anodic mobility in buffer, pH 8.6 (Figure 5.14).

The methods and procedures employed for the separation and quantitation of isoenzymes have been comprehensively described by many authors.[14,15] Location of the activities after electrophoresis is usually performed by direct application to the support medium of histochemical reagents designed to produce intense colours at the areas of enzyme activity. The gel is then scanned either visually or densitometrically in order to quantitate the relative abundance of each isoenzyme.

5.3.2   Structures and Properties of Some Isoenzymes*

5.3.2.1   *Lactate Dehydrogenase* (EC 1.1.1.27)

Lactate dehydrogenase isoenzymes are probably the most extensively studied of the 300 or so enzymes found to possess multiple forms. On electrophoresis five separable isoenzyme bands, each of which exhibits lactate dehydrogenase activity, are given by most vertebrates (Figure 5.14). All are tetramers of MW about 140,000 daltons requiring four molecules of NADH for activity. The equal spacing of the isoenzymes on the gel indicates that they differ by equal increments of charge, a conclusion supported by amino acid analysis of the extreme isoenzymes. LD1 contains approximately twenty more aspartic acid residues and ten fewer lysine side chains than LD5. These values and the molecular weights of the

* See also Chapter 6.

isoenzymes are reduced to one-quarter after denaturation. Taking these observations into account, Markert[14] proposed that LD1 and LD5 were homopolymers and that the intermediate isoenzymes were mixtures containing different proportions of the two different subunits. This proposal has since been amply verified. Repeated freeze-thaw hybridisation of a mixture of LD1 and LD5, for example, gives a mixture of all the five isoenzymes.

The four identical polypeptide chains of LD1 have been abbreviated H (Heart) and the four LD5, M (Muscle).* Thus LD1 can be written H_4 and LD5, M_4. It is now thought that these two polypeptide chains are under the control of two distinct genes located on different chromosomes. By analogy with the suggested evolution of the α and β chains of haemoglobin, polypeptides H and M may also have arisen by gene duplication followed by mutations at the two loci. The activities of the two gene apparati in the particular tissues will then determine the concentrations of the two chains and hence the proportions of the isoenzymes.

Typical lactate dehydrogenase isoenzyme distribution patterns found in various vertebrate organs and cells are described in Table 5.12. Although the quantities reported from individual sources may differ slightly, these average percentages indicate the general, characteristic pattern, elaborated by most animals. Heart muscle and kidney cells contain mainly the anodic isoenzymes LD1 and LD2, while a larger concentration of the slower moving LD4 and LD5 is

Table 5.12: Percentage Distribution of Lactate Dehydrogenase Isoenzymes in various Vertebrate Tissues

Isoenzyme	Heart	Erythrocyte	Skeletal muscle	Liver
LD1	60	40	—	—
LD2	30	50	—	5
LD3	10	10	10	10
LD4	—	—	25	30
LD5	—	—	65	60

Source: Reference 17.

* Some authors have labelled these A and B respectively.

possessed by liver cells and skeletal muscle. The lack of LD5 in mature erythrocytes probably reflects the lower stability of this isoenzyme.

Besides variations in electrophoretic mobility, the different chemical structures of the two polypeptides elicit distinct kinetic behaviour. Some of the differences in substrate specificity, apparent Michaelis constants and pH maxima are listed in Table 5.13.

Table 5.13: Kinetic Behaviour of Lactate Dehydrogenase Isoenzymes 1 and 5

Substrate	Assay pH	LD1 $K_{m(app)}$(M)	pH Optimum	LD5 $K_{m(app)}$(M)	pH Optimum
Pyruvate	7.4	1.2×10^{-4}	6.5–8	8×10^{-4}	6.8
Hydroxy-butyrate	7.4	6×10^{-4}	—	3×10^{-3}	—
Lactate	8.7	1×10^{-2}	8.5	2×10^{-2}	8.5

Source: Reference 15.

With all three substrates the heart-type isoenzyme exhibits lower $K_{m(app)}$ values. Although, as argued in Chapter 2, $K_{m(app)}$ cannot always be equated to K_s (the enzyme-substrate dissociation constant), for lactate dehydrogenase this assumption appears valid. Thus LD1 manifests a slightly greater affinity for its substrates. This isoenzyme is also more susceptible to inhibition by high pyruvate levels than is LD5. The metabolic significance of this *in vitro* observation has yet to be fully explained, but one proposal is that inhibition of the heart enzyme by high substrate levels prevents the accumulation of lactate in this tissue, which of course would be detrimental to the individual, and allows the more complete utilisation of glucose for energy. The physiological roles of the LD isoenzymes will be discussed in Section 5.3.3.

The two isoenzymes may also be distinguished by their reactivity towards 2-hydroxybutyrate. This substrate shows greater affinity for the heart isoenzymes and accordingly these are often named hydroxybutyrate dehydrogenases, particularly in clinical laboratories where this reaction is used for the differentiation of the heart and muscle isoenzymes (Chapter 6).

5.3.2.2   *Glucose-6-phosphate Dehydrogenase* (EC 1.1.1.49)

D-Glucose-6-phosphate + NADP$^+$

$$= \text{D-Glucono-}\delta\text{-lactone-6-phosphate} + \text{NADPH} \quad (5.54)$$

Glucose-6-phosphate dehydrogenase is the first directional enzyme to the pentose phosphate pathway* and is thus widely distributed in nature. Glucose-6-phosphate dehydrogenase from human erythrocytes has been studied extensively as a result of its involvement in drug induced anaemias and in congenital disorders resulting from its decreased activity.

Normal erythrocyte glucose-6-phosphate dehydrogenase is dimeric with subunits of molecular weight around 10^5, each of which binds one molecule of nicotinamide adenine dinucleotide phosphate.

Multiple forms of the enzyme have been examined electrophoretically. The number of stainable bands has been variously reported between two and six, a variation attributed to the extent of coenzyme binding. Most reports agree however with the identification of one major and other minor bands. The major band in normal human red cells is described as B +, to distinguish it from that of a faster migrating band found in about one-fifth of normal American negro males, labelled A + (the positive sign indicates enzyme sufficiency). The increased electrophoretic mobility of A + is caused by an asparagine side chain in place of aspartic acid in B +. At the nucleic acid level this corresponds to a mutation in the first base of the aspartic acid codon triplet from G to A. The resulting genes are then allelic variants, and thus the origin of G6PDH multiplicity is similar to that responsible for variations in the haemoglobin family, and is another example of enzyme polymorphism.

From a study of the distribution of glucose-6-phosphate dehydrogenase polymorphism amongst American negro families, it has been concluded that inheritance of the A + phenotype is carried by the sex chromosomes rather than the autosomal** chromosomes.

---

* In distinction from the glycolytic pathway this is linked to the production of NADPH and via glutathione to the maintenance of haemoglobin in the active reduced state.

** An autosome is a chromosome other than a sex chromosome.

Table 5.14: Glucose-6-phosphate Dehydrogenase Allelozymes

Allelozyme	Erythrocyte proportion (%)	Occurrence	Electrophoretic mobility[a]	Enzymopathy[b,c]
B+	100	Most common	N	none
B−	10	Mediterranean	N	DI, Favism
A+	80	American negro males	F	none
A−	20	American negro males	F	DI
Baltimore (C+)	75	Negroes	S	none
Barbieri	50	Italy	F	DI
Canton	25	Orient	F	DI
Ibadan	70	Negroes	S	none
Kerala	50	Asia	S	none
Madison	100	Scandinavia	S	none
Markham	10	New Guinea	N	none
Oklahoma	10	W. Europe	N	HA
Seattle	20	Celts	S	HA

Source: Reference 20.
[a] F, fast; N, normal; S, slow.
[b] Pathological consequence of an enzyme variant.
[c] DI, drug induced haemolysis; HA, haemolytic anaemia.

This applies also to the other genetically determined G6PDH variations, of which approximately twenty-five have been recorded. Some of these are listed in Table 5.14.

Deficiencies of the enzyme or the presence of its unstable variants cause reduced NADPH levels, and these lead to decreased concentrations of reduced haemoglobin which in turn destabilises the erythrocyte. Haemolysis may then take place which if prolonged may cause anaemia. This pathological response of the stressed cell may be induced by sulphonamides, antimalarials, analgesics, fava beans, or by stress caused by disease.

5.3.2.3   Carbonic Anhydrase (EC 4.2.1.1)

$$HCO_3^- + H_3O^+ = CO_2 + H_2O \qquad (5.55)$$

The simplest way in which isoenzymes can differ is in the amino acid sequences of their polypeptide chains. Such an explanation holds for the multiplicity of human erythrocyte carbonic anhydrase. This activity is resolvable into three forms A, B and C which constitute respectively about 5, 80 and 15 percent of the total weight of the

Figure 5.15: The Primary Structures of HCAB and HCAC (The sequences are numbered to match exactly homologous regions)

```
              1                5                  10               15                20               25              30
HCAB:Ac-Ala-Ser-Pro-Asp-Trp-Gly-Tyr-Asp-Asp-Lys-Asn-Gly-Pro-Glu-Gln-Trp-Ser-Lys-Leu-Tyr-Pro-Ile-Ala-Asn-Gly-Asn-Gln-Ser-Pro-
HCAC:   Ac-Ser-His-His-Trp-Gly-Tyr-Gly-Lys-His-Asp-Gly-Pro-Glu-His-Trp-His-Lys-Asp-Phe-Pro-Ile-Ala-Lys-Gly-Glu-Arg-Gln-Ser-Pro-

                          35                40                 45               50                55               60
     Val-Asp-Ile-Lys-Thr-Ser-Glu-Thr-Lys-His-Asp-Thr-Ser-Leu-Lys-Pro-Ile-Ser-Val-Ser-Tyr-Asn-Pro-Ala-Thr-Ala-Lys-Glu-Ile-Ile-
     Val-Asp-Ile-Asp-Thr-His-Thr-Ala-Lys-Tyr-Asp-Pro-Ser-Leu-Lys-Pro-Leu-Ser-Val-Ser-Tyr-Asp-Gln-Ala-Thr-Ser-Leu-Arg-Ile-Leu-

                          65                70                 75               80                85               90
     Asn-Val-Gly-His-Ser-Phe-His-Val-Asn-Phe-Glu-Asp-Asn-Asp-Asn-Arg-Ser-Val-Leu-Lys-Gly-Gly-Pro-Phe-Ser-Asp-Ser-Tyr-Arg-Leu-
     Asx-Asx-Gly-His-Ala-Phe-Asn-Val-Glu-Phe-Asp-Asp-Ser-Glx-Asx-Lys-Ala-Val-Leu-Lys-Gly-Gly-Pro-Leu-Asp-Gly-Thr-Tyr-Arg-Leu-

                          95               100                105              110               115              120
     Phe-Gln-Phe-His-Phe-His-Trp-Gly-Ser-Thr-Asn-Glu-His-Gly-Ser-Glu-His-Thr-Val-Asp-Gly-Val-Lys-Tyr-Ser-Ala-Glu-Leu-His-Val-
     Ile-Gln-Phe-His-Phe-His-Trp-Gly-Ser-Leu-Asp-Gly-Gln-Gly-Ser-Glu-His-Thr-Val-Asp-Lys-Lys-Lys-Tyr-Ala-Ala-Glu-Leu-His-Leu-

                         125               130                135              140               145              150
     Ala-His-Trp-Asn-Ser-Ala-Lys-Tyr-Ser-Ser-Leu-Ala-Glu-Ala-Ala-Ser-Lys-Ala-Asp-Gly-Leu-Ala-Val-Ile-Gly-Val-Leu-Met-Lys-Val-
     Val-His-Trp-Asn-Thr------Lys-Tyr-Gly-Asp-Phe-Gly-Lys-Ala-Val-Gln-Gln-Pro-Asp-Gly-Leu-Ala-Val-Leu-Gly-Ile-Phe-Leu-Lys-Val-

                         155               160                165              170               175              180
     Gly-Glu-Ala-Asn-Pro-Lys-Leu-Gln-Lys-Val-Leu-Asp-Ala-Leu-Gln-Ala-Ile-Lys-Thr-Lys-Gly-Lys-Arg-Ala-Pro-Phe-Thr-Asn-Phe-Asp-
     Gly-Ser-Ala-Lys-Pro-Gly-Leu-Gln-Lys-Val-Val-Asp-Val-Leu-Asp-Ser-Ile-Lys-Thr-Lys-Gly-Lys-Ser-Ala-Asp-Phe-Thr-Asn-Phe-Asp-

                         185               190                195              200               205              210
     Pro-Ser-Thr-Leu-Leu-Pro-Ser-Ser-Leu-Asp-Phe-Trp-Thr-Tyr-Pro-Gly-Ser-Leu-Thr-His-Pro-Pro-Leu-Tyr-Glu-Ser-Val-Thr-Trp-Ile-
     Pro-Arg-Gly-Leu-Leu-Pro-Glu-Ser-Leu-Asp-Tyr-Trp-Thr-Tyr-Pro-Gly-Ser-Leu-Thr-Pro-Pro-Leu-Leu-Glu-Cys-Val-Thr-Trp-Ile-

                         215               220                225              230               235              240
     Ile-Cys-Lys-Lys-Glu-Ser-Ile-Ser-Val-Ser-Ser-Glu-Gln-Leu-Ala-Gln-Phe-Arg-Ser-Leu-Leu-Ser-Asn-Val-Glu-Gly-Asp-Asn-Ala-Val-Pro-
     Val-Leu-Lys-Glu-Pro-Ile-Ser-Val-Ser-Ser-Glu-Gln-Val-Leu-Lys-Phe-Arg-Lys-Leu-Asn-Phe-Asn-Gly-Glu-Gly-Glu-Pro-Glu-Glu-Leu-

                         245               250                255              260
     Met-Gln-His-Asn-Asn-Arg-Pro-Thr-Gln-Pro-Leu-Lys-Gly-Arg-Thr-Val-Arg-Ala-Ser-Phe
     Met-Val-Asp-Asn-Trp-Arg-Pro-Ala-Gln-Pro-Leu-Lys-Asn-Arg-Gln-Ile-Lys-Ala-Ser-Phe-Lys
```

enzyme, about 1 g per litre of human red cells. Each isoenzyme is a single chain of MW about 30,000 daltons containing one zinc atom per molecule. The amino acid sequences of the B and C isoenzymes (HCAB and HCAC) are shown in Figure 5.15 and have been delineated to show their extensive (60 percent) homology. The proteins are probably functionally independent, the amounts of each vary independently in disease, and are encoded by two separate genes. Sequence homology may however have arisen early, in evolutionary time, by gene duplication, with the resultant genes then undergoing independent evolution. In a study of the primate isoenzymes, Tashian[19] found that HCAB possessed the greater variability and concluded that this isoenzyme has evolved more rapidly, whereas HCAC was possibly an older evolutionary form.

Homology is also evident in the crystal structures and three-dimensional configurations of the two isoenzymes, probably reflecting their polypeptide similarities. The folding of HCAC is shown schematically in Figure 5.16, and although for sequence comparison, two deletions are needed, these occur on the surface of the molecule with no ramifications transmitted elsewhere; many of

Figure 5.16: A Schematic Drawing of the Main Chain of the Carbonic Anhydrase Molecule. Cylinders represent helices and arrows represent β-structure. The dark ball in the middle is the essential zinc ion bonded to the protein by three histidyl ligands.

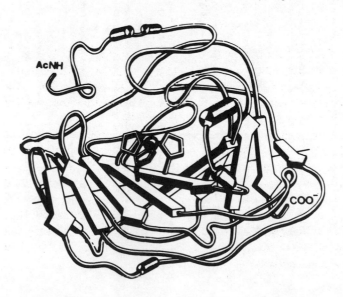

the homologous residues possess similar micro-environments. The molecule is roughly ellipsoidal in shape with dimensions 41 × 45 × 50 Å. Approximately 20 percent of the amino acid complement is organised into seven helices, found mainly on the surface and about 35 percent is folded into the internal sheet structures which bisect the molecule. Most of the hydrophilic side chains appear on the surface wrapped around a central hydrophobic core comprised of a majority of the leucyl, isoleucyl, valyl and aromatic side chains.

Depressed into one half of the molecule is the active site with its incumbent zinc(II) ion. Structurally the active site is comprised of a non-polar half and a polar half. Particularly important in the latter are three histidines which provide ligands to stabilise the Zn(II) ion attachment. The metal ion forms a distorted tetrahedral arrangement of bonds with a solvent molecule as the fourth ligand. This arrangement, participating in an extensive network of hydrogen bonds (VII), is found in the active site of both HCAC and HCAB,

$$
\begin{array}{c}
\text{Thr 199} \\
|\\
\text{CH·CH}_3 \\
|\\
\text{Glu 106} \diagup\!\!\!\!\diagdown \!\! \overset{O}{\underset{O}{}} \cdots H - O \cdots H - O \cdots Zn \cdots His \\
\end{array}
$$

VII

but the HCAC active site is more open compared to HCAB and is inactivated by alkylation at his 64 with bromopyruvate. The corresponding group in HCAB is shielded by his 67 and his 200, both more bulky than asn 67 and thr 200 of HCAC, and consequently only his 200 is modified.

Both isoenzymes exhibit Michaelis-Menten kinetics in the hydration and dehydration directions with similar pH dependencies of K_m and V_{max}. K_m is constant between pH 6 and pH 8 and V_{max} has a sigmoidal dependency with an inflexion at pH 8.0. The two enzymes do however differ considerably in their kinetic activities, probably as a result of the active site commutations. The isoenzymes were originally classified with respect to their total catalytic activities, but although HCAB is in a greater molecular proportion in blood, it possesses lower activity, as shown in Table 5.15.

Table 5.15: Rate Constants for Human Carbonic Anhydrase (CA) Catalysis (Units, $s^{-1} \times 10^{-6}$)

$$H_2O + CO_2 \underset{k'_{cat}}{\overset{k_{cat}}{\rightleftharpoons}} H^+ + HCO_3^-$$

	k_{cat}	k'_{cat}
HCAB	0.2	0.003
HCAC	1.4	0.081

Source: Reference 18.

5.3.3 Physiological Roles

Isoenzymes contribute to the functioning of their host cells by their individual responses to varying environmental conditions. In this respect, two features of their physiological roles will be considered, their involvement with the integration of carbon skeleton flux through divergent metabolic pathways and their functions in maintaining certain intercompartmental concentration gradients.

The basic features of enzyme regulation were treated in Chapter 3, where a common regulatory device, negative feedback inhibition, was described in which the product of a metabolic pathway inhibits an enzyme involved earlier in its biosynthesis. For linear unbranched pathways the logistics involved are relatively straightforward, but the fluxional balancing of branched, multifunctional pathways such as those of amino acid biosynthesis, which involve interdependent reaction sequences, requires more complex integration.

A mechanism evolved by *E. coli* for the regulation of aspartic acid conversion into threonine, lysine, isoleucine and methionine involves the participation of isoenzymes early in the sequences which are inhibited and repressed* by specific combinations of the products.[22] The overall sequence is illustrated in Figure 5.17. The common step from aspartic acid to aspartyl phosphate is catalysed by three ATP dependent aspartokinases, designated AK I, II and III. AK I is inhibited by threonine and repressed by the combined action of threonine and isoleucine, AK II is sensitive to methionine and AK III is both inhibited and repressed by lysine and inhibited by lysine in conjunction with isoleucine.

---

* Repression refers to the restriction of enzyme biosynthesis at the operon level.

Figure 5.17: Regulation of Amino Acid Biosynthesis through Control of Enzyme and Isoenzyme Activity

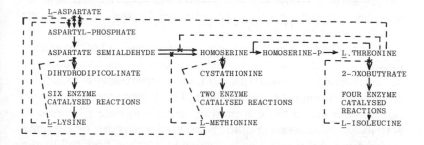

Isoenzymic control of carbon flow is also exercised further along the chain. Homoserine dehydrogenase (HDH) which catalyses the formation of homoserine from aspartate-β-semialdehyde, has been found in two forms. HDH I is inhibited and repressed by threonine and repressed by isoleucine while HDH II is repressed by methionine. The response of HDH I to these amino acids parallels that of AK I, and interestingly in *E. coli* (K12), the two different catalytic activities are carried by the same bifunctional enzyme molecule. Likewise the AK II and HDH II activities are also exhibited by one assemblage. The molecular weight of the aspartokinase I-homoserine dehydrogenase I complex is approximately 350,000 daltons and the molecule is a hexamer of equal size subunits; threonine binds in a co-operative manner. AK II—HDH II is a tetramer of molecular weight 170,000 daltons. Being in the same complex and possibly on the same polypeptide chain the partners are then simultaneously repressed and depressed at the operon by the same combination and concentration of amino acid end-products. In addition each activity in the bicephalic complexes is inhibited by the substrate of its partner. One consequence of this is the reciprocal dependence of the flux along the pathway on the relative concentrations of the intermediates.

In the pathways for biosynthesis of these amino acids, it appears that there have evolved three isodynamic proteins to catalyse the phosphorylation of aspartic acid, corresponding to the three end products lysine, methionine and threonine. In addition, at the aspartate semialdehyde branch point there are two homoserine dehydrogenases corresponding to the two eventual products methionine and threonine, but only one dihydrodipicolinate synthetase, towards L-lysine. Similarly, after homoserine, multiple forms of homoserine kinase and cystathionine synthetase would be unnecessary as only single end products are formed, but the enzymes that have evolved are repressed by their respective terminal products.

An efficient and sensitive regulation of amino acid biosynthesis is thus present. Excess of one end product will decrease the activity of its corresponding isoenzyme and so proportionally reduce the amount of a common intermediate. The lower concentration will not be directed to the end product already in excess because the activity or concentration of the respective isoenzyme or enzyme at each branch point will also be regulated. Carbon flow will then be directed towards the amino acids in deficient proportions.

Metabolic regulation through feedback control of specific isoenzymes is also found in the biosynthesis of the aromatic amino acids tyrosine, phenylalanine and tryptophan in *E. coli* and *Aerobacter aerogenes*. The first committed step common to all, is the formation of 3-deoxy-D-arabino-heptulonic acid (DAHP) from erythrose-4-phosphate and phosphoenol pyruvate. This reaction is catalysed by three DAHP synthetases, one of which is inhibited and repressed by tyrosine, a second is inhibited by phenylalanine and bivalently repressed by phenylalanine plus tryptophan and the third is inhibited and repressed by tryptophan.

The same chemical reactions occurring in different biochemical pathways in an organism, and hence possessing dissimilar physiological roles, are often catalysed by enzymic variants. An example is the threonine dehydrase catalysed deamination of threonine to 2-oxobutyric acid.[23] In coliform bacteria, two classes of this pyridoxal phosphate dependent activity are present, one performs a catabolic and the other an anabolic role. The kinetic behaviours of both are compatible with the mechanism of pyridoxal phosphate catalysed β-elimination outlined in Chapter 4 and in both the cofactor is bound to a lysine ε-amino group. However the two isoenzymes differ in the numbers they bind. The anabolic isoenzyme from *Subnoxella typhimurium* binds two molecules per MW 194,000 daltons, whereas the catabolic form from *E. coli* binds four per MW 150,000 daltons. Both enzymes are tetramers however and the subunits of that from *S. typhimurium* are probably identical. The two enzymes are also dissimilar in their responses to modifiers. The biosynthetic enzyme (Figure 5.16) is negative feedback inhibited by isoleucine, which elicits the characteristic sigmoidal binding curves and the isoenzyme is multivalently repressed when isoleucine, valine and leucine are all present in excess. The biodegradative enzyme on the other hand is insensitive to isoleucine but is activated by AMP. The nucleotide appears to effect a monomer to tetramer association at low enzyme levels and induce a conformational change in the subunits which increases their substrate affinities. The catabolic enzyme is not constitutive, being induced by anaerobic growth of the microorganism on a glucose free medium containing threonine plus serine.

Both the threonine dehydratases act in the same, deaminating direction, but from studies on the multiple forms of other enzymes such as malate dehydrogenase and aspartate transaminase, it appears that isoenzymic catalysis cannot in general be assumed to be unidirectional. Distinct molecular species of malate dehydrogenase, localized in the cell supernatant and mitochondria, both catalyse the interconversion:

$$\text{L-Malate} + \text{NAD} = \text{Oxaloacetate} + \text{NADH} \qquad (5.56)$$

The two forms are electrophoretically separatable and exhibit dissimilar kinetic properties (Table 5.16).

Although both isoenzymes exhibit higher turnover constants for oxaloacetate than L-malate, the soluble form shows a more pronounced capacity. In addition the cytoplasmic isoenzyme is inhibited by excess malate whereas the mitochondrial form is inhibited by high levels of oxaloacetate and to a lesser extent by malate. From these observations it has been proposed that the cytosol isoenzyme preferentially catalyses the reaction shown in eq. 5.56 in the direction of oxaloacetate reduction while the mitochondrial enzyme catalyses the reverse, malate oxidation. Acting in these opposite ways,

Table 5.16: Comparison of the Mitochondrial and Soluble Bovine Malate Dehydrogenases

	Relative concentration	Electrophoretic mobility	L-Malate $K_{m(app)}$ (mM)	Turnover no.	Oxaloacetate $K_{m(app)}$ (mM)	Turnover no.
Mitochondrial	70	Cathodic	1.0	35000	0.04	60000
Soluble	30	Anodic	0.54	20000	0.05	70000

it can be seen that they could promote the transfer of reducing equivalents into the mitochondria, as outlined in Figure 5.4. In order for this shuttle system to be effective however, the active participation of organelle specific oppositely acting aspartate amino-transferases would be necessary, because the mitochondrial membrane is limited in its permeability to oxaloacetate.

Two different forms of aspartate aminotransferase have been found specifically localised in the cytoplasm and mitochondria. Like the malate dehydrogenases, these are isoenzymes and both require pyridoxal phosphate as cofactor. They can be separated by electrophoresis; the supernatant enzyme has anionic and the mitochondrial cationic mobilities. Their description as isoenzymes is based on the marked differences found in their immunological properties, kinetic parameters and particularly their primary structures.

The isoenzymes discussed above are all enclosed within the one cell type, but different cells in an organism often contain different and distinctive isoenzyme patterns. This is most prominent for lactate dehydrogenase (Section 5.3.2). The differences observed in its tissue distribution could arise from varying rates of catabolism in different tissues, which is the probable explanation for the observed erythrocyte complement. For most other tissues a mechanism of this type is not considered significant, and Kaplan et al.[24] in order to rationalise the observed distributions and the known kinetic properties of the different isoenzymes have propounded the 'aerobic-anaerobic' hypothesis.

This hypothesis suggests that the pyruvate sensitive LD1 predominates in aerobic tissues such as heart muscle whereas the pyruvate insensitive LD5 isoenzyme occurs mainly in anaerobic tissues such as skeletal muscle. In skeletal muscle the immediate and rapid supply of energy is provided by anaerobic glycolysis, the essential irreversibility of which is maintained by lactate formation. In this tissue therefore LD5 is acting as a pyruvate reductase. In heart, LD1 is inhibited by pyruvate preventing the overproduction and accumulation of lactate. In this tissue when energy is needed, lactate, provided by the blood stream, is oxidised to pyruvate which then enters the mitochondria to provide energy via the citric acid cycle; LD1 is therefore acting as a lactate oxidase.

Several pieces of evidence support this proposal. No lactate has been found accumulated in heart muscle, but lactate does dissociate an abortive LD1-NAD-pyruvate ternary complex in which the enol form of pyruvate is covalently coupled to the enzyme (Figure 4.12). NADH has no effect on the dissociation. The proposal is that this complex acts as a one way valve, allowing the oxidation of lactate but not its formation, and workers have reported the presence of heart muscle LD1 in this inhibited form.[24]

This hypothesis has been criticised on several accounts.[25] Some tissues such as liver have aerobic metabolisms with high mitochondrial content but also have significant LD5 levels. The importance of substrate inhibition at physiological temperatures, which is reduced compared with that *in vitro*, has been questioned, as have the concentrations of enzyme, coenzyme and substrate available *in vivo* to form the abortive ternary complex at a substantial rate compared to that of catalysis. Finally it has been pointed out that on the anaerobic-aerobic basis only two isoenzymes would be needed whereas each tissue contains several. Thus the theory must be expanded to take into account the functional significances of LD2, LD3 and LD4.[25]

5.4 REFERENCES

1. M. A. Hayat, *Electron Microscopy of Enzymes* (Van Nostrand, New York, 1973).
2. G. Birnie (ed.), *Subcellular Components: Preparation and Fractionation* (Butterworth, London, 1972).
3. J. W. De Pierre and L. Ernster, *Ann. Rev. Biochem.*, vol. 46 (1977), p. 201.
4. P. D. Boyer, B. Chance, L. Ernster, P. Mitchell, E. Racker and E. C. Slater, *Ann. Rev. Biochem.*, vol. 46 (1977), p. 955.
5. H. L. Kornberg, *Essays in Biochemistry*, vol. 2 (1966), p. 2.
6. V. Z. Gorkin, *Adv. Pharmacol. Chemother.*, vol. 11 (1973), p. 2.
7. J. I. Dingle (ed.), *Lysosomes in Biology and Pathology* (North Holland, Amsterdam, 1969–76), vols. 1–5.
8. B. S. Cohen and R. W. Estabrook, *Arch. Biochem. Biophys.*, vol. 143 (1971), p. 54.
9. S. H. Kim, *Adv. Enzym.*, vol. 46 (1978), p. 279.
10. A. R. Fersht and M. Kaethner, *Biochemistry*, vol. 15 (1976), p. 3342; *Biochemistry*, vol. 16 (1977), p. 1025; D. Soll and P. R. Schimmel, *The Enzymes*, vol. 10 (1974), p. 489.
11. K. Block and D. Vance, *Ann. Rev. Biochem.*, vol. 46 (1977), p. 263.
12. F. Lynen, *Fed. Proc.*, vol. 20 (1961), p. 941.
13. H. Kacser and J. A. Burns, *Symp. Soc. Expl Biol.*, vol. 27 (1973), p. 65.
14. C. L. Markert (ed.), *Isozymes* (Academic Press, New York, 1975), vols. 1–4.
15. J. H. Wilkinson, *Isoenzymes* (Chapman and Hall, London, 1970).
16. C. L. Markert and F. Moller, *Proc. Natl Acad. Sci.*, vol. 45 (1959), p. 753.
17. R. L. Foster, *Clin. Chem.*, vol. 21 (1975), p. 1845.
18. R. G. Klalifah, *J. Biol. Chem.*, vol. 246 (1971), p. 2561; E. Christiansen, *Biochem. Biophys. Acta.*, vol. 220 (1970), p. 630.
19. R. E. Tashian, *Biochem. Genet.*, vol. 5 (1971), p. 183.

20. I. Szorady, *Pharmacogenetics, Principles and Paediatric Aspects* (Akademia; Kiado, Budapest, 1973).
21. B. Nostrand, I. Vaara and K. K. Kannan, in reference 14, vol. 1 (1975), p. 578.
22. G. N. Cohen, *Curr. Top. Cell. Reg.*, vol. 1 (1969), p. 183.
23. H. E. Umbarger, *Adv. Enzym.*, vol. 37 (1973), p. 349.
24. J. Everse and N. O. Kaplan, in reference 14, vol. 2 (1975), p. 29.
25. E. S. Vessell in reference 14, vol. 2 (1975), p. 1.
26. L. L. Kisseler and O. O. Favorova, *Adv. Enzym.*, vol. 40 (1974), p. 141.

6 Medical Enzymology

6.1 DIAGNOSTIC ENZYMOLOGY

The central role of the hospital clinical laboratory is to investigate and quantitatively analyse the biochemical alterations which underlie or accompany human disease. As part of this function the assay of enzyme levels in the extracellular body fluids (blood plasma and serum, urine, digestive juice, cerebrospinal fluid, amniotic fluid, and synovial fluid) are important aids to the clinical diagnosis and management of disease.

Although the majority of enzyme catalysed reactions take place within living cells, whenever an energy imbalance occurs in the cells as a result of exposure to infective agents, bacterial toxins, etc., enzymes 'leak' out through the membranes and into the circulation. This causes their fluid levels to be augmented above those present as a result of normal cellular metabolism and turnover. Measurements of the type, extent and duration of these raised enzyme activities can then provide information on the identity of the damaged cell and indicate the degree of injury and its response to corrective treatment.

Enzyme assays can make important contributions to the diagnosis of disease because very small changes in the concentration of an enzyme activity can be easily measured, as explained in Chapter 2, and in serum, several thousand fold increases, such as occur on severe intoxication, have been found compatible with the living state. Because the concentration gradient between cells and their external fluids are large (Table 6.1), even small alterations in tissue composition can elicit quite pronounced changes in fluid enzyme levels. Measurements of the changes in enzyme levels therefore offer a greater degree of organ and disease differentiation than is possible using the other clinico-chemical parameters such as albumin or gamma-globulin levels.

In addition to these facilities, enzyme assays fulfil many of the clinical criteria required of a diagnostic test. An unhealthy state is indicated with a high degree of sensitivity, the practical procedures are reliable, specific, fairly reproduceable, and cause little discomfort to the patient.

Table 6.1: Enzyme Concentration Gradients between Hepatocytes and Blood Plasma

Enzyme*	GOT	GPT	LD	ALD
Gradient ($\times 10^{-3}$)	12	10	3	3

* See Table 6.2 for abbreviations.

At present however, the diagnostic specificity of enzyme tests is such that they are restricted mainly to confirming diagnoses, providing data to be weighed with other clinical evidence. One reason for this limitation is the lack of disease-specific enzymes.

The mechanisms of enzyme release and their subsequent metabolism and elimination from the circulation are currently areas of intense research but the empirical observations of changes in their activities, particularly those that occur in the blood plasma and serum have practical significance and form an established part of clinical practice. This is adduced by the generality of clinical enzyme measurements, which are performed in most hospitals and medical institutes in Europe and the United States of America, and by the number of assays carried out. In a hospital containing around 500 beds, between 10,000 and 20,000 analyses could be requested annually.

Scales of this order have necessitated the design, development and generalised use of automated enzyme analysers, some of which can now process up to 300 patients' specimens per hour. The mechanical and chemical concepts involved and their history, from the simple modified colorimeters to the centrifugal fast analysers have been extensively reviewed.[1]

6.1.1 Historical Development

The age of clinical enzymology started approximately at the beginning of this century. In 1908 Wohlgemuth[2] recorded an increase in amylase activity in the sera and urine of patients suffering from acute pancreatitis. Occasional literature reports then followed, also principally concerning gastrointestinal disorders. Two decades were to elapse however before the next significant advance, the demonstration of raised levels of alkaline phosphatase activity accompanying bone and liver disease.[3] Soon afterwards the practicality of routine measurements of enzyme activities were enhanced by the introduction of spectrophotometric equipment and Warburg's optical test.[4] Warburg and his collaborators also showed that certain enzymes of tissue metabolism were additionally found in serum and that glycolytic enzymes were present in sera of rats and humans bearing malignant tumours—abnormal cells which exhibit high rates of glycolysis.

Most significant for the development of diagnostic enzymology were the studies of La Due *et al.*,[5] and de Ritis *et al.*,[6] on the transaminases, particularly aspartate transaminase. The activity of this enzyme was found to be elevated in serum after myocardial infarction and liver disease. These reports, showing that injury to a cell caused the release of intracellular enzymes into the serum, catalysed investigations into the causal relationship between disease and the concomitant alterations in the enzyme levels of body fluids.

Altered levels of activity were then reported for a wide range of enzymes associated with muscle, gastrointestinal, hepatic, renal, malignant and congenital diseases. Many however proved to be unsuitable for accurate and routine diagnosis but about eighteen have been demonstrated to exhibit various degrees of diagnostic reliability. These are listed in Table 6.2. Of these probably GPT, GOT, OCT, AP, SP, CPK, amylase and aldolase are most frequently assayed in hospital laboratories. Even so, not all of these are assayed within the one clinical laboratory, different biochemistry departments preferring various combinations of tests depending on their individual specialities and experience.

6.1.2 The Practical Significance of Raised Enzyme Levels

Discussion will be centred mainly on the changes in enzyme levels which occur in blood serum since these constitute the majority of enzyme assays performed clinically.

The enzymes in blood plasma have been classified by Bücher[7] into three main categories, depending not on their catalytic activities, but their biological function and source. One group comprises the secreted *plasma specific* enzymes, cholinesterase and the enzymes of blood coagulation, all of which have specific roles in blood. A second group includes the *secreted* enzymes, such as those of pancreatic origin. These do not contribute to plasma function and are often catalytically ineffectual through combination with the specific plasma inhibitors, e.g. α-antitrypsin. Classified into a third group are the enzymes of *cellular and tissue metabolism*. The concentrations of some of these catabolic enzymes, lactate dehydrogenase,

Table 6.2: Diagnostically Important Enzymes

EC Numbers	Enzyme	Abbreviation	Tissue source*	Reaction
1.1.1.27	Lactate dehydrogenase	LD	HLMK	Lactate $\overset{NAD}{=}$ Pyruvate
	(Hydroxybutyrate dehydrogenase)	HBD(LD1)	H	2-Hydroxybutyrate $\overset{NAD}{=}$ 2-Oxobutyrate)
1.1.1.42	Isocitrate dehydrogenase	ICD	L	Isocitrate = 2-Oxoglutarate + CO_2
2.1.3.3	Ornithine carbamoyltransferase	OCT	L	Carbamoyl-P + Ornithine = Citrulline + P_i
2.3.2.2	γ-Glutamyl transferase	GGT	KL	γ-Glutamylpeptide + Amino acid = γ-Glutamylamino acid
2.6.1.1	Aspartate aminotransferase	GOT(AST)	HLMKB	Aspartate + 2-Oxoglutarate = Glutamate + Oxaloacetate
2.6.1.2	Alanine aminotransferase	GPT(AAT)	L	Alanine + 2-Oxoglutarate $\overset{ATP}{=}$ Glutamate + Pyruvate
2.7.3.2	Creatine kinase	CPK	MHB	Creatine = Creatine phosphate
3.1.1.3	Triacylglycerol lipase	Lipase	Pa	Triacylglycerol = Diacylglycerol + Fatty acid
3.1.1.7	Acetylcholinesterase	ACHE	BE	Acetylcholine = Acetate + Choline
3.1.1.8	Cholinesterase	CHE	L	Acylcholine = Fatty acid + Choline
3.1.3.1	Alkaline phosphatase	AP	BILPIK	Phosphate monoester = Alcohol + P_i (pH 8–10)
3.1.3.2	Acid phosphatase	SP	Pr E	Phosphate monoester = Alcohol + P_i (pH 5)
3.1.3.5	5'-Nucleosidase	5.N	Ht Pa	5'-Ribonucleotide = Ribonucleoside + P_i
3.2.1.1	α-Amylase	Amylase	Pa S	Starch = Maltose + Maltotriose
3.4.21.1, 3.4.21.4	Chymotrypsin, trypsin	CT, T	Pa	Proteins = Polypeptides
4.1.2.13	Fructose-bisphosphate aldolase	ALD	MH	Fructose-1,6-bisphosphate = Triose phosphates

* B, brain; E, erythrocytes; H, heart muscle; Ht, hepatobiliary tract; I, intestinal mucosa; K, kidney; L, liver; M, skeletal muscle; Pa, pancreas; Pl, placenta; Pr, prostate gland; S, saliva.

aldolase and aspartate aminotransferase in particular, are the sum of individual enzymic forms released from many different cells in contact with the circulating fluid, whereas others arise from particular organs. Urea cycle enzymes come from the hepatocytes and creatine phosphokinase is essentially derived from striated muscle. In the blood, the activities of the third group of enzymes are latent due to the absence of their cofactors and substrates, but it is largely from their assays in serum that inferences are drawn concerning the states of health of the cells and tissues of their origin.

Interpretations of the changing enzyme levels can only be made if the *normal ranges* of enzyme activities in serum are known. By this term is meant 'the numerical values obtained from a particular test when applied to individuals free from clinical conditions likely to lead to elevated values'.[8] The results from many persons, categorised with respect to age, sex, mobility and condition of health are statistically assessed to give the normal range. Numerous pitfalls can occur in the setting up of this normal scale not least of which is the uniqueness of the pattern of enzyme activities for each person. Large differences between apparently healthy individuals are observed and significant changes in the values often occur after ingestion of commonplace drugs such as aspirin or alcohol. Some ranges of enzyme activities in sera considered to be normal for healthy adults are given in Table 6.3. The factors controlling these levels in the extracellular fluids are discussed in Section 6.2.

Deviations from the normal patterns of these values in serum are diagnostically important for identification of their source and the nature of the disease. For these purposes, three features of the

Table 6.3: Normal Ranges of Enzyme Activities in the Sera of Healthy Adults

Enzyme	LD	GGT	GOT	GPT	CPK	AP
Activity range* (mU per ml)	< 250[a]	< 20[a]	< 15[a]	< 15[a]	< 1[b]	20–50[b]

Enzyme	SP (Total)	SP (Prostatic)	Amylase	Aldolase
Activity range* (mU per ml)	5–15[b]	1–5[b]	50–250[b]	1–6[b]

* Assay Temperature: a, 25°C; b, 37°C.

enzyme activities in serum are utilised: (1) the levels of organ specific enzymes, (2) the isoenzyme composition, and (3) the pattern of enzyme activities.

It was demonstrated in the last chapter that organs and tissues possess quantitative differences in their enzyme complements. In some specialist tissues these differences are large, giving rise to the notion of 'organ-specific enzymes'. Thus when these are elevated in the serum, their measurement can lead to identification of the parent tissue. Examples are creatine phosphokinase which is associated mainly with striated muscle and so is useful in the diagnosis of muscle disorders, especially progressive muscular dystrophies, and ornithine carbamyltransferase, a urea cycle enzyme restricted mainly to liver cells.

Less distinct, but equally valuable are changes in the isoenzymic complement of the serum sample. Many enzymes in serum exhibit heterogeneity but as yet only three have been shown to possess broad diagnostic significance, those with lactate dehydrogenase, alkaline phosphatase and acid phosphatase activities. Isoenzymes are isodynamic proteins found in different tissues of the same species. Thus identification of the isoenzyme responsible for an elevated catalytic activity could prove valuable for the determination of the site of disease. Experimental methods have therefore been devised for the automated and routine practical differentiation of isoenzymes in serum.

Total lactate dehydrogenase activity in healthy serum is comprised of LD1, LD2, LD3, LD4 and LD5 in the approximate percentage ratios 35 : 40 : 15 : 5 : 5. LD1 and LD2, the anodic isoenzymes, are mainly of heart tissue origin and their levels increase in certain heart conditions. Conversely, the cathodic LD4 and LD5 isoenzymes predominate in liver and skeletal muscle and are substantially raised in acute hepatitis or obstructive jaundice. In order to determine the source of raised LD activity several methods have been devised, including electrophoretic separation, heat stability, substrate inhibition and, in particular, the hydroxybutyrate test (HB). LD1 and LD2 reduce 2-oxobutyrate as readily as pyruvate, unlike the other isoenzymes which are specific for pyruvate. Thus the augmentation of HB dehydrogenase over lactate dehydrogenase activity is used as a confirmation of myocardial infarction.

Elevated levels of acid phosphatase in serum point to carcinoma of the prostate gland and changes in its levels during treatment are used to monitor the progress of the disease. Measured values in serum however represent the sum of the prostatic and erythrocyte activities and thus several methods, using inhibitors, have been developed for their resolution. Of these 0.5 percent formaldehyde, which has little effect on prostatic acid phosphatase but almost completely inhibits the red cell enzyme, and L-tartaric acid (0.01 M) which has the reverse effect probably draw the clearest distinctions.

Heterogeneous variants with alkaline phosphatase activity are found localised in the membranes of the intestinal mucosa, placenta, bone osteoblasts and the biliary canalicula of the liver. The exact molecular and genetic relationships between these forms have yet to be determined, as have their biological functions, but their measurement in the diagnosis of bone and liver dysfunction is one of the oldest and most reliable clinical tests. Normal serum contains mainly liver alkaline phosphatase and this isoenzyme is found elevated in hepatobiliary diseases. In bone disease, AP activity is also raised but in this case resulting from an increased synthesis of the specific osteoblast isoenzyme. These two forms are separable by electrophoresis but a clear distinction is not always possible and therefore kinetic methods (for automated assays) have been introduced. Heat inactivation, urea inhibition and L-phenylalanine inhibition have all been applied; the different responses of each variant to these influences are shown in Table 6.4.

Only a few organs can be distinguished by means of organ-specific enzymes and a third feature of enzymes in serum—their pattern of activities—can be used to determine the site of injury. Here the activity pattern of several enzymes in the unhealthy serum is compared to the patterns of the same enzymes in the tissues.

Table 6.4: Physico-chemical Differences between Alkaline Phosphatases*

	Liver	Bone	Placenta	Intestine
Anodic mobility	1	2	3	4
Thermal stability (70°, 0.5 h)	+	+	−	+
Urea (0.5 M)	+	+	−	+
L-Phenylalanine	−	−	+	+

* Notation: + = affected, − = unaffected.

Table 6.5: Distribution of Clinically Important Enzymes in the Tissues*

Enzyme	Duodenum	Prostate	Liver	Bone	Heart muscle	Skeletal muscle
LD			150		120	150
GOT			60		50	35
GPT			35		3	4
CPK					30	200
AP	50		10	30	40	
SP		1000	2		1	
ALD			6		5	50

	Brain	Kidney	Pancreas	Erythrocytes
LD		100	50	40
GOT	15	10	5	1
GPT		1	1	0.1
CPK	70			
AP		30		
SP		2		3
ALD		1		1

* Activities, Units per gram of organ.

In Table 6.5 are given the organ distributions of several enzymes of clinical interest. As is seen, each organ has a characteristic pattern from which in principle it could be recognised and differentiated. In practice however this is rarely possible or needed, for two main reasons. First, measurement of all these enzymes, as well as their isoenzyme ratios would be too costly in time and expense. Moreover such a multitude of assays would be unnecessary since other clinical and medical information is available from which a diagnosis can be made, and enzyme studies have greatest advantage when contributing to or confirming the diagnosis. Secondly, in diseases other than those involving severe cell damage such as acute hepatitis, in which the liver may be easily recognised from the typical enzyme pattern in serum, the picture will appear distorted. Several factors cause this distortion, (1) different locations of the enzymes in the cell, (2) differences in the rates of diffusion from the cell, (3) variations in the rates of inactivation and elimination of the enzymes from the plasma, (4) different rates of intracellular enzyme synthesised on damage, and (5) superposition of the enzyme patterns from more than one organ.

The influence of different cellular locations is illustrated by the behaviour of the transaminases. GPT is localised mainly in the cytoplasm whereas GOT activity is distributed fairly evenly between the cytoplasm and mitochondria. In severe acute liver damage (toxic poisoning) serum GOT and GPT levels parallel those of the organ, indicating that both the cellular and mitochondrial membranes have been damaged. On slight damage in which the mitochondria remain intact, the amount of GPT activity released will be approximately twice that of GOT. From the GOT : GPT ratios therefore it might be possible to determine the extent and severity of the cell damage. However their interpretation is complicated by the different rates of activity loss for the two transaminases in serum; the half-lives (time for loss of 50 percent initial activity) are approximately 20 and 50 h for GOT and GPT respectively. Aspartate aminotransferase (GOT) is thus more readily cleared, leading to relatively higher alanine aminotransferase (GPT) levels, even on extensive damage. (One important aspect of these rapid rates of inactivation however is the continuous and up-to-date pattern of enzyme activities that is obtained, Figure 6.2.)

Despite the above drawbacks significant diagnostic information is provided by the changes in activity of certain enzymes accompanying disease processes. For the practising clinical enzymologist the subject has been discussed in considerable detail by Wilkinson.[8]

6.1.3 Enzymes and Diseases of the Liver and Biliary System

With the diseases of the gastrointestinal tract, liver diseases were among the first to which serum enzyme tests were applied. Diagnostically they have since proved to be the most successful, possibly because of the large size of the organ and the wide variety and abundance of its enzymes, the activities of which in sera are raised even with only slight damage to the cells. The liver-function enzymes GOT, GPT and AP are assayed in the majority of clinical laboratories and in addition some or all of LD, GGT, OCT and CHE (see Table 6.2) are monitored to determine the site and nature of the liver disease and to follow its course.

Hepatobiliary disorders have been classified[8] into:
1. Hepatocellular diseases, such as acute hepatitis, chronic hepatitis and cirrhosis which affect the liver cells.
2. Cholestatic diseases of the biliary tract.

3. Neoplastic diseases of the interstitial tissue.

In all these liver dysfunctions, the alanine and aspartate amino-transferase levels are increased in serum, the extents giving a useful differential index of the type of dysfunction. Gamma-glutamyl transferase (GGT) activity is also raised and this enzyme is gaining acceptance as a more sensitive indicator of liver disease, particularly as serum GGT is not raised in bone diseases, whereas the cholestasis indicator alkaline phosphatase is raised. GGT is also found to be elevated earlier than AP in liver diseases.

6.1.3.1   Acute Hepatitis

High values of both GOT and GPT always accompany acute viral hepatitis, attaining 20 to 50 times their normal values in the prodromal phase before the onset of icterus (jaundice). Even damage to 0.1 percent of the liver cells gives measurable amino-transferase elevation in serum and some of the highest values have been found associated with acute hepatitis. Measurement of their values thus constitutes a powerful diagnostic tool. The occurrence of enhanced activities is about a week before clinical symptoms appear and the higher the elevation the greater is the likelihood of acute hepatitis. Even in the 50 percent of non-icteric cases, the tests are useful in following its progress and in monitoring the effectiveness of prophylactic measures.

The time courses of the enzyme levels in serum are shown schematically in Figure 6.1. Early in acute hepatitis serum GOT activity exceeds GPT as the former is at a higher concentration, but since the latter is less rapidly inactivated, its relative amount gradually increases. This is reflected in the de Ritis ratio, GOT : GPT which decreases to about half its normal value of 1.1–1.3 during this period.* The quotient then remains fairly constant until the recovery phase when the aminotransferase values decrease to the normal range. The GPT levels fall more slowly than those of GOT and parallel the clinical recovery, which is usually complete within about six weeks. During the later course of the disease the amino-transferase activities tend to parallel the bilirubin levels but again are more sensitive indices.

* The quotient GOT : GPT was proposed by de Ritis[6] as a means of differentiating inflammatory lesions from necrotic processes. GPT is confined to the liver cell sap, whereas GOT activities are found in both the cytoplasmic (cGOT) and mitochondrial (mGOT) fractions. In acute hepatitis and myocardial infarction, both characterised by extensive cell necrosis, significant amounts of mGOT appear in the serum.

Figure 6.1: Time Course of Aminotransferase Activities in Acute Hepatitis

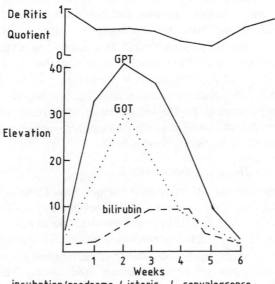

Aminotransferase assays are important for determining the extent of recovery. Relapses are accompanied by new peaks in their activities, and continuous, moderately elevated levels indicate the onset of cirrhosis. In viral hepatitis, in which a few cells at a time are attacked over a long period, the aminotransferase levels become moderately elevated and the de Ritis ratio remains at intermediate values. Treatment then consists of rest plus a light diet and the prognosis is generally favourable. Very high aminotransferase levels and a high de Ritis quotient however indicate more extensive hepatocyte necrosis requiring therapeutic treatment and the prognosis is less favourable. Alkaline phosphatase is only moderately increased in the majority of uncomplicated acute hepatitis cases but cholestatic diseases cause protracted increases in both AP and the aminotransferase activities.

Serum levels of the aminotransferases but not liver alkaline phosphatase are also increased in drug induced toxicity, and thus this condition cannot be enzymically distinguished from hepatitis. The liver is a major site of drug metabolism and increases in the serum levels of GPT and GOT serve as early biochemical indicators of drug associated liver damage and drug overdose; the de Ritis ratios are usually less than unity. The clinical picture accounting for release of the enzymes is varied and depends on the drug type and dosage but elevations have been induced by administration of tetracycline, iproniazid, testosterone, paracetamol, salicylate and the monoamine oxidase inhibitors used in psychiatric therapy.

In acute hepatitis, the serum levels of the enzymes secreted by the liver, the blood coagulation enzymes and cholinesterase, remain fairly constant, except in protracted forms of the disease, liver cell necrosis or recurrent damage when decreases in activity are observed as a result of impaired biosynthesis.

6.1.3.2   Chronic Hepatitis and Cirrhosis

Elevated transaminase activities of the order 3 to 12 fold accompanying hepatomegaly point to chronic hepatitis—acute inflammation of the liver—or to cirrhosis—hepatic fibrosis accompanied by parenchymal cell regeneration nodules. The aetiology of each includes viral infection and alcohol consumption, although in some, cryptogenic cases the cause remains unidentified. The levels of the aminotransferases in the serum tends to parallel the inflammatory activity and the main applications of their measurements are as indices to the courses of these chronic liver diseases. Their advantages over liver biopsy are the simple manipulations involved which are more comfortable for the patient. Acute episodes of these diseases are accompanied by deterioration of the patient's condition, dysproteinemia and augmented serum bilirubin levels, but these symptoms are occasionally absent, in which case the enzyme tests become additionally important.

Increases in the aminotransferases always occur but the relative increases in GOT and GPT vary between individual patients and a de Ritis ratio below unity in chronic hepatitis and above unity in cirrhosis is applied reservedly only as a rough guide. Progressive stages of the diseases however are fairly reliably indicated by the levels of the two aminotransferases. Relapses, whether icteric or anicteric always lead to increases and a moderately high GOT activity with a de Ritis quotient greater than 1.5 is unfavourable. This is the enzyme pattern for decompensated cirrhosis. Also unpropitious is a decrease in the levels of aminotransferase activity accompanied by an increase in GGT and LD, which attend hepatic coma. Some authors have suggested that the activities of these enzymes in serum can be used to distinguish between 'true' hepatic coma and porto-caval encephalopathy. 'Typical' values (U per l) for the GOT : GPT : LD ratios accompanying these two conditions have been reported as 550 : 250 : 250 and 40 : 20 : 150 respectively. Blood serum levels of cholinesterase (biosynthesised by the parenchymal liver cells) are also of diagnostic value, since in most chronic liver diseases of long duration or with severe cell damage as in decompensated cirrhosis, its levels are depressed below the normal range in serum.

The rapid inactivation of enzymes released into the serum provides an ongoing picture of the effectiveness of chemotherapy. This is illustrated by Figure 6.2 which shows the influence of corticosteroids and immune-suppressants on the measured transaminase levels in chronic hepatitis. From the responses of their activities, the dosages, their effectiveness and the duration over which therapy must be prolonged can be gauged.

Figure 6.2: *Effects of Corticosteroid and Immunosuppressant Therapy on Aminotransferase Activities in Highly Active Chronic Hepatitis*

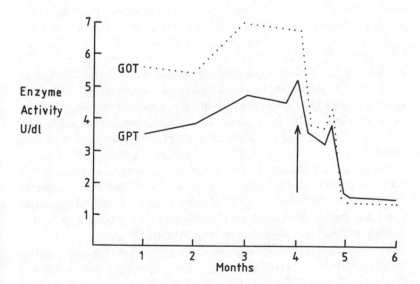

6.1.3.3   Toxic Liver Damage

Fatty Liver.   Fatty infiltration of the liver (steatosis) is a common liver disorder of alcoholics and diabetics. 'Social' drinking causes only slight and reversible histological changes in the liver together with small changes in aminotransferase levels in the serum. But alcoholism may cause steatosis attended by GPT levels raised to approximately twice the mid-normal range. In acute-on-chronic alcoholism, steatonecrosis may arise causing jaundice and elevating the transaminase levels up to five times the normal. Liver failure can then result. GGT has a microsomal location in the liver and its activity in serum is particularly sensitive to alcohol consumption, but whether the enzyme is induced or released under its influence is at present unclear. Like the transaminases however, this enzyme is only slightly elevated in the serum of social drinkers.

Solvent Poisoning.   The most pronounced changes in the enzyme activities in serum occur as a result of organic solvent poisoning, particularly by carbon tetrachloride. Often the serum enzyme pattern reflects that of the liver, testifying to the extensive damage to the hepatocytes caused by the solvents, and is very different from those of acute or chronic hepatitis. The GOT : GPT : LD levels are around 400 : 500 : 250 (U per l) for hepatitis but at least an order greater in solvent poisoning, 6,500 : 3000 : 10,000 respectively.

6.1.3.4   Hepatobiliary Diseases

In extra-hepatic biliary obstructive jaundice, both the GOT and GPT levels in serum are elevated by five to ten fold with GPT raised slightly higher than GOT, as in hepatitis. Unlike acute hepatitis the levels are lower and in contrast to both uncomplicated acute and chronic hepatitis and cirrhosis alkaline phosphatase is also raised. During the early stages of the biliary diseases, the initial large increase in the aminotransferases is frequently attributed to hepatitis but after a few days their reduction to more intermediate values and significant increases (up to three times) in the delayed biliary alkaline phosphatase activity leads to confirmation of an intra- or extra-hepatic impairment of bile flow (cholestasis).

More recently, several workers have found GGT to be a more sensitive indicator, arriving earlier in the serum than alkaline phosphatase and giving more pronounced elevations, up to fifty times its normal average values.

6.1.4 Enzymes and Heart Disease

Applications of enzyme assays in serum for the diagnosis of the heart diseases are less general than for liver diseases and are restricted mainly to those causing extensive damage to the heart muscle. Of these most success has been achieved in the diagnosis of myocardial infarction, enzyme tests for which are now routine in most hospital laboratories. No one enzyme has yet been found entirely specific for myocardial damage and a combination of results from assays of CPK, GOT, and HBD, each of which has been shown to be elevated in more than 90 percent of cases, is used for diagnostic purposes. Of these, CPK levels appear to exhibit the highest degree of specificity; they are also elevated the earliest. Even so enzymic differentiation between infarction and cardiac ischaemia is far from straightforward.

The time course of elevated enzyme activities in serum after acute myocardial infarction is shown in Figure 6.3. The rise in CPK

Figure 6.3: Serum Enzyme Activities after an Episode of Myocardial Infarction

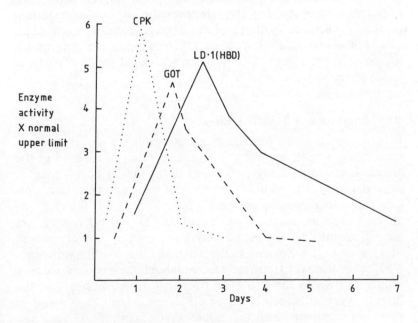

Source: Redrawn with permission from Reference 12.

activity begins three to four hours after the initial onset of pain, followed in order by GOT and HBD which appear after an elapse of approximately eight hours. Maximum activity is reached in the same sequence, CPK after 24 hours, GOT after 36 hours and HBD after about two days. HBD is also the last to return to normal, after 10 to 14 days, although the total LD activity may fall earlier. During these progressions the aminotransferase levels remain normal with a de Ritis ratio slightly less than unity unless a secondary liver involvement complicates the enzyme pattern as a result of the heart failure causing oxygen or substrate deficiency.

The observed increments in the enzyme levels are fairly moderate, around four to ten fold for GOT and CPK and two to five times for HBD, reflecting the small volume of cellular damage. The sizes of the elevations together with their durations have been found to be reliable guides to the damage, correlating well with the extent of the infarct as judged morphologically, and with mortality. The latter has been found to increase substantially when the elevations approach the upper limits of the ranges or when CPK activity is continually raised for more than four days.

The clinical data provided by enzyme tests in heart conditions are important adjuncts to the electrocardiographic information, especially when the cardiographical displays are not typical of an infarct or when reinfarction occurs. Often an infarct pattern remains for some time in the ECG whereas additional peaks of enzyme levels are readily discernable.

6.1.5 Enzymes and Muscle Disease

Disorders of skeletal muscle, which constitutes about 30 percent total body weight, are the consequences either of diseases of the muscle fibre (entitled myopathies) or of the muscle nerves (neurogenic diseases). The clinical symptoms of these two types are similar, pain, weakness, tenderness and atrophy. In the sera of afflicted patients, the CPK, LD, ALD, GOT and GPT levels are more frequently raised in the first type of condition and less often in those of neurogenic origin. In the latter, and the hereditary diseases, CPK is occasionally transiently raised though the increases are only moderate (two to three fold) and occur in the early stages of the disease. Very infrequently raised CPK levels have been found in some of the other neurogenic diseases such as multiple sclerosis and infective poliomyelitis.

Damage to the muscle can be caused by muscular exercise, drugs, physical trauma, inflammatory diseases, microbial infection or metabolic dysfunction or it may be genetically predisposed. In all these myopathies, CPK is the enzyme elevated in the serum with the highest frequency and consequently is the one most frequently assayed in the diagnosis of muscle disorders.

6.1.5.1 Genetically Determined Myopathies— Progressive Muscular Dystrophies

The activities of all the five enzymes CPK, ALD, LD, GPT and GOT are elevated in the Duchenne, limb-girdle and facio-scapulo-humeral types of degenerative muscular dystrophy. Of these the largest changes occur in the first type, the most common form of muscular dystrophy affecting male persons. Unfortunately, the wide range of enzyme elevations in patients with this atrophic disorder overlap so no clear differential diagnosis between the three types may be made on this basis. Typical elevations are shown in Table 6.6.

Table 6.6: Average Elevations (above the Upper Limit of the Normal Range) in the Progressive Muscular Dystrophies

	ALD	CPK	GOT, GPT	LD
Duchenne	2–7	10	5	2–3
Limb-girdle	1–2	2–10	1–2	2
Facio-scapulo-humeral	1–2	1–2	1–2	1–2

In common with the diseases of heart and liver, the extent and severity of the muscle diseases may be estimated from the enzyme activity values in the sera. High levels, particularly of CPK are found at a very early age before the advent of clinical symptoms and they often correspond with histological findings after muscle biopsy. Such observations have also been found to correlate with family history. As a corollary, from the absence of a raised CPK activity in the serum, a 95 percent probability of no dystrophy may be inferred. For those affected, as the disease progresses, an increasing amount of muscle fibre is replaced by fat and connective tissue, causing a time dependent lowering of the muscle specific enzymes in the serum. Eventually, in chair-bound patients, the activities become within the normal range. Large individual variations in the enzyme activities are found at each stage of the disease, which depend in part on the ambulatory aspect of the patient. Partly for this reason and because at present there are no effective treatments available, enzyme based prognoses and therapy assessment are less fruitful than for the liver and heart diseases.

Duchenne muscular dystrophy is an X-chromosome-linked recessive disease carried by female persons, half of whose sons will be afflicted and half of whose daughters will be heterozygous carriers. Carriers have few clinical symptoms although occasionally changes in muscle histology have been found. Those who possess a family history of the disease often prefer to have their state defined prior to pregnancy. For this purpose measurement of serum creatine kinase activities have some value. Several studies have reported that in 'definite' carriers, those with a history plus one affected son, the incidence of raised CPK activity is approximately 70 percent, although others have found a lower frequency and recommend combining CPK with muscle biopsy investigations for a clearer indication. For carriers with affected sons but no family history CPK activity has been found elevated in 35 to 50 percent of instances. In known pregnant carriers, amniocentesis can be used to sex the foetus before possible therapeutic abortion.

6.1.5.2   Traumatic Myopathies

Creatine kinase and LD5, both isoenzymes of muscle fibre origin have been found discharged in the serum following direct traumatic damage to the muscle fibre during surgery and intramuscular injection. Surprisingly these enzymes are also raised in the serum following muscular exercise. The levels in untrained persons undergoing heavy or prolonged strenuous exercise have been observed to approach pathological values, with CPK values often raised 20 to 30 times during the first twelve hours. The activities usually return to normal within two or three days.

A probable cause of the observed elevations is increased membrane permeability resulting from anoxia or changes in the metabolite concentrations during exercise. This explanation is consistent with observations from the experimental study of leakage of enzymes from cells discussed in the next section.

6.2 FACTORS GOVERNING ENZYME LEVELS
IN BLOOD PLASMA

Three main factors, the same as those considered in the clinical interpretation of raised enzyme levels, determine the concentration of an enzyme in blood plasma: (1) the turnover and proliferation of cells with access to the blood plasma, (2) their total enzyme complement and (3) their intracellular enzyme distribution. The tissue dependent level of activity in the blood is then a function of two vectors, the flux of extrication from the cell and the fluxes of inactivation, degradation and elimination from the plasma. Both of these depend on the state of health of the individual.

6.2.1 Factors Contributing to Normal Enzyme Levels

In healthy individuals the concentrations of cellular enzymes in the extracellular fluids are fairly low (Table 6.3), and although daily variations occur, the basal enzyme level of a healthy individual stays reasonably constant. This circumstance represents a balance between the two fluxes described above.

The constant, low level of enzyme release into plasma is the result of normal cell turnover, supporting the concept of a highly coordinated process regulating cell turnover and metabolism in the healthy state, but the identity of the cells and the factors controlling the process have yet to be clearly identified. One possible source of cellular enzymes in the plasma of healthy persons is their red blood cells. Young[9] has argued that the turnover of 5×10^{10} erythrocytes would be sufficient to yield the daily plasma lactate dehydrogenase concentrations and is within the normal decay rate of the cells. An alternative source, the turnover of two grams of liver cells, which contain a similar concentration of enzymes, was considered unlikely since even in acute hepatitis only a few grams of this tissue are destroyed. Red blood cells are the main source of acid phosphatase in normal plasma and increased turnover of these cells, as happens for example in haemolytic anaemia, causes the associated plasma enzyme activities to be increased. But it is unlikely that the total complement of enzymes in serum is provided by the erythrocytes. The lactate dehydrogenase isoenzyme patterns of red blood cells and blood plasma are different. From a consideration of the various isoenzyme ratios in plasma the enzymes are considered to issue from those hepatocytes, osteoblasts, spleen and endothelial cells with direct access to the blood.

6.2.2 Factors Affecting the Appearance of Cellular Enzymes in Blood

Though exercise, haemoconcentration, hormonal changes and nutritional balance can all contribute to alterations in blood plasma enzyme levels, their effects, like the diurnal variations, are generally smaller than those arising from physiological dysfunction. Damage to the cellular tissues from virus infection, chemical or mechanical trauma, and the overproduction of enzymes in cells (e.g. the enhanced alkaline phosphatase biosynthesis in osteoblasts on hypoparathyroidism or in liver on biliary obstruction), can all lead to substantial elevations in the plasma non-specific enzyme levels.

Originally it was thought that the increases in enzyme activities in serum accompanying pathophysiological conditions were the result of an equalisation of enzyme concentrations between the cellular source and the extracellular fluid. In patients suffering from hyperthyroidism for example, the creatine phosphokinase (CPK) levels in serum were found to be raised as a consequence of increased turnover of muscle tissue. But although many of the neuromuscular diseases are attended by extensive tissue damage they are not accompanied by significant changes in CPK activity in the serum. Thus cellular constituents are not necessarily liberated on tissue damage and some other explanation is necessary.

Key metabolic enzymes are retained within their intracellular compartments by means of the membrane systems. As was discussed in the last chapter, membranes perform crucial roles maintaining the integrity and functional character of the cell by restricting ion and enzyme permeability. Consequently, they sustain the large concentration gradients between the intracellular and extracellular environments (Table 6.1). These membraneous systems are in dynamic states and require continuous supplies of energy for their maintenance and to preserve the concentration gradients. Therefore if their energy supplies are interrupted, extensive leakage of cell constituents, including enzymes, could result possibly without the immediate appearance of structural damage. This proposal is supported by the investigations of Wilkinson and his colleagues[10] who studied the release of enzymes from human and rat lymphocytes and erythrocytes and by Schmidt[11] and coworkers using perfused rat livers. These groups found that cellular integrity and membrane permeability were directly related to the intracellular adenosine triphosphate (ATP) content. If the anabolic-catabolic

Figure 6.4: Suggested Mechanism for the Release of Intracellular Enzymes from Tissues Damaged by a Pathological Process

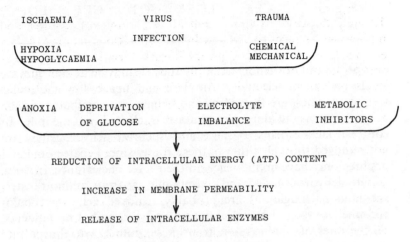

REDUCTION OF INTRACELLULAR ENERGY (ATP) CONTENT

INCREASE IN MEMBRANE PERMEABILITY

RELEASE OF INTRACELLULAR ENZYMES

Source: Redrawn with permission from Reference 12.

equilibrium which maintains this energy balance was perturbed by anoxia, hypoglycemia or starvation then the intracellular osmolality rose, membrane permeability increased and enzyme leakage resulted. In the experimental systems escape of the enzymes could be prevented by addition of ATP or its precursors to restore the energy balance. The mechanism proposed by Wilkinson[12] consistent with the observations is shown in Figure 6.4.

6.2.3 Factors Governing the Disappearance of Proteins from Blood Plasma

Different enzymes have long been known to disappear from the blood circulation at different rates. For example in Figure 6.1 the raised aspartate aminotransferase activity falls at a slower rate than alanine aminotransferase activity and in Figure 6.3 creatine kinase activity returns to normal about three fold faster than that of lactate dehydrogenase. Some quantitative values for the half lives of enzymes in serum are listed in Table 6.7.

Experimentally, the time courses of the clearances often depart considerably from simple first order kinetics, from which it may be inferred that complex mechanisms exist for the clearance of enzymes from plasma. That active mechanisms for enzyme removal

Table 6.7: Disappearance Rates of Enzymes from Plasma

Enzyme	CPK	LD5	GOT	GPT	LD1
$t_{1/2}$ (days)	1.4	1.5	2	6.3	6.8

do operate can be demonstrated by their saturation at high enzyme levels and their selectivity. Amylase and lipase, low molecular weight enzymes, are filtered by the kidney and eliminated via the urine, whereas alkaline phosphatase is cleared via the bile. In contrast, lactate dehydrogenase and isocitrate dehydrogenase are not removed through either of these routes since hepatectomy and nephrectomy have no influence on their rates of clearance. In dogs, lactate dehydrogenase-5 and cytoplasmic aspartate aminotransferase (and haemoglobin) are probably removed via the reticulo-endothelial system but its inhibition by zymosan has no influence on the rates of alanine aminotransferase, mitochondrial aspartate aminotransferase or LD1 clearance.

Posen[13] has put forward the idea that enzymes are first inactivated in the blood circulation before their elimination rather than being removed catalytically intact. Support for this proposal has come from the results of Qureshi and Wilkinson[14] who found that the rate of inactivation of LD5 in rabbit plasma was faster than the loss of the enzyme protein. Rabbit muscle lactate dehydrogenase was radioactively labelled with ^{125}I and injected intravenously into rabbits. After an initial distribution period, radioactivity was lost from the circulation at a slower rate than enzymic activity. During the elimination phase, relatively high concentrations of radioactivity were found in the intestinal juices, leading to the suggestion that discharge via the small intestine is a major route whereby inactivated enzyme fragments are removed from the circulation. In the intestine proteolysis most likely converts the polypeptides into low molecular weight peptides and amino acids. Three-quarters of the radioactivity was recovered in the urine and part identified as labelled amino acids and small peptides, thus the intestinal breakdown is followed by absorption of the products and their urinary excretion.

Circulating enzymes may therefore undergo intravascular inactivation with the major proportion of the products removed via the small intestine. But the actual inactivation mechanisms involved await elucidation.

6.3 ENZYMES AS DRUGS[15,16]

Enzymes possess many of the characteristics required of therapeutic agents. They are specific, capable of producing the desired effect without eliciting side reactions, they are water soluble, making formulation and administration of their preparations relatively easy, and they are highly efficient in the biological environment. For these reasons, enzyme preparations have long been used internally as digestive aids and topically for the treatment of burns and superficial infections. More recently, their efficacies as selective chemotherapeutic agents for the treatment of thrombo-embolic, inherited and neoplastic diseases are under active investigation.

Enzymes are limited in this application however. The large molecular structures exclude them from the intracellular domain and being proteins, in most instances foreign to the host, they are also antigenic. Hypersensitivity reactions preclude their administration over long periods and the induced formation of antibodies lowers their effective catalytic activities, an effect which exacerbates the normal rapid clearance of proteins from the blood plasma. Consequently, the administration of large quantities is usually necessary in order to maintain an adequately high level of activity, substantially raising the cost of enzyme therapy. For parental use the preparations must be extensively purified free from pyrogens and toxins, and as outlined in Chapter 1, enzyme purification on a large scale is a lengthy, tedious and expensive process.

Nevertheless, considerable relief has been afforded to patients with digestive and absorption problems and although the parental administration of enzyme preparations of specific functionality is still at a relatively early developmental stage, this type of therapeutic application is one of the exciting new areas of enzymology (Table 6.8).

6.3.1 Treatment of Digestive Disorders and Inflammation[17]

Microbial, plant and animal enzyme extracts have been used over many years to supplement enzyme deficiencies of the pancreas and small intestine. Pancreatin, obtained as an off-white powder after alcohol extraction of animal pancreas is given buccally to assist the enzymatic digestion of dietary starch and proteins in patients with pancreatitis or fibrocystic disease of the pancreas. Children with cystic fibrosis respond well to pancreatin therapy although favourable responses in adults have also been obtained. A lipase enriched

Table 6.8: Therapeutic Applications of Enzyme Preparations

Enzyme preparation (activity)	Source	Therapeutic application
Asparaginase	*E. coli*, guinea pig serum	Cytotoxic agent
Bromelain (protease)	*Ananas comosus*	Inflammation, oedema
Chymotrypsin (protease)	Bovine pancreas	As for bromelain plus ophthalmology and upper respiratory tract diseases
Deoxyribonuclease (DNA hydrolysis)	Bovine pancreas	Reduces viscosity of pulmonary secretions
Dextranase (dextran hydrolysis)	*Penicillium funiculosum*	Dental plaque restriction
Diastase (starch hydrolysis)	Malt	Amylaceous dyspepsia
Galactosidase (lactose hydrolysis)	*Aspergillus niger*	Inherited β-galactosidase deficiency
Hyaluronidase (mucopolysaccharide hydrolysis)	Animal testes	Increase absorption rate, increase effectiveness of local anaesthetics
Pancreatin	Animal pancreas	Pancreatitis
Papain (protease)	*Carica papaya*	Dyspepsia, gastritis
Penicillinase	*Bacillus cereus*	Penicillin allergy
Plasmin (protease)	Plasminogen	Thrombotic disorders, anticoagulation
Streptodornase (DNA-ase)	*Streptococci*	Depolymerisation of DNA in purulent exudates
Streptokinase (protease)	*Streptococci*	Thrombo-embolic diseases
Trypsin (protease)	Animal pancreas	Cleaning necrotic tissue
Urokinase (protease)	Human urine	Thrombo-embolic disease

pancreatin has been used to treat patients eliminating fatty stools. Other hydrolytic enzyme preparations such as papain and extracts from the fungi *Aspergillus niger* and *A. oryzae* are also used to aid absorption from the small intestine. These extracts contain cellulases as well as amylases and proteases which together assist degradation of the indigestible fibres of cabbages etc., so reducing dyspepsia and flatulence.

Micro-organisms are valuable as large scale sources of therapeutic enzymes; the organisms *Saccharomyces cerevisiae, S. fragilis, Bacillus subtilis* as well as the two Aspergillus species referred to above are 'generally regarded as safe' (GRAS) by the Food and Drug Administration of the USA to be taken orally. That is, they contain no toxins harmful to humans. For example, administration of an augmented β-galactosidase preparation from *A. oryzae* is of benefit to patients suffering from the inherited intestinal disease lactase deficiency. Children with this genetic trait are unable to effectively digest milk lactose and consequently suffer gastrointestinal disorders. β-Galactosidase degrades the lactose to glucose plus galactose, both of which are readily absorbed by the intestine. Another useful therapeutic enzyme derived from a microorganism is penicillinase from *Bacillus cereus*. Penicillinase is used to treat the small proportion of the population who undergo a hypersensitivity reaction when treated with penicillin antibiotics. The enzyme catalyses the hydrolysis of penicillin to penicilloic acid, which is not antigenic nor, unfortunately, biologically active.

Non-specific microbial and plant hydrolases are also used occasionally to reduce inflammation and oedema. Bromelain, trypsin, chymotrypsin, papain and streptokinase-streptodornase (Varidase) have all been subjected to clinical trials with varying degrees of success. All are given orally and have been demonstrated to elicit measurable proteolytic activity in the serum. The argument for the use of these proteolytic-nucleolytic mixtures is that catalytic dissolution of the soft fibrin clots would increase tissue permeability and reabsorption of the oedema. An added advantage, it is argued, is the possible debridement of the inflammatory exudates. Varidase in particular has been shown to relieve pain on systemic injection. The preparations have also been used to clean dirty wounds and necrotic tissue and to remove the debris from second and third degree burns.

6.3.2 Fibrinolytic Enzymes

Blood circulates through every tissue and participates in every major activity in the body. Consisting of a plasma suspension of non-nucleated red cells (erythrocytes), nucleated white cells (leukocytes) and platelets (thrombocytes), this extremely important biological fluid performs three major functions (1) the transport of oxygen from the lungs to the tissues, and of carbon dioxide in the opposite direction and the conveyance of carbohydrates, hormones etc. from their production sites to their sites of action; (2) the regulation of salt-water levels and acid-base balance, and (3) defence. In its defensive role blood carries antibodies to counteract invading antigens and white cells to ingest bacteria, but one of its most intricate functions is the prevention of fatal blood loss after blood vessel injury.

In the normal, healthy circulation, blood remains fluid, but when shed a 'haemostatic' mechanism operates to arrest the bleeding and aid healing. On injury, damage to the endothelial lining of the blood vessel exposes the collagen fibres and the first stage in the defence mechanism is attachment to these protein fibres of the platelets. On adherence the platelets swell, releasing serotonin plus ADP. ADP increases further thrombocyte aggregation and eventually a plug or 'thrombus' forms which seals the ruptured blood vessel. The platelets simultaneously release 'factor 3' which initiates the second stage of haemostasis—clot formation or coagulation. The clot, an insoluble network of fibrin polypeptide strands in which red cells and other blood components are trapped, is formed on the initial thrombus.

Even though many of the details are still incomplete, elucidation of the main steps in the complex blood coagulation mechanism is a major triumph of biochemical and haematological investigation. The cascade mediating clot formation is shown in Figure 6.5.*

Figure 6.5: Mechanism of Blood Coagulation

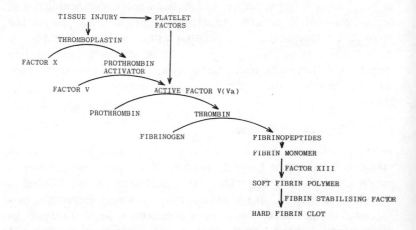

* The principles and physiological roles of 'cascades' are explained in Chapter 3.

Damage to the tissues causes release of thromboplastin plus calcium ions. Thromboplastin catalyses the formation of the active enzyme thrombin from the plasma zymogen prothrombin. Thrombin then promotes the conversion of soluble plasma fibrinogen into insoluble fibrin. Fibrinogen is a dimer of molecular weight 340,000 daltons containing three pairs of dissimilar chains α (or A), β (B) and γ, disulphide bond linked. Thrombin splits the monomers at specific arg-gly peptide bonds, releasing two fibrinogen peptides A and B plus sialic acid to give fibrin monomers containing four new amino terminal glycines. The fibrin molecules then polymerise by end-to-end and side-by-side alignment into a loose aggregate or 'soft' clot. In the final stage, fibrinase (fibrin stabilisation factor) hardens the clot by catalysing disulphide interchange.

After a few days, the blood clot is digested by the proteolytic enzyme plasmin (fibrinolysin). This trypsin-like enzyme is released from plasminogen by activator(s) catalysed hydrolysis of an arg-val bond. These activators are released from the blood cell walls on vascular injury, exercise, emotional stress or raised epinephrine levels. The mechanism of fibrin dissolution is shown in Figure 6.6.

In health, abnormal blood clot formation in the blood vessels never occurs, partly because of the nature of the endothelium lining the blood vessels and the presence of circulating anticoagulants like heparin, and partly because of the delicate balance maintained between the regulation mechanisms described above. Formation of intravascular clots (thrombosis*) may occur however if the endothelium is damaged by injury or inflammation, if blood flow is slow, in varicose veins or in a recumbent person for example, or if the chemical constitution of the blood alters. Many aspects of the

Figure 6.6: Mechanism of Blood Clot Dissolution

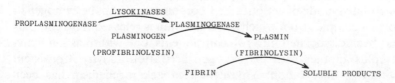

* A thrombus is formed in moving blood from agglutinated platelets, white cells, red cells and fibrin. An embolus is a piece of thrombus broken off and carried along in the blood stream. If it originates in the systemic veins it can lodge in the pulmonary artery causing pulmonary embolism. The problem of circulating thrombi becomes particularly acute if they lodge in the blood vessels serving the brain.

aetiology of arterial thrombosis have been established from epidemiological studies and obesity, cigarette smoking, genetic influences and certain contraceptive pills have been identified as high risk factors. Thrombo-embolic diseases, which include strokes, heart attacks and pulmonary embolism are in fact the primary health hazards of Caucasian adults.

The procedures for therapeutic treatment of these diseases can be divided into: (1) prevention of platelet aggregation; (2) prevention of fibrin formation (anticoagulation) and (3) induction of fibrin removal. Enzymes have been tested clinically as anticoagulants and fibrinolytic drugs, but because intravenous administration of heparin or anticoagulant drugs, such as coumarin or indanedione, are more immediately effective, their general clinical usage is the promotion of fibrinolysis. The rationale of enzymic fibrinolysis rests on the supposition that increasing the levels of circulating proteases should more rapidly dissolve the blood clot or preformed thrombus. For this purpose infusion of two types of enzymes can be considered, those with direct and general proteolytic activity such as plasmin or trypsin, or those such as the plasminogen activators which increase the levels of native plasmin. Clinical evidence supports the second alternative. Promotion of natural lysis utilises the avalanche effect of the cascade, does not require a prior infusion to neutralise circulating anti-proteases and is more specific. In this respect, two lysokinase plasminogen activators have received the most attention, streptokinase and urokinase.

Streptokinase is a protease of molecular weight approximately 47,600, and is derived from the filtrates of haemolytic Streptococci. The protein molecule contains no cysteine or cystine and has an isoelectric point at pH 4.7. The protein molecule possesses thrombolytic capability, catalysing the transformation of human plasminogen to plasmin by hydrolytic cleavage of the arg-val bond. Its postulated mode of action is 1 : 1 combination with the zymogen to form a complex from which active plasmin is released. This complex has been dissociated into two components, one with plasmin activity plus another with plasminogen activation activity. Consistent with this observation, a positive feedback mechanism has been proposed[18] (eq. 6.1), in which complexation first takes place between streptokinase and the plasmin contaminant of plasminogen. This complex or a lighter derivative then activates plasminogen to plasmin, simultaneously releasing the activator complex which is then free to transform a second plasminogen molecule.

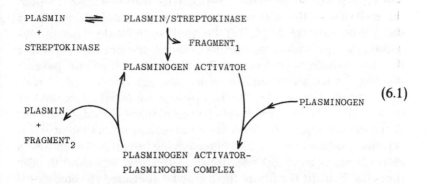

$$(6.1)$$

Streptokinase was the first enzymic activator of fibrinolysis to be clinically investigated. In humans, dogs and rabbits, the enzyme is fairly well tolerated. Some success has been achieved in the lysis of artificially induced thrombi in veins after systemic infusion and in deep vein thrombosis; between 50 and 70 percent of patients showed radiographic clearance within a few days of the clinical onset of symptoms. A success level of 60 to 70 percent lysis was also achieved in artificially induced pulmonary embolism in animals and the superiority of streptokinase over heparin in reducing mortality in myocardial infarction was confirmed by a European Working Party.[15] However because the enzyme is bacterial in origin it is antigenic and sufficient must first be given to bind antibodies and non-specific inhibitors. The immune response also limits the number of dosages that can be administrated and thus therapy can only be carried out over a short time period. Streptokinase effectivity is reduced further by its short lifespan in the circulation, its clearance behaviour being biphasic with half-lives of 18 and 80 minutes. For these reasons attention has been directed to urokinase.

It has been known for many years that urine can digest blood clots because of the presence of the proteolytic enzyme urokinase. Human urokinase can act as a plasminogen activator, its specificity is to cleave peptide bonds adjacent to lysine and arginine side chains, and being of human origin it is non-antigenic. Synthesised in the kidney, the protein is a single polypeptide of molecular weight 54,500 daltons and is very stable to temperature and extremes of pH. Its enzymic action is similar to that of trypsin, catalysing the activation of plasminogen in a first order reaction with cleavage of the arg-val bond.

A large scale and controlled clinical trial, the Urokinase Pulmonary Embolism Trial was launched in 1968 by the National Heart and Lung Institute, USA.[19] Their findings indicated that thrombolytic therapy with urokinase may become the treatment of choice in massive pulmonary embolism with shock in high risk patients although patients with smaller emboli also responded well to treatment. Its effectiveness for the therapy of myocardial infarction was less certain however. Compared to streptokinase, clinical experience of urokinase is less comprehensive because of its availability. It is extremely dilute in urine, 2300 litres having been reported to yield a mere 29 mg of pure enzyme. Thus treatment is expensive in both time and cost. In the future these may be alleviated to some extent by developments in the tissue culture of human kidney cells and affinity chromatography techniques.

6.3.3   Enzyme Therapy of Neoplastic Diseases[20]

The therapeutic treatment of cancer* with enzymes is based on the same principle as that of other drug treatments, that is, by depriving the abnormal cells of their essential metabolic precursors (nucleic acids, amino acids and folates) their growth may be checked or terminated. To be considered suitable as drugs in this context, enzymes must satisfy several thermodynamic and physiological conditions: (a) the reactions catalysed must be irreversible so that the substrate concentrations are effectively depleted; (b) they must exhibit high affinities and considerable specificity for the metabolite; (c) the rate of substrate catalysis must be higher than the rate of its tissue repletion; (d) the enzyme should be optimally active under physiological conditions; (e) an adequate cofactor concentration must be present; (f) they should be non-destructive towards the host's normal cell metabolism and (g) the products must be non-toxic. The main biological limitations to enzyme therapy arise principally from their protein natures. Systemic injection of non-human proteins leads to immunological responses which preclude prolonged treatment unless combination therapy with immunosuppressants is also given. Induced formation of antibodies will also increase the rate of clearance of the administered enzyme. Proteins are rapidly cleared from the circulation and thus the effective life-

---

* Cancer is the excessive and uncontrolled proliferation (neoplasia) of living cells. Leukemia is neoplasia of the blood forming tissue, one characteristic of which is an increase in the white blood cells in the circulation. Cancer of the bone marrow is named myelocytic and that of the lymphoid tissues is named lymphocytic.

span of each dose is severely limited. It has been possible in mice to decrease the rate of enzyme disappearance by co-injection of Riley lactate dehydrogenase elevating virus but its effect on other species is limited. (However as explained in Chapter 7 the possibility exists for the stabilisation of enzyme activity by immobilisation or microencapsulation and this approach is an exciting and promising possibility for the future.) Proteins and their conjugates of molecular weights greater than about 6×10^5 daltons are limited in their vascular permeability and thus their potential is mainly restricted to the intravascular space; neoplastic cells growing adjacently to normal cells supplying essential metabolites will then remain relatively inert to enzymic action.

Although the above criteria provide definite boundaries, several enzymes have been tested and proved suitable as anti-tumour agents. L-Asparaginase, L-arginase, carboxypeptidase G (folate depletion), L-glutaminase, L-methioninase, L-phenylalanine ammonia lyase, L-serine dehydratase, L-tyrosinase and xanthine oxidase, have all been investigated for their capacity to deprive rapidly growing cells of their metabolites. But of these, the first to be identified and which stands as the prototype for enzyme therapy is the protein that catalyses the hydrolysis of L-asparagine to aspartic acid plus ammonia.

6.3.3.1 L-*Asparaginase*[21] (L-*Asparagine Amidohydrolase*, EC 3.5.1.1)

The history of the identification of asparaginase as an anti-tumour agent is quite involved. In 1953, Kidd[22] found that only adult guinea pig serum, but none of juvenile guinea pig, rabbit, horse or human sera induced regression of lymphosarcoma strain 6C3HED in C3H mice and lymphoma III in albino mice. It had no effect however on the growth of sarcoma 180 or mammary carcinoma. Although Kidd correctly identified the active agent as a protein, ten years were to elapse before Broome[23] showed that the antilymphoma capability of guinea pig serum fractions paralleled their asparaginase activities. This advance was followed by the successful treatment by Dolowy[24] of a girl suffering from acute lymphatic leukemia with guinea pig serum. A scientific basis for these empirical observations was provided also by Broome who demonstrated that the tumour cells most sensitive to asparaginase, unlike normal or resistant abnormal cells, required an exogenous supply of L-asparagine for growth. The conclusion that asparaginase was responsible for 'Kidd's phenomenon' was confirmed by Mashburn and Wriston[25] who showed that several other proteins with asparaginase activity were also anti-neoplastic, particularly those isolated from the bacteria *Escherichia coli* and *Erwinia carotovora*. Clinical investigations of L-asparaginase were impelled by this discovery which meant the enzyme could be obtained in much larger quantities.

In *E. coli*, two isoenzymes with asparaginase activity, labelled EC-1 (or I) and EC-2 (or II) are present within the cell proper and in the periplasmic space respectively. Asparaginase-I is constitutive and asparaginase-II is induced by anaerobiosis but only the latter has anti-lymphoma activity. Isoenzyme II has been purified to homogeneity and the amino acid sequence of that from *E. coli* strain A.1.3 reported.[26] The enzyme from a related strain, *E. coli* A.1.3 KY3598, has also been crystallised[27] and preliminary X-ray studies support a tetrameric quaternary structure with the four subunits related by 222 symmetry, most probably in a tetrahedral arrangement. The subunits are identical with molecular weights 34,080. From circular dichroistic data, it is estimated that 10 percent of its amino acid complement is arranged into α-helices and 40 percent into β-sheets. No free cysteine groups are apparent, judged by its insensitivity to iodoacetate or *p*-mercuribenzoate, but the monomers each contain one intra-chain disulphide bond.

Each subunit can be labelled and inactivated by titration with one equivalent of the aspartate analog 5-diazo-4-oxo-L-norvaline (DONV). Since this compound also behaves as a poor substrate, liberating nitrogen, the existence of one active site per subunit is supported. The active sites are specific for asparaginase, catalysing the hydrolysis of D-asparagine and L-glutamine at 30–50 fold slower rates. Also the enzyme is not inactivated by the glutamine analog 6-diazo-5-oxo-L-norleucine (DON). The presence of an α-carboxyl group in the substrate is necessary for catalysis to occur but a free α-amino group appears to be required mainly for attachment since its elimination or substitution decreases the stability of the corresponding enzyme-substrate complex. The active sub-site complementary to the carboxamido group is also restricted in its steric capacity, limiting the substituents to short, linear aliphatic chains.

E. coli asparaginase-II obeys Michaelis-Menten kinetics and has maximum activity between pH 6 and 8. Its catalysed hydrolyses of asparagine and L-aspartyl-β-hydroxamate are consistent with a 'ping-pong' mechanism, i.e. a catalytic pathway involving initial release of ammonia to form an acyl-enzyme intermediate, which then breaks down to form aspartate plus native enzyme, eq. 6.2.

$$E + R \cdot CO \cdot NH_2 \rightleftharpoons E \cdots R \cdot CO \cdot NH_2 \rightleftharpoons E \cdot CO \cdot R \xrightarrow{H_2O} E + R \cdot CO_2H$$

$$R = -CH_2 \cdot CH(NH_2)CO_2H \qquad NH_3 \qquad (6.2)$$

This mechanism has obvious similarities to that proposed for proteolytic enzyme action. A similar mechanism is also proposed for the action of the enzyme from *Erwinia carotovora*. The vegetable bacterium has however a much higher specific activity, its value of 700 IU/mg is approximately twice that of EC-2, which is interesting since the two enzymes appear to possess similar molecular weights, specificities and quaternary structures.

The normal plasma asparagine concentration is maintained fairly constant at $4 \times 10^{-5}M$ by liver metabolism, and although a wide variety of mammalian livers contain asparaginase, no measurable sera activity is found in the same animals except for adult members of the rodent family Cavioidea. Injection into humans of 100 IU of asparaginase lowers the plasma concentrations to an undetectable level and maintains it at this level for several hours.

L-Asparaginase therapy has proved of value for the treatment of three diseases, acute lymphoblastic leukemia, leukemic lymphosarcoma and myeloblastic leukemia. High levels, between 100–1000 IU/kg/day for up to one month can be tolerated and the documented remission rates from acute lymphoblastic leukemia, the most sensitive malignancy of the three, range from 25 to 70 percent depending on the institution. The rates appear to be independent of dosage within the above range. The therapeutic index for L-asparaginase against acute lymphoblastic leukemia is about 500 which is the highest found for an anti-tumour agent. However the remission is not prolonged and the patients are resistant to further treatment. A total kill of all leukemic cells thus appears to be unachievable with the enzyme alone and combination therapy with prednisone and vincristine has proven superior in inducing remission in children. Asparaginase is less effective (10 to 20 percent) against myeloblastic leukemia, lymphosarcoma and other solid tumours.

The simplest rationalisation of the biological action of asparaginase is deprivation of the sensitive tumours of their nutritional asparaginase supply. In resistant tumours high levels of the enzyme asparagine synthetase* (EC 6.3.1.1) which catalyses the transamination, eq. 6.3, have been detected.

$$\text{L-Aspartate} + NH_3 + ATP \rightleftharpoons \text{L-Asparagine} + PP_i + AMP \quad (6.3)$$

Asparagine is needed by the cell for protein synthesis, at least two specific activating enzymes having been identified in *E. coli* extracts and a number of cognate tRNA molecules noted in mammalian systems. Its depletion from cells has also been recorded to synergistically impede the incorporation of valine and glycine. Decreased protein synthesis leads to a fall in DNA and RNA synthesis. There is also evidence to support disorganised glycoprotein synthesis and hence disruption of the cell wall structure as the major consequence of asparagine depletion. One function of asparagine in the protein of the glycoproteins is to serve as the link, via the carboxamido group, to the *N*-acetylglucosamine units of the oligosaccharide component.

Impaired protein synthesis is also probably responsible for the immunosuppressant and toxic repercussions of asparaginase therapy. Both the humoral and cellular immune responses are inhibited and IgG immunoglobulin (a glycoprotein) production deteriorates. Restricted synthesis of the oligopeptide insulin and the blood coagulation proteins are also probably responsible for the observed clinical hypoglycaemia and clotting defects.

From early studies it was evident that not all asparaginases were equally effective as anti-lymphoma drugs, the yeast enzyme for example having no measurable efficacy. Several possible reasons for this have been considered, differences in K_m values for substrate, isoelectric points, and clearance rates having received the most attention (Table 6.9). Of these the lifespan of the enzyme in the circulation appears to be the most important factor. The half lives for clearance of guinea pig serum activity in mice, eleven and nineteen hours, and of *E. coli*-II and *E. carotovora* enzymes, about four hours, can be compared to the one hour for the ineffective yeast or *E. coli*-I derived enzymes. These are empirical observations however and very little is known at present concerning the mechanism of

---

* After initial sensitivity some lymphoid tumours become resistant to asparaginase action, probably as a result of de-repression of asparagine synthetase. In these cases, combination therapy with asparagine synthetase inhibitors such as the transition state analog homoserine adenylate has been attempted.

Table 6.9: L-Asparaginases

Source	Anti-tumour activity	MW (daltons)	Approximate isoelectric points	K_m Asparagine (M^{-1})	Mucine clearance rates (h)
Guinea pig serum	Yes	135,000	3.6–4.5	7.2×10^{-5}	11–19
E. coli K-12, I	No	130,000	5.4	1.2×10^{-5}	Very rapid
E. coli K-12, II	Yes	130,000	4.9		4
Serratia marcescens	Yes	140,000	6.0	1×10^{-4}	3–6
Erwinia carotovora	Yes	135,000	8.6–8.9	1×10^{-5}	4
Baker's yeast	No	800,000			Very rapid

plasma asparaginase clearance. The reticulo-endothelial system (RES) may be involved but the RES inhibitor zymosan has no effect on the rates of clearance. A role in plasma clearance for the liver has been suggested from comparison of the high (15 percent) asparagine/glutamine content of asparaginase with other proteins known to be cleared by this organ, but not proved conclusively.

The future of enzymes in anti-cancer and other therapy looks fairly bright provided the problems of antigenicity and short circulatory lifespans can be overcome. Research attention is accordingly devoted partly to suitable chemical modification of enzymes and immobilisation in suitable matrices for stabilisation, and partly to attempted direction of enzymes to particular interstitial localities or sanctuaries, a technique known as 'address labelling'.

6.4 REFERENCES

1. D. W. Moss, Adv. Clin. Chem., vol. 19 (1977), p. 1.
2. J. Wohlgemuth, Biochem. Zeit., vol. 9 (1908), p. 1.
3. A. Bodansky, J. Biol. Chem., vol. 101 (1933), p. 93.
4. O. Warburg and W. Christian, Biochem. Zeit., vol. 314 (1943), p. 399.
5. J. S. La Due, F. Wroblewski and A. Karmen, Science, vol. 120 (1954), p. 497.
6. F. de Ritis, M. Coltorti and G. Giusti, Minerva Med., vol. 46 (1955), p. 1207.
7. Th. Bücher and P. Baum, Dtsch. Kong. f. Ärztl. Fortbildung Berlin (1958).
8. J. H. Wilkinson, Principles and Practice of Diagnostic Enzymology (Arnold, London, 1975).
9. D. Young, in Blume and Frier (ed.), Enzymes in the Practice of Laboratory Medicine (Academic Press, New York, 1974), p. 253.

10. J. H. Wilkinson, J. M. Robinson and K. P. Johnson, *Ann. Clin. Biochem.*, vol. 12 (1975), p. 58; *Clin. Chem.*, vol. 20 (1974), p. 1331.
11. E. Schmidt, *Lab.*, vol. 1 (1974), p. 258.
12. J. H. Wilkinson, 'Clinical Enzymology,' Proc. Symp. Inst. Med. Lab. Sci. Liverpool (1975).
13. S. Posen, *Clin. Chem.*, vol. 16 (1970), p. 71.
14. A. R. Qureshi and J. H. Wilkinson, *Clin. Sci. Mol. Med.*, vol. 50 (1976), p. 1.
15. J. S. Holcenberg and J. Roberts, *Ann. Rev. Pharmacol. Toxicol.*, vol. 17 (1977), p. 97.
16. D. A. Cooney and R. J. Rosenbluth, *Adv. Pharmacol. Chemother.*, vol. 12 (1975), p. 185.
17. Martindale, *The Extra Pharmacopoeia*, 26th ed. (Pharmaceutical Press, London, 1972).
18. F. B. Taylor and J. G. Beisswanger, *J. Biol. Chem.*, vol. 248 (1973), p. 1127.
19. Urokinase Pulmonary Embolism Trial Study Group, *J. Am. Med. Ass.*, vol. 214 (1970), p. 2163.
20. J. R. Uren and R. E. Hanschumacher, 'Cancer. A Comprehensive Treatise,' in Becker (ed.) *Chemotherapy*, vol. 5 (Plenum Press, New York, 1977), p. 457.
21. J. C. Wriston and T. O. Yellin, *Adv. Enzym.*, vol. 39 (1973), p. 185.
22. J. G. Kidd, *J. Exp. Med.*, vol. 98 (1953), pp. 565, 583.
23. J. D. Broome, *J. Exp. Med.*, vol. 118 (1963), pp. 99, 121.
24. W. C. Dolowy, D. Henson, J. Cornet and H. Sellin, *Cancer*, vol. 19 (1966), p. 1813.
25. L. Mashburn and J. Wriston, *Arch. Biochem. Biophys.*, vol. 105 (1964), p. 450.
26. T. Maita, K. Morokawa and G. Matsuda, *J. Biochem.*, vol. 76 (1974), p. 1351.
27. M. Yonei, Y. Mitsui and Y. Iitaka, *J. Mol. Biol.*, vol. 110 (1977), p. 179.

7   Enzyme Technology

Over the last decade, 'enzyme technology' or 'enzyme engineering' has rapidly become a specialism, the result of a successful marriage between the principles of enzymology, as outlined in earlier chapters, and chemical engineering techniques. The many different spheres of expertise and knowledge that have been combined to form this applied aspect of enzymology are briefly summarized in Figure 7.1. Two areas in particular have contributed to its progress, the improvements made in procedures for the large scale production and isolation of enzymes from microbial sources and the practical potentiality of enzymes immobilised on insoluble matrices.

In distinction to procedures used for the purification of enzymes in quantities of only a few grams, those required for the purification of amounts ten or one hundred fold larger introduce their own, different requirements of handling, yield optimisation, plant design, and process control. These are the traditional realms of chemical engineering and thus one aim of enzyme technology is to use this expertise to supply the large quantities of enzymes required by the food, pharmaceutical and textile industries to improve the quality and economic supply of their products.*

The bulk of the enzymes used in these industries are the relatively cheap extracellular enzymes secreted by non-toxic microorganisms. They are inexpensive because they can be recovered in a partially pure form from a fermentation vessel by simple filtration without expensive cell disruption and purification procedures. Once used these crude, bulk enzymes are then either discarded, or incorporated into the product if they are found to contribute additional flavour characteristics or visual attractiveness.**

In most cases these bulk enzymes are preferred, on economic grounds, to the purer preparations of higher specific activity. But to some extent, the application of the latter has been expanded and the handling of enzymes made easier by attachment to insoluble supports. Insolubilisation of an enzyme confers easier removal from a

---

* The utilisation in the textile industry is however gradually diminishing as synthetic (man-made) fibres become more predominant.

** Being protein they are, of course, nutritious.

Figure 7.1: *Enzyme Engineering*

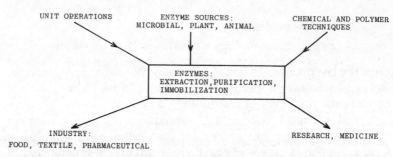

reaction mixture and thus its more frequent reuse, especially if its stability is also promoted. The present size of the enzyme industry can be gauged from the annual world market sales which are of the order £20 to 40 million,[1] ($40–80 million). Sales of the various enzyme classes, and the different industrial users are shown in Table 7.1.

By far the most widespread industrially used bulk enzymes are the proteolytic enzymes, especially the neutral and alkaline proteases. Of these approximately half are used in the food industry and the rest are divided between the pharmaceutical, tanning and brewing industries. In the late 1960s, the detergent industry accounted for the largest share of the worldwide volume of proteolytic enzymes, but over recent years this has gradually waned as hazards to the users' health became apparent.

Table 7.1: Approximate Percentage Worldwide Sales of Enzymes, Classified as to Type (a) and Application (b)

(a)	Glycosidases	15
	Proteases	65
	Medical, diagnostic research	15
	Other (glucose oxidase, invertase)	5
(b)	Dairy	35
	Detergents	25
	Pharmaceutical	25
	Other	15

7.1 ENZYMES IN INDUSTRY

7.1.1 Food Enzymology[2,3]

Food science and enzymology have always been intimately related. Over countless generations, alterations in the consistency of foodstuffs have been effected by the application of enzyme preparations; for example, one of our most staple foods, milk, has been preserved as cheese by the application of calf stomach rennin. All foods are derived from naturally occurring materials, which are themselves produced by enzyme action. In other words, the constituents of food are enzyme products and substrates, one consequence of which is the spoilage of over-ripe food by endogenous enzymes.

Enzymes can be considered to have two main influences on the foodstuff industry (1) as inherent parts of the food (Table 7.2) and (2) as exogenous agents that can be added to induce a change in its constituency during manufacture (Table 7.3).

Although commercial emphasis has been placed on the latter type of food related enzyme, detailed studies of the activity of endogenous enzymes have provided valuable information regarding the treatment, analysis and history of food products. For example, the loss of alkaline phosphatase activity in milk is used as a guide to the success of its pasteurisation, and measurements of lysolecithinase activities have been employed to indicate fish quality.

The physical appearance and flavours of some food products are functions of their endogenous enzyme activities. One of the most common experiences is the darkening of apple tissue on its exposure to air. In molecular terms this has been explained as the result of oxidation of phenolic substances, flavones and flavonyl glycosides by atmospheric oxygen catalysed by the polyphenol oxidases present in the tissue. The fundamental step in the browning process appears to be oxidation of these diphenols to chromogenic *o*-quinones, eq. 7.1.

$$\text{(7.1)}$$

Table 7.2: Food Endogenous Enzymes

Enzyme	Source	Reaction	Effect
Alliin lyase	Onion, garlic	S-Alkyl-cysteine sulphoxide = 2-Amino acrylate + Alkyl sulphenate	Activates flavour
Amylase	Potatoes	Hydrolysis of 1,4-α-glycosidic links	Starch metabolism
Esterases	Fruit	Esters = Acid + Alcohol	Flavour activation
Invertase	Honey	β-Fructofuranoside hydrolysis	Invert sugar
Lipase	Butter	Triacylglycerol = Diacylglycerol + Fatty acid	Rancidity
Monophenol mono-oxygenase	Fruit, vegetables	Tyrosine + O_2 = Dioxophenylalanine	Aerobic darkening on tissue damage
Nucleosidases	Meat, fish, poultry	N-Ribosyl-purine = Ribose + Purine	Purine levels indicate freshness
Peroxidase/catalase	Vegetables	$H_2O_2 = H_2O + O_2$	Causes off-flavour
Thioglucosidase	Brassicas	Thioglucoside = Thiol + Sugar	Flavour and sugar formation plus toxic isothiocyanates

Table 7.3: Enzyme Additives in Food

Enzyme	Food	Use
Amylase	Beer, bread, corn syrup	Starch degradation
Catalase	Eggs, cheese	Peroxide removal after pasteurisation
Exo-1,4-α-glucosidase (glucamylase)	Corn syrup	Glucose production
β-Galactosidase (lactase)	Ice cream, desserts	Prevents graininess due to lactose crystals
Glucose isomerase	Sugar syrup	Promotes fructose levels— a sweetener
Papain	Meat	Tenderiser
Pectinases	Fruit juices	Dehazer, prevents gelling
Pepsin	Milk	Cheese manufacture
Proteases	Protein hydrolysates	Flavouring
Rennin (chymosin)	Milk	Cheese manufacture

Another visual phenomenon due to enzyme action is the dark purple pigmentation of vacuum packed beef. The colour is that of reduced myoglobin, which because of the paucity of oxygen caused by locally active oxidases is not oxidised to the red oxygenated form. Some of these oxidases, which act on unsaturated lipids to give epoxides, are partly responsible for rancidity in meat. Other enzymes, the hydrolases, released after the death of the animal, enhance the flavour by acting as natural tenderisers.

The taste and texture of fruit and vegetables are also dependent on their cellular and extracellular enzymes. The natural flavour substances are, of course, enzymes synthesised during the growth of the plant but after ripeness has been attained endogenous proteases and glycosidases hydrolyse the macromolecular cellulose and pectin structures, leading to disruption of the internal architecture, post-ripeness softening and decline of flavour.

The food processing enzymes can be divided into three broad groups: (1) integral components of the manufacturing process, for example, rennin and cheese production; (2) enzyme preparations applied to increase the rate and economy of production, for example, amyloglucosidase and sugar production, and (3) enzymes employed to improve the quality of food and food products, for example, pectinases and fruit and vegetables.

7.1.1.1   Cheese Production

Rennin (EC 3.4.4.3) is prepared from the fourth stomach of weaning calves. Its action in cheese formation is to specifically hydrolyse the kappa-casein of milk to para-casein. Calcium ions then cause the latter to clot and release the whey. By careful adjustment of pH and temperature the clot is made to form cheese, but in an unripe state. Mature cheese is produced by a process of ageing, during which time rennin probably continues its action. In recent years, because of meat shortages, calves have been allowed to reach maturity, thereby causing an insufficiency of rennin. This has necessitated a search for suitable substitutes. All proteases will coagulate milk but in general their wide specificities usually cause extensive proteolysis of the casein protein and thus they are unsuitable for cheese manufacture. Mixtures containing equal proportions of rennin and pepsin have been found to alleviate the deficiency, to some extent, and furthermore to reduce the cost. Also successful in experimental trials have been microbial proteases isolated from Mucor and Rhizopus species.

7.1.1.2   Starch and Sugar Industries

Amyloglucosidase (glucamylase) (EC 3.2.1.3) converts starch into glucose by hydrolysing the 1,4-α- and 1,6-α-glycosidic bonds in the polysaccharide. The enzyme is produced by deep fermentation from fungi such as *Aspergillus niger*, *A. oryzae* and *A. faetidus*, microorganisms generally regarded as safe (GRAS) by the Food and Drug Administration of the United States of America. It is prepared and used for sugar production in an impure form and as such exhibits rather broad specificity, acting on a wide variety of glucose polymers, including amylose, amylopectin, dextrins and glycogen. Partly for this reason and because of the large quantity of sugar that is consumed, glucamylase is the most important bulk enzyme currently used commercially. It gives higher yields of glucose than the time honoured acid catalysed hydrolysis of starch plus a less astringent mother liquor. Pretreatment of the starch with acid is still carried out although this is also gradually being displaced by technologies based on enzymes. Heat stable α-amylases have been found suitable.

α-Amylase (EC 3.2.1.1) hydrolyses the 1,4-α links in starch and glycogen to produce glucose, maltose and higher oligosaccharides. *Bacillus subtilis* is the main source of this enzyme, from which organism it is obtained by submerged fermentation and isolated by alcohol or acetone precipitation. Again, little expenditure or effort is used to further purify the enzyme. The micro-organism produces two enzymes with amylase activity, a starch saccharifying protein with maximum activity at 60°C and a liquefying enzyme with a temperature optimum ten degrees higher. These amylases can thus be designated as thermophilic enzymes.

As well as their use in the confectionary industry these amylases are also applied to textile designing and starch liquefaction in the paper industry.

An enzyme preparation also exhibiting amylase activity and from the fungus *A. oryzae*, 'Takadiastase', was the first enzyme to be sold commercially in western markets (1895), but as a digestive aid.* Produced by surface fermentation and extracted by solvent precipitation, Takadiastase is a complex mixture containing proteases, esterases, ribonucleases, glucamylases, glycosidases and phosphatases. Nowadays, its major use in the European market is to enrich flour, but in the United States of America it plays a major role in the production of corn syrup and bread. In baking its partial solubilisation of the starch has also been found to retard the staling processes and to impart good crust colour.

Approximately ten percent of total sucrose consumption is in the confectionary industry either as invert sugar for artificial honey or mixed as fondants for soft centred sweets and chocolates. Invert sugar is an equal mixture of D-fructose and D-glucose produced from sucrose by the action of invertase (EC 3.2.1.26). The main commercial source of this enzyme is the yeast *Saccharomyces cerevisae*.

7.1.1.3 Fruit and Vegetables

Pectins are the gelling substances of jam, purees and dressings. In these preparations they perform structural and textural roles which aid foodstuff stability and lifetime. In fruit and vegetable juices however, colloidal pectins retain particulate matter in suspension and give rise to a 'haze'. To produce a commercially acceptable product these juices must first be clarified, to which task pectinolytic enzymes are employed. These enzymes de-esterify and partially depolymerise the polymethyl galacturonates, giving products which, in the presence of calcium ions, coagulate and sediment,

* It is still available for this application in both oriental and occidental countries.

322   *Enzyme Technology*

simultaneously also absorbing some of the brown colour produced by the monophenol mono-oxygenases. The main pectinases can be listed as (a) EC 3.1.1.11, pectin methyl esterase, which demethylates pectin to pectic acid plus methanol, (b) EC 3.2.1.4, polymethylgalacturonidase, which catalytically hydrolyses the 1,4-α bonds in pectin, (c) EC 3.2.1.15, and EC 3.2.1.40, *endo-* and *exo-*polygalacturonases, which hydrolytically cleave the 1,4-α bonds of pectic acid, (d) EC 4.2.2.1 and EC 4.2.2.2, *endo-* and *exo-*polygalacturonic acid lyases, which rupture the 1,4-α bonds in pectic acid by trans-elimination, and (e) EC 4.2.2.3, polymethyl galacturonate lyase, which also cleaves the 1,4-α bonds of pectin by trans-elimination.

7.2 IMMOBILISED ENZYMES[4]

In the previous chapters, considerations have been given to the properties of enzymes and enzyme systems translationally and conformationally free in homogeneous solution. And although the modification of enzyme activity has occasionally been raised, for example, as a means of determining active site structure, no drastic changes in the molecular weight of the enzyme were envisaged. On the contrary, immobilisation restricts possible conformational movements in the enzyme and maintains it within a defined space. This is achieved either by conjugating the enzyme with another polymer or by covalently crosslinking the protein molecule, but always retaining as high an activity as possible. Immobilisation often results in the precipitation of the enzymic activity, the enzyme then being *insolubilized*. In manufacturing industry, immobilised enzymes have much potential.

The specificity and catalytic efficiencies of enzymes would make them ideal as industrial catalysts but their instability and the practical difficulties of removal, limit their effective utilisation, except in those cases already discussed. Normally, removal of soluble enzymes is a lengthy, tedious and hence costly process, whereas insolubilisation will obviously facilitate their separation. For some enzymes, but not all, insolubilisation has also conferred thermal or storage stability, enabling several reuses of one batch of enzyme material and reducing the process cost. Against these two main advantages have to be weighed several drawbacks. These include leaching of the enzyme from the stationary phase, causing loss of activity and the contamination of product, lower activities of the attached enzyme, and high costs of synthesis of the conjugates compared to those of the bulk enzymes discussed in the previous section. Some enzymes are stabilised by immobilisation but this is a generalisation true mainly for proteolytic enzymes and the stabilisation of other enzymes has generally been arrived at by empirical procedures. In addition the efficiency of conversion of high molecular weight polymeric substrates by immobilised enzymes is often reduced as a result of steric interactions with the polymeric support.

Some of these limitations however are gradually being overcome as the present extensive research programmes undertaken by many university and industrial laboratories, bear fruit. This work has so far resulted in many novel operations and principles[1,4,5] some of which will be outlined in the following sections.

The heterogeneity of immobilised enzymes makes them suitable as models for enzymes localised at naturally occurring interfaces. It is apparent from studies such as those described in Chapter 5 that cellular and subcellular organisation is complex. Intracellular enzymes function within the organised structures of the living cell, some are attached to membranes and associated with lipids, perhaps essentially removed from extensive contact with water, whereas others are free in solution but subject to concentration gradients. These micro-heterogeneous systems will therefore be very different from the homogeneous, aqueous and buffered systems used in the standard determinations of enzyme kinetic properties. Other significant properties of these microenvironments include surface effects, dielectric changes, electric fields, hydrophobic effects, all of which will affect the protein molecule and thereby its mode of reaction. Although some chemical insights into these effects are gradually emerging, methods for analysis of the integrated enzyme systems are lacking. One current approach to this problem is the synthesis of artificial enzyme systems, modelled on the heterogeneous membrane structures, in which the physico-chemical factors influencing enzymatic behaviour may be ascertained and evaluated.

Figure 7.2 gives a general classification scheme of immobilised enzymes, divided according to the solubilities of the conjugates and the practical techniques used for their formation. Four enzyme immobilisation procedures are used:

1. Covalent coupling to a polymeric matrix,
2. Crosslinking,
3. Adsorption on a polymeric matrix,
4. Entrapment within a polymeric matrix.

Figure 7.2: Classification of Immobilised Enzymes

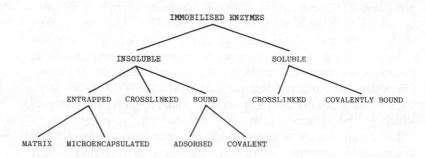

7.2.1 Covalent Attachment

The enzyme is coupled to the polymer through functional groups in the former non-essential to biological activity. It is usually difficult to restrict the modification to amino acid side chains remote from the active site. Few procedures are selective for these particular residues since by definition those at the active site possess enhanced reactivity. Coupling is therefore very often carried out in the presence of a competitive inhibitor to protect the active site. This procedure has the added advantage of fixing the enzyme in an active conformation. Reactive functional groups in the enzyme molecule available for attachment to the matrix are: (1) amino groups, the N-terminal α-amine, the ε-amino group of lysine and the guanidine group of arginine; (2) the α, β, and γ carboxylic acid groupings of the C-terminal amino acid residue, aspartic acid and glutamic acid residues respectively; (3) the tyrosine phenol and (4) sulphydryl groups. Mild methods have therefore been devised to couple the enzyme via one or more of these groups to the polymeric matrix without incurring major disruption of the protein molecule. A representative list of such coupling reactions, together with examples of polymers and enzymes tied together is given in Table 7.4.

Table 7.4: Examples of Enzymes Immobilised on Water Insoluble Matrices

Polymer	Activator	Enzymes*
Ala-glu copolymer	Woodward's reagent K	Acetylcholinesterase (45)
p-Aminophe-leu copolymer	Polydiazonium salt	Urease (80), streptokinase (5)
Cellulose	Titanium chloride	Amyloglucosidase (46)
Cellulose, *p*-aminobenzyl-	Polydiazonium salt	Trypsin (60), thrombin (a)
Cellulose, bromoacetyl-		Pronase (70), trypsin (2)
Cellulose, carboxymethyl-	Azide	Ribonuclease (40), ficin (13)
Cellulose, DEAE-	Gluteraldehyde	Glyceraldehyde-3-phosphate DH(1)
	Cyanuric chloride	Lactate DH (40)
Ethylene-maleic anhydride	Copolymer	Trypsin (70), papain (a), streptokinase (a)
Polyacrylamide	Acyl azide	Amylase (a), trypsin (35)
Polyaminostyrene	Polydiazonium salt	Lipase (a), glucose oxidase (a)
Polyaspartic acid	Woodward's reagent K	Cholinesterase (35)
Porous glass	Isothiocyanato-alkylsilane	Papain (75)
	Aminoarylsilane, diazonium salt	Alkaline phosphatase (a)
Sepharose	Cyanogen bromide	Glucose-6-phosphate DH (a)

* Figures in brackets are percentage activities retained; a = active.

Enzymes have been attached to both electrically neutral carriers, such as acrylic polymers and nylon, both of which contain amide groups, polysaccharides such as cellulose, dextran gels and agarose gels and also to polyelectrolytes, for example, polyamino acids, ethylene-maleic anhydride copolymers and the ion-exchange resins, carboxy-methyl cellulose and aminoethyl cellulose. These hydrophilic materials have been found to induce a smaller degree of enzyme destabilisation than the lipophilic types, such as polystyrene, although the majority of enzymes so far studied have been either extracellular or cytoplasmic in origin, and this may become less a generalisation as more organellular enzymes are studied.

The carriers are available in many different forms, beads, granules, sheets, membranes and tubes, which enable the attached enzyme to be used for a variety of purposes. Porous glass for example has good mechanical properties and also suffers no bacterial attack, unlike the natural supports, dextran and agarose, but leaching of the enzyme from the support becomes more of a problem with the former. Nylon and polystyrene, obtainable in tubular form have been internally coated with enzyme and incorporated into continuous flow systems. In such a manner several enzymes may be linked in series, the product of one serving as the substrate for the next in line. For example a system for the routine, continuous analysis of sucrose has been described in which the test samples are first passed through an invertase coated tube, which converts the sucrose to fructose plus glucose, followed by continuous passage of the monosaccharide solution through a glucose oxidase tube. The hydrogen peroxide produced by the second immobilised enzyme is determined colorimetrically and is proportional to the concentration of sucrose in the original sample. Controlled pore size and bead sizes of polyacrylamide, agarose and dextran gels and granules of starch and cellulose are used as the packed stationary phases for chromatography. The properties required of the preparations for their application are then the same as those for affinity chromatography as discussed in Chapter 1.

Normally, the majority of these polymers are unreactive towards protein amino acid side chains and therefore either the polymer or the enzyme or both must be activated prior to reaction with its partner, or as an alternative a third, intermolecular, crosslinking agent must be included. The last choice is usually least preferred because of the greater probability of side reactions which could decrease the yield of attached enzyme. Preactivation of the polymer to a form capable of direct one-step attachment of the enzyme is usually favoured since the polymer is often in greater supply and any adverse effect on the protein is minimised. Most of the reactions utilised to activate the polymers fall into the following types:

1. Acylation or alkylation with acyl azides, acyl halides, anhydrides, isocyanates, isothiocyanates, alkyl halides, aldehydes, iminocarbonates, carbonates—all of which give products susceptible to nucleophilic attack by enzyme amino groups.

2. Reaction of aromatic diazonium salts with tyrosine and histidine.

3. Amide formation between a carboxylic acid and an amine, promoted by a water soluble carbodiimide or Woodward's reagent, K.

4. Sulphydryl exchange.

5. Reaction with heavy metal salts.

7.2.1.1   *Acylation and Alkylation*

Activation of polyhydroxyl polymers by acylating or alkylating reagents, followed by nucleophilic attack by a non-essential enzyme amino group on the acylated or alkylated carrier constitutes one of the earliest and still used methods of immobilisation. One of the first patented procedures for the insolubilisation of trypsin, chymotrypsin and ribonuclease utilised bromoacetylcellulose, synthesised from cellulose and bromoacetyl bromide, eq. 7.2.

$$
\text{[}-\text{OH} \xrightarrow{\text{Br·CO·CH}_2\text{·Br}} \text{[}-\text{O·CO·CH}_2\text{·Br}
$$

$$
\xrightarrow{\text{E·NH}_2} \text{[}-\text{O·CO·CH}_2\overset{\text{H}}{\underset{\centerdot}{\text{N}}}\text{·E}
$$

$$(7.2)$$

Reactions of this mechanistic type were introduced to activate the otherwise relatively unreactive polysaccharide hydroxyl groups. Another versatile reagent, cyanogen bromide has already been outlined in Chapter 1.

One of the earliest reports on enzyme immobilisation described the application of the Curtius azide reaction to cellulose activation,[6] as shown in eq. 7.3:

$$
\text{[}-\text{OH} \xrightarrow{\text{Cl·CH}_2\text{·CO}_2\text{H}} \text{[}-\text{O·CH}_2\text{·CO}_2\text{·H}
$$

$$
\xrightarrow[\text{HCl}]{\text{CH}_3\text{·OH}} \text{[}-\text{O·CH}_2\text{·CO}_2\text{·CH}_3
$$

$$
\xrightarrow{\text{NH}_2\text{·NH}_2} \text{[}-\text{O·CH}_2\text{·CO·NH·NH}_2
$$

$$
\xrightarrow{\text{HNO}_2} \text{[}-\text{O·CH}_2\text{·CO·N}_3 \xrightarrow{\text{E·NH}_2} \text{[}-\text{O·CH}_2\text{·CO·}\overset{\text{H}}{\text{N}}\text{·E}
$$

$$(7.3)$$

Considerable activity was retained by several proteolytic enzymes after attachment and their autolytic stability was much improved. Consequently the Curtius reaction has been used to activate poly-acrylamide gels, eq. 7.4, and the cationic exchange resin carboxy-methyl cellulose (eq. 7.3).

$$\boxed{\quad}\!\!-CO \cdot NH_2 \longrightarrow \boxed{\quad}\!\!-CO \cdot NH \cdot NH_2$$

$$\longrightarrow \boxed{\quad}\!\!-CO \cdot N_3 \xrightarrow{\text{E} \cdot NH_2} \boxed{\quad}\!\!-CO \cdot \overset{H}{N} \cdot E$$

$$(7.4)$$

Nucleophilic attack by enzymic amino groups on the carboxyl carbon of the polymeric acylating agent is a feature of this type of activation reaction. Another much investigated support is the synthetic macromolecular acylating reagent, ethylene-maleic anhydride co-polymer, which reacts as in eq. 7.5.

$$-CH_2 \cdot CH_2 \cdot CH \cdot CH \cdot CH_2 \cdot CH_2-$$

$$O = C \quad C = O$$
$$\underset{O}{\diagdown\diagup}$$

$$\xrightarrow{\text{E} \cdot NH_2} \quad -CH_2 \cdot CH_2 \cdot CH \cdot CH \cdot CH_2 \cdot CH_2-$$
$$\underset{HO_2C}{\qquad} \quad \underset{CO \cdot \overset{H}{N} \cdot E}{\qquad}$$

$$(7.5)$$

Other common polymeric materials are those containing amino groups, for instance the anionic exchange resin, aminoethyl cellu-lose. These are convertible into enzyme retention supports by the action of phosgene or thiophosgene. The respective isocyanates or isothiocyanates formed couple to the enzyme with good yields and with few side products. One interesting example of their use is the activation of porous glass. The silica is pre-activated by treatment with γ-aminopropyltriethoxysilane to substitute the free silica hydroxides with alkyl amines, these are then converted to the iso-cyanate or isothiocyanate derivatives, eq. 7.6.

$$-O-\overset{\displaystyle \overset{|}{O}}{\underset{\displaystyle \underset{|}{O}}{Si}}-OH + H_2N(CH_2)_3 \cdot Si \cdot (OCH_2CH_3)_3$$

$$(7.6)$$

$$\longrightarrow -O-\overset{\displaystyle \overset{|}{O}}{\underset{\displaystyle \underset{|}{O}}{Si}}-O-\overset{\displaystyle \overset{|}{O}}{\underset{\displaystyle \underset{|}{O}}{Si}}-(CH_2)_3-NH_2$$

$$\xrightarrow[\substack{X = O \text{ or } S}]{Cl_2C:X} -O-\overset{\displaystyle \overset{|}{O}}{\underset{\displaystyle \underset{|}{O}}{Si}}-O-\overset{\displaystyle \overset{|}{O}}{\underset{\displaystyle \underset{|}{O}}{Si}}-(CH_2)_3-NCX \xrightarrow{E \cdot NH_2}$$

$$-O-\overset{\displaystyle \overset{|}{O}}{\underset{\displaystyle \underset{|}{O}}{Si}}-O-\overset{\displaystyle \overset{|}{O}}{\underset{\displaystyle \underset{|}{O}}{Si}}-(CH_2)_3-NH-CX-NH-E$$

Treatment of the amino groups with dialdehyde reagents, of which the most common is glutaraldehyde, also provides materials suitable for enzyme attachment. β-Galactosidase, amongst many enzymes, has been coupled with a 25 percent yield of activity to aminoethyl cellulose preactivated by this pentanedial, eq. 7.7.

$$\boxed{}-NH_2 + OHC \cdot (CH_2)_3 \cdot CHO \longrightarrow \boxed{}-N=C \cdot (CH_2)_3 \cdot CHO$$

$$(7.7)$$

$$\xrightarrow{E \cdot NH_2} \boxed{}-N=C \cdot (CH_2)_3 \cdot C=N \cdot E$$

The disadvantage of this reagent however is its extensive self-polymerisation and protein crosslinking. Thus the support needs to be well washed prior to enzyme addition.

By the use of these various reagents enzymes can be attached to most support materials, and used for a variety of purposes; for example, lactate dehydrogenase[4] has been attached to filter paper and to DEAE cellulose first modified with cyanuric chloride (trichloro-S-triazine), eq. 7.8, and then incorporated into a two enzyme reactor with pyruvate kinase, similarly insolubilised.

$$(7.8)$$

7.2.1.2 Diazonium Coupling

Diazonium salts react rapidly and almost quantitatively with phenols and imidazole, and also to some extent with the amino and guanidine groups present in protein molecules. The main reaction however is with the enzyme tyrosine groups, eq. 7.9.

$$(7.9)$$

This type of reaction enables the interposition of a spacer arm between the two polymers to reduce steric interactions and yield higher enzyme : carrier conjugation ratios than is often possible using the other coupling procedures. For example trypsin could be directly coupled to dialdehyde starch but greater yields were obtained by first reacting the oxidised polysaccharide with bis-amino-benzidine followed by diazotization of the free aromatic amino group.

7.2.1.3 Amide Formation

Carbodiimides are the reagents of choice for amide formation between an organic acid and an amine. The reactions are easily performed at room temperature and water soluble derivatives are available which facilitate the removal of the elements of water between the two reactants at moderate pH, eq. 7.10.

$$\boxed{-}\text{CO}_2\text{H} + \text{R}_1-\text{N}=\text{C}=\text{N}-\text{R}_2 \longrightarrow \boxed{-}\text{CO}\cdot\text{O}\cdot\text{C} \begin{array}{l} \text{NH}\cdot\text{R}_1 \\ \\ \text{NH}-\text{R}_2 \\ + \end{array}$$

$$\xrightarrow{\text{E}\cdot\text{NH}_2} \boxed{-}\text{CO}\cdot\text{NH}\cdot\text{E} + \text{R}_1\cdot\text{NH}\cdot\text{CO}\cdot\text{NH}\cdot\text{R}_2$$

$$\text{R}_1 = \text{ethyl-} \qquad \text{R}_2 = \text{3-dimethyl aminopropyl-} \qquad (7.10)$$

This method can therefore be used for coupling enzymes to macromolecules containing either carboxylic acids or amines. Unlike the foregoing conjugation reactions the polymer is not pre-activated but all three reagents are simultaneously present.

Similar remarks apply to the other commonly used amide forming reagent, Woodward's reagent K, N-thyl-5-phenylisooxazolium-3'-sulphonate, eq. 7.11.

$$\boxed{-}\text{COOH} + \text{R}\overset{O}{\underset{\text{N}-\text{R}'}{\diamond}} \rightarrow \boxed{-}\text{COO-}\overset{O}{\underset{\text{R'NHCO·CH}}{\text{C-R}}} \xrightarrow{\text{E·NH}_2} \boxed{-}\text{CO·NH·E} + \text{R·}\overset{O^-}{\text{C}}\text{=CH·CONH·R'} \qquad (7.11)$$

$$\text{R} = \langle O \rangle\text{-SO}_3^- \ ; \ \text{R}' = -\text{C}_2\text{H}_5$$

7.2.1.4 Sulphydryl Exchange

Enzymes which possess a cysteine non-essential to activity and in an accessible region on the protein, can be immobilised at acid pH by oxidation in the presence of a mercapto-polymer. A crosslinked polyacrylic copolymer containing free sulphydryl groups which reacts in this way has been prepared by co-polymerisation of acrylamide and N-acryloyl-cysteine. This reacts with sulphydryl containing enzymes according to eq. 7.12.

$$\boxed{-}\text{CH}\begin{array}{l}\text{CO}_2\text{H}\\ \\ \text{CH}_2-\text{SH}\end{array} + \text{E·SH} \longrightarrow \boxed{-}\text{CH}\begin{array}{l}\text{CO}_2\text{H}\\ \\ \text{CH}_2-\text{S}-\text{S}-\text{E}\end{array}$$

$$(7.12)$$

This reaction is reversed by reducing or sulphydryl reagents. Inactivated or expired enzyme can then be displaced and the polymer recharged with active enzyme.

7.2.1.5 Metal-linked Enzymes

In distinction to some heavy metals, which inactivate enzymes, some authors[4] have found that transition metal salts can be used to transform many materials such as cellulose, nylon, borosilicate glass, soda glass, filter paper and yeast cells, into enzyme supports. The solid is steeped in a solution of the metal salt, filtered, washed and the enzyme added. Transition metal salts tried have been $TiCl_4$, $TiCl_3$, $SnCl_4$, $ZnCl_4$, VCl_3, $FeCl_2$ and $FeCl_3$; the exact reaction mechanism awaits elucidation but can be represented simply as:

$$\text{[}\text{-} OH + MCl_2 \longrightarrow \text{[}\text{-} O-M-Cl$$

$$\xrightarrow{\text{E·NH}_2} \text{[}\text{-} O-M-NH·E$$

$$(7.13)$$

7.2.2 Crosslinking[7]

Immobilisation and insolubilisation of enzymes is also induced by their subjection to polyfunctional crosslinking reagents (Table 7.5). Both intramolecular and intermolecular modification will result but the relative proportions of these can be controlled by careful adjustment of protein concentrations. At low concentrations intramolecular crosslinking predominates in which case the enzyme will most probably remain in solution, whereas at higher concentrations, polymerisation and possible insolubilisation will result. Most popular has been the dialdehyde, gluteraldehyde. This also self polymerises in aqueous solution to form an extensive crosslinked network in which the enzyme molecules are embedded as well as covalently modified, but the precise nature of the resultant matrix is far from clear at present. Gluteraldehyde is entitled a homofunctional crosslinking reagent because it contains identical reactive groupings, whereas those possessing groups of differing reactivities are classed as heterofunctional reagents (Table 7.5).

Table 7.5: Polyfunctional Crosslinking Reagents

Reagent	Structure
Gluteraldehyde	$CHO(CH_2)_3CHO$
1,5-Difluoro-2,4-dinitro-benzene	
Bis-diazobenzidine-2,2'-disulphonic acid	
Phenol-2,4-disulphonyl chloride	
Toluene-2-isocyanate-4-isothiocyanate	
Trichloro-S-triazine	
Dimethyl adipimidate	

7.2.3 Adsorption

Immobilisation of an enzyme by adsorption onto a matrix is the simplest method to carry out, requiring little prior activation of the matrix and no third crosslinking component. Adsorption is described as the adhesion or condensation of the protein to the surface of a carrier which has not been specifically functionalised for covalent attachment. It is thus a non-specific phenomenon and always accompanies the other techniques of immobilisation on solid surfaces, and for this reason covalently bound enzyme-polymer conjugates are exhaustively washed prior to use. Desorption of the enzyme especially in high salt or high substrate containing media is a greater inconvenience during usage and thus limits the applications of enzymes immobilised by this method. For

Table 7.6: Enzymes Immobilised by Adsorption

Support material	Enzyme*
Activated clay	Amylase (a)
Activated charcoal	Hexokinase (a), phosphatases (80)
Alumina	Fructofuranosidase (a)
Amberlite	Lipase (a)
Barium stearate	Urease (5)
Cellulose	Deoxyribonuclease
Cellulose, carboxymethyl-	Asparaginase (a), trypsin (a)
Cellulose, citrate-	Chymotrypsin (a)
Cellulose, DEAE-	β-Galactosidase (a), acylase (a)
Cellulose nitrate (membrane)	Lactate dehydrogenase (0)
Cellulose phosphate	Chymotrypsin (a)
Cellophane sheets (+ gluteraldehyde)	Glucose oxidase (10)
Cephalin coated carbon	Phosphoglucomutase (50)
Ceramics, titania	Glucose oxidase (1)
Collagen (+ gluteraldehyde)	β-Galactosidase (70)
	glucose oxidase (30)
Collodion membrane (+ gluteraldehyde)	Alkaline phosphatase (a)
Dowex-50	Ribonuclease (a)
Glass	Ribonuclease (a), trypsin (a)
Hydroxy-apatite	NAD pyrophosphorylase (a)
Kaolinite	Chymotrypsin (a)
Polyaminostyrene	Catalase (a)
Sephadex, DEAE-	Acylase (a)
Silica gel	Catalase (300),
	acid phosphatase (90)

* Figures in brackets are percentage retention of activity; a = active.

use in such media the enzyme is usually further fixed after adsorption by covalent crosslinking. Table 7.6 gives a short list of the support matrices used and examples of enzymes immobilised by adsorption.

Both charged and neutral materials, including controlled pore glass, collagen membranes, titania, charcoal, clay, dialysis tubing, Millipore filters and silica gels are able to adsorb a wide variety of proteins. The enzymes are held predominantly by electrostatic and hydrogen bonds and thus, because of the lack of directional selectivity a statistical proportion of the attached molecules will possess altered activities as a result of interference with the active site. Exceptions at both activity extremes have been observed however. Catalase adsorbed on silica gel is considerably activated by the surface, whereas lactate dehydrogenase is totally inactive on cellulose nitrate membranes (Table 7.6). Also consequent on the nature and types of these bonds are the activities of holoenzymes. If the prosthetic groups of such enzymes bind more strongly to the matrix than to the apoenzyme, dissociation and hence inactivation will result. For example, the haem group rather than the globular protein of haemoglobin is preferentially adsorbed by silica glass.

Table 7.7: Some Enzymes Immobilised by Occlusion

Support material	Enzyme*
Microcapsules:	
Collodion	Catalase (25), β-fructofuranosidase (a)
Nylon	Urease (a), trypsin (a), carbonic anhydrase (a)
Polyacrylamide gel	Alcohol dehydrogenase (5), amylase (2), ATPase (45), enolase (90), phosphoglycerate mutase (60) *Arthrobacter simplex, Curvularia lunata*
Silastic resin (silicon polymer)	Cholinesterase (40), glucose oxidase (50), trypsin (60)
Starch gel	Acetylcholinesterase (a)

* Figures in brackets are percentage retention of activity, a = active.

7.2.4   Occlusion and Entrapment

Polymerisation of unsaturated monomers in the presence of an enzyme often results in its occlusion within the interstitial spaces of the gel (Table 7.7). A popular gel is that produced from acrylamide plus *N,N*-methylene-bisacrylamide (I). By varying their relative

$$
\begin{array}{c}
\overset{\displaystyle |}{\text{CH}_2} \\
\overset{\displaystyle |}{\underset{}{}} \\
\text{NH}_2 \qquad \text{NH} \\
\overset{\displaystyle |}{} \qquad \overset{\displaystyle |}{} \\
\text{CO} \qquad \text{CO} \\
\overset{\displaystyle |}{} \qquad \overset{\displaystyle |}{} \\
-\text{CH}_2-\text{CH}-\text{CH}_2-\text{CH}-\text{CH}_2-\text{CH}-\text{CH}_2- \\
\overset{\displaystyle |}{\text{CO}} \\
\overset{\displaystyle |}{\text{NH}} \\
\overset{\displaystyle |}{\text{CH}_2} \\
\overset{\displaystyle |}{\text{NH}} \\
\overset{\displaystyle |}{\text{CO}} \\
-\text{CH}_2-\text{CH}-\text{CH}_2-\text{CH}-\text{CH}_2-\text{CH}-\text{CH}_2- \\
\overset{\displaystyle |}{\text{CO}} \\
\overset{\displaystyle |}{\text{NH}} \\
\overset{\displaystyle |}{\text{CH}_2}
\end{array}
\qquad \text{I}
$$

concentrations, these can be co-polymerised to yield polyacrylamide gels of controlled pore size. Moreover the gels can be dispersed mechanically to give particles of defined size ranges.

An alternative method entraps the enzyme within a three-dimensional network based on a preformed polymer. Cholinesterase for example can be enclosed by dissolution in a warm, soluble starch solution, to which gluteraldehyde is then added. The dialdehyde immediately both self polymerises and crosslinks the polymer mixture to form an insoluble network which occludes the enzyme.

No covalent bonds are formed between the occluded enzyme and its support. Thus no constraints are placed on the enzyme molecule which could affect its catalytic behaviour. Free radicals produced during the polymerisation process and the exothermic formation of polyacrylamide which can cause localised heating effects, are however potential sources of enzyme inactivation. In addition many polymerases have been rendered latent by occlusion, the diffusion of high molecular weight substrates to the enzyme being restricted by the crosslinked network.

In distinction to the enzymes immobilised by the other methods, several advantages result from occlusion. The enzyme is more resistant to bacterial attack and the mechanical strength of the polymeric matrix can be controlled. In addition, the possibility exists of co-immobilising several enzymes within the one framework. A model for mitochondrial oxaloacetate metabolism was synthesised in this way by co-entrapping malate dehydrogenase, citrate synthase and lactate dehydrogenase within a single polyacrylamide matrix. In the gel the following reactions take place:

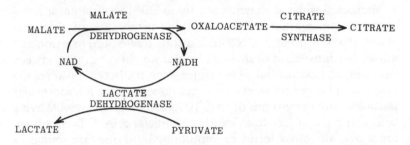

The rate of citrate production in this three-enzyme system was found to be four fold faster than that of the corresponding reaction in solution. This was attributed by the authors to the production of higher local concentrations of the intermediates.[8]

The gel entrapment technique has also been applied by Mosbach and his colleagues[4] to the immobilisation of microbial cells. The microorganisms, *Curvularia lunata* and *Arthrobacter simplex*, which contain an 11-β-hydroxylase and a $\Delta^{1,2}$-dehydrogenase respectively were treated as enzyme reservoirs by encasement within the polyacrylamide gel and used to transform the steroid, Reichstein S into Prednisolone via the intermediate formation of cortisol. Immobilisation facilitated separation of the catalysts and retrieval of the steroid products.

Since occluded enzymes are not fixed to the matrices and because of the wide distribution of pore sizes in the gels, leaching of the enzymes can occur, but this is generally less a problem than with adsorbed enzymes. It has been possible to reduce this inconvenience in some instances by co-polymerising a polymerisable derivative of the enzyme; e.g. glucose oxidase has been vinylated with a heterobifunctional electrophilic acylating agent containing a double bond and then co-polymerised with the unsaturated monomers,[4] eq. 7.14.

$$E + nCH_2\!:\!CH\cdot CO\cdot Cl$$

$$\longrightarrow E(-CO\cdot CH\!=\!CH_2)_n \tag{7.14a}$$

$$E(-CO\cdot CH\!=\!CH_2)_n +$$

$$Acrylamide + Bisacrylamide \longrightarrow \tag{7.14b}$$

Microencapsulated enzymes are those enveloped within a semipermeable membrane.[9] The ultrathin membrane, of thickness about 250 Å, is usually of cellulose nitrate (collodion) or nylon and allows free movement of small molecular weight substrates between the intra- and extrastitial compartments but restricts diffusion of the enzyme and other polymers. In some respects these microcapsular particles, with dimensions of 10 to 100 μm diameter, resemble living cells and have in fact been entitled 'artificial cells'.[9] Their advantages over the other forms of immobilised enzymes are similar to those afforded by the other occlusion methods but in addition loss of enzyme through leakage is reduced by the semipermeable membrane. The large surface to volume ratio allows rapid substrate and product exchange. Therefore the enzymes exhibit unaltered catalytic properties and *in vivo* the inert membrane elicits no immunological reaction (Chapter 6).

Enzymes have also been encapsulated within liposomes. These are mechanically stable macrovesicular assemblages formed when the lipidaceous mixtures phosphatidyl-ethanolamine, choline-serine and lecithin-cholesterol are suspended in water. The technique is to first spread these substances as a film inside a rotating flask and then add an aqueous solution of the enzyme. When the two phases are rapidly mixed, liposomes form spontaneously, dispersing the enzyme molecules throughout the ordered structures. Some are occluded within the liposome, others occupy the lipidaceous bilayers while others are adsorbed on the liposomal surface.

7.3 IMMOBILISED ENZYME ACTIVITY

The particulate nature of insoluble enzymes makes their handling easier than that of soluble derivatives, facilitating their assay by the discontinuous methods outlined in Chapter 2. Aliquots can be periodically removed from bulk solution, filtered to remove enzyme activity and analysed. Continuous assays based on spectrophotometric measurements are of limited use since they suffer particularly from light scattering effects which limit their accuracy, but those based on electrode methods are not affected by this disadvantage and are therefore preferred.

One main factor governing the reproduceability of the assays is the adequacy of stirring. This must be at a rate sufficient to prevent restricted substrate and product diffusion around the enzyme and to reduce clumping of the insoluble material. The thickness of the diffusional layer (Nernst layer) around the particle governs the movement and thus the mass transfer of substrates and products from the external bulk solution into the matrix. The numerical values of the measured catalytic parameters depend on the availability of substrate at the enzyme, and therefore its Michaelis parameters will be a function of the stirring or flow rate. Diffusion restrictions within the particle will also be present when very high catalytic activities per volume are attached. Thus for kinetic investigations, preparations of low specific activity have been suggested.

7.4 PROPERTIES OF IMMOBILISED ENZYMES

The conformational, kinetic and stability properties of an immobilised enzyme are consequent upon its transfer from homogeneous solution to the heterogeneous biphasic environment. In this alien microenvironment the chemical characteristics of the support matrix are superimposed upon the intrinsic properties of the enzyme. An idea of the effect of insolubilisation on enzymic activity can be gauged from a comparison of the catalytic activity of soluble poly(L-tyrosyl) trypsin, which exhibits the activity of the native protease, and the insoluble derivatives formed by block copolymerisation of this soluble form with poly(*p*-diazo-D,L-phenylalanine-L-leucine) and diazobenzyl-cellulose. Only 30 percent and 50 percent activities respectively were retained against low molecular weight substrates.[10]

The influence of the carrier on the catalytic behaviour of the active immobilised enzyme can be considered under four broad headings: (1) type of immobilisation; (2) steric effects; (3) microenvironmental effects and (4) diffusion restrictions.

7.4.1 Type of Immobilisation

The kinetic properties exhibited by the insolubilised enzyme will depend on the type and number of constraining forces to the carrier. Entrapped in microcapsules where no additional bonds are formed and where the enzyme resides in an inert proteinaceous matrix, the matrix would be expected to have little effect on activity, provided diffusion limitations are absent, whereas covalent modification or adsorption could impose more significant influences.

Coupling of the enzyme to matrices is achieved through the more reactive groups on the enzyme, namely amino, carboxyl, tyrosyl and sulphydryl residues, all of which are charged or capable of hydrogen bonding. Covalent modification and adsorption will cause electronic rearrangements in these groups, which, if they are transmitted to active or allosteric sites, will affect the activity.

Some workers however report only apparent modification of the enzyme activity by the carrier. Effects have on occasion been attributed directly without the appropriate control experiments having been performed. For example, attachment of trypsin via its ε-amino groups to polymeric carriers has been reported to induce an alkaline shift in the pH maximum of activity, but in fact similar shifts have been observed after simple *N*-acetylation of the enzyme.[11]

7.4.2 Steric Effects

The approach of high molecular weight substrates to immobilised enzymes is generally found to be sterically hindered by the support. This behaviour is particularly common with immobilised proteases, nucleases and glycosidases all of which act on polymers, respectively polypeptides, nucleic acids and oligosaccharides. For this reason the attached enzymes are usually assayed with low molecular weight substrates in order to ascertain the retention of activity, e.g. papain attached to collagen is 50 percent less reactive towards the protein casein than the amino acid derivative *N*-benzoyl-L-arginine ethyl ester, and trypsin attached to soluble dextran has its activity towards haemolobin decreased twenty-five fold relative to *N*-benzoyl-D,L-arginine, *p*-nitranilide.[11]

7.4.3 Matrix Microenvironment

Charged groups present on the matrix and on adjacent protein molecules, changes in ion concentrations caused by ionised macromolecules and alterations in the water structure around the conjugate caused by the hydrophobic groups present, will all affect the revealed enzyme properties. Of these effects, the electrostatic influence of charged carriers has been studied in most detail.[12]

Polyanionic carriers, ethylene maleic anhydride, polyglutamic acid, etc., have been observed to elicit a shift in the pH dependent maximum activity of the attached enzymes towards more alkaline values compared to the native enzymes. Conversely, enzymes fixed to polycationic supports such as polyornithine exhibit a displacement to more acidic values (Figure 7.3).[13] However this behaviour was found to predominate only at low ionic strengths, and as the salt concentration of the buffers was progressively increased, the displacement was incrementally reduced towards native pH values.

Also found to be dependent on ionic strength is the Michaelis constant of some enzyme conjugates. For example in dilute buffers, the Michaelis parameter for a trypsin/ethylene maleic anhydride (a polyanionic matrix) conjugate is thirty fold lower than that of the free enzyme for the positively charged substrate *N*-benzoyl-D,L-arginine amide. This probably reflects the electrostatic attraction between the two, favouring an increased local concentration of substrate in the enzyme vicinity. In more concentrated buffers the measured value returns to that typical of the unattached enzyme.

Such effects have been attributed by Goldstein *et al.*[10] to an unequal distribution of charged molecules and ions between the immediate enzyme neighbourhood and bulk solution, Figure 7.4. A partitioning which they proposed would be reflected in the concentration dependent properties of the enzyme.

If a_H^i and a_H^0 are respectively the activities of hydrogen ions in the immobilised phase and outer solution, their relative concentrations in the two phases will be given by the Maxwell-Boltzmann distribution:

$$a_H^i / a_H^0 = \exp(-eU/kT) \qquad (7.15)$$

or
$$pH_i - pH_0 = 0.43 \cdot eU/kT \qquad (7.16)$$

Figure 7.3: pH-*Activity Curves at Low Ionic Strength for Chymotrypsin* (○), *a Polyanionic Derivature, Ethylene-maleic Anhydride-chymotrypsin* (●) *and a Polycationic Derivative* (△), *using Acetyl-L-tyrosine Ethyl Ester as Substrate*

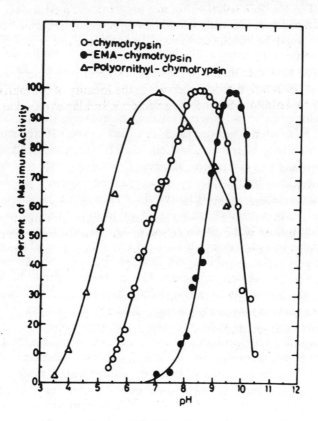

Figure 7.4: *Immobilised Enzyme*

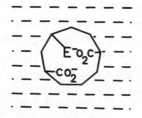

where e is the electronic charge, U the electrostatic potential, k the Boltzmann constant and T the absolute temperature.

The hydrogen ion concentration experienced by the enzyme in the locality of a polyanionic matrix ($U < 0$) is predicted to be higher than that of the bulk solution but in a polycationic medium ($U > 0$) it would be lower. On this basis the pH maximum exhibited by an enzyme would be displaced towards higher and lower pH values respectively.

Charged substrate molecules will also experience the partitioning effect and the relative concentrations in the locality of the matrix, S^i and in bulk solution, S^0, will be given by a similar expression

$$S^i/S^0 = \exp(-ZeU/kT) \tag{7.17}$$

where Z is the charge on the substrate molecule.

If the matrix and substrate are oppositely charged, Ze and U will have opposite signs, giving $S^i > S^0$; if both have the same sign the equation predicts $S^0 > S^i$. When the first situation prevails the immobilised enzyme will experience a saturating substrate concentration and hence reach the maximum limiting rate at a lower bulk concentration of substrate than the free enzyme. The opposite will apply in the second case, when $S^0 > S^i$.

In the absence of diffusion limitations and assuming the enzyme obeys Michaelis-Menten kinetics, an equation for the influence of the substrate concentration on rate can be obtained by combining the Michaelis equation and eq. 7.17 to give:

$$v = \frac{V \cdot S^i}{K'_m + S^i} = \frac{V \cdot S^0 \cdot \exp(-ZeU/kT)}{K_m + S^0 \cdot \exp(-ZeU/kT)} \tag{7.18}$$

$$\therefore \ K'_m = K_m \cdot \exp(ZeU/kT) \tag{7.19}$$

Therefore, when Ze and U have the same sign, $K'_m > K_m$, but for oppositely charged species, $K'_m < K_m$.

However, these equations apply only to the kinetic behaviour of immobilised enzymes in media of low ionic strength where $\mu \ll 0.1$. To explore the dependencies on ionic strength more fully, Wharton et al.[14] considered further the unequal distribution of charged particles between the bulk solution and matrix, i.e. the Donnan effect. Because of the charge on an ionised macromolecule, the concentration of small ions of like charge in its environment is decreased but compensation takes place with an increase in the concentration of oppositely charged particles. This Donnan effect can be suppressed by high concentrations of neutral salts or by adjusting the pH to the isoelectric point of the polyelectrolyte.

The influence of ionic strength on the apparent Michaelis constant, K'_m, was derived to be,

$$(K'_m)^2 = \gamma^2 \cdot K_m(K_m - K'_m \cdot Zm_c/\mu) \tag{7.20}$$

where Zm_c is the effective concentration of charged groups in the immobilised phase, γ the activity coefficient equals $\gamma^i_\pm / \gamma^0_\pm$ where $\gamma^i_\pm$ and $\gamma^0_\pm$ are the mean ion activity coefficients of matrix and bulk solution respectively. From eq. 7.20,

$$K'_m = \gamma \cdot K_m(1 - K'_m Zm_c/\mu K_m)^{1/2} \tag{7.21}$$

thus $$K'_m = \gamma \cdot K_m(1 - K'_m Zm_c/2K_m \mu) \tag{7.22}$$

which is obtained from eq. 7.21 by binomial expansion and elimination of terms after the first, assuming $0 < (K'_m Zm_c/2K_m \mu) < 1$. Rearrangement yields,

$$(K'_m)^{-1} = (\gamma K_m)^{-1} + Zm_c/2K_m \cdot (\mu)^{-1} \tag{7.23}$$

which predicts a linear relationship between the reciprocals of the observed Michaelis constant and the ionic strength of the medium, of slope $Zm_c/2K_m$ and intercept $(\gamma K_m)^{-1}$. Wharton et al.[14] showed that this equation was applicable to the observed kinetics of carboxymethylcellulosebromelain catalysed hydrolysis of benzoyl-L-arginine ethyl ester.

7.4.4   Diffusion Restrictions

In order for the substrate to be converted it must diffuse from the outer bulk solution to the enzyme within the support; if the rate of the substrate diffusion is slower than its enzymic conversion the observed catalytic rate will be lower than that obtained for the free enzyme. The diffusion controlled process consists of two events— transfer from bulk solution to the conjugate across a boundary layer, and an internal transport within the matrix. In the first, increasing the diffusion rate by more rapid stirring or faster flow down a column will increase the observed enzymic rate, since reaction occurs after the substrate has reached the catalyst. In the second, reaction and diffusion occur together, no equilibrium is then possible between internal and bulk solution and thus local gradients in the concentrations of substrates and products will arise. The same phenomena will also be present *in vivo* in the membrane linked enzymic reactions of the cell.

Synthetic enzyme-containing membranes are prepared by adsorbing the enzyme molecules onto a preformed membrane of collodion, nylon, etc., followed by fixation with gluteraldehyde or bisdiazobenzidine-2,2'-disulphonic acid. Alternatively the enzyme may be occluded within an inert polymeric film formed on a flat surface.

In order to analyse the kinetic behaviour of a membrane, it is subjected to differing boundary conditions, that is, to varying substrate and product concentrations on its two interfaces. These concentrations determine the velocity with which the stationary state is attained, i.e. the state in which the local concentrations of substrate and product are time invariant.

For a membrane, of width x, Fick's Law of diffusion is applicable, and if a boundary condition of zero product concentration ($P = 0$) is set, then

$$D'_s \cdot \frac{d^2 S}{dx^2} = f(S) \qquad (7.24)$$

where $f(S)$ is the local reaction rate and D'_s is an apparent diffusion coefficient independent of substrate or product concentration. In place of $f(S)$ can be substituted the enzymic rate, given by the Michaelis-Menten relationship,

$$D'_s \cdot \frac{d^2 S}{dx^2} = \frac{V_{max} \cdot S}{S + K_m} \qquad (7.25)$$

Integration gives the concentration gradient between any point, x and the midpoint ($L/2$) of the membrane,

$$\frac{dS}{dx} = \left| \frac{2V_{max}}{D'_s} \left| S_0 - S + K_m \cdot \ln\left(\frac{K_m + S}{K_m - S}\right) \right| \right|^{1/2} \qquad (7.26)$$

where S is the substrate concentration at $L/2$. A second integration then enables the positional substrate concentration at any value of

S_0 to be obtained, and the rate at each point can then be derived by insertion into the equation:

$$\text{Rate} = -2D'_s \frac{dS}{dx} \qquad (7.27)$$

These two equations predict that for a bound enzyme the substrate concentration necessary to give half maximum rate is raised. In other words, the apparent Michaelis constant of a membrane attached enzyme is increased relative to that of the native enzyme, its numerical value depending on the membrane thickness and on the rate of diffusion. At very high substrate concentrations however the effects of internal diffusion will be overcome and an unaltered maximum rate will be approached.

Around the membraneous or particulate enzyme unstirred layers of solution are present, the 'Nernst diffusion layers', of thickness 10 to 100 μm, depending on the rate at which the bulk solution is stirred. Substrate and product concentrations will be different on either side of these layers and the sizes of their gradients will determine the rates of metabolite flow back and forth from the solution to the conjugate. If d is the thickness of the unstirred layer and J_S is the rate of substrate flow across the gradient, then,

$$J_S = \frac{D_s}{d}(S_0 - S) \qquad (7.28)$$

The rate of flow is given by the Michaelis equation, thus:

$$\frac{V_{max} \cdot S}{K_m + S} = \frac{D_s}{d}(S_0 - S) \qquad (7.29)$$

and on expansion,

$$S^2 + S\left(K_m - S_0 + V_{max}\frac{d}{D_s}\right) - K_m S_0 = 0 \qquad (7.30)$$

Binomial expansion of the quadratic root of eq. 7.29 and truncation of terms yields the effective substrate concentration (S) at the conjugate,

$$S = \frac{K_m \cdot S_0}{K_m - S_0 + \dfrac{V_{max} \cdot d}{D_s}} \qquad (7.31)$$

and the rate

$$v = \frac{S_0 V_{max}}{K_m + \dfrac{d \cdot V_{max}}{D_s}} \qquad (7.32)$$

At low substrate concentrations therefore the apparent Michaelis constant, $K_{m(app)}$ equals $(K_m + V_{max} d/D_s)$. Thus when $D_s > V_{max}$ diffusion is not rate limiting and the chemical reaction is kinetically controlled, and when $V_{max} > D_s$, $K_{m(app)} > K_m$ and the reaction becomes dependent on mass transport.

These equations have been verified by Goldstein *et al.*,[10] who found that the apparent Michaelis constants for three alkaline phosphatase-collodion membranes of thickness 1.6, 2.6 and 8.8 × 10^{-4} cm were 0.85, 2.9 and 12 mM, respectively 25, 100 and 350 times larger than those of unattached alkaline phosphatase. These increments they attributed mainly to the existence of unstirred layers around the particle although partly to additional concentration gradients within the membrane itself.

7.5 APPLICATIONS OF IMMOBILISED ENZYMES

Immobilised enzymes are useable in particulate form, in suspension, as column packings, embedded in membranes, on the inside of tubes and on the outside of electrodes. The applications of this variety of forms are just as varied, and Table 7.8 gives a short representative list.

To all of these uses, two main properties of immobilised enzyme preparations apply. First in an insoluble form, enzyme activity is easily separated and retrieved from its reaction mixture. This lessens possible contamination of the product and allows reusage of the enzyme. The latter is an economic necessity since insoluble enzymes are more costly to prepare than the bulk enzymes described earlier. Secondly, enzyme recycling depends on the thermal and storage stability of the conjugates. Many of the enzymes listed are in use because of their stabilisation on immobilisation. In contrast to non-enzymic catalysts enzymes loose more activity by lying idle than by

Table 7.8: Some Applications of Insolubilised Enzymes

Insoluble enzyme	Use
Aminoacyl-tRNA synthetase	Affinity chromatography of tRNAs
Aminoacylase	D- and L-Methionine resolution
Amylase	Glucose production
Amyloglucosidase	Starch conversion
Asparaginase	D- and L-Asparagine resolution
Carbonic anhydrase	Membrane model for CO_2 transport
Carboxypeptidase	D- and L-Alanine resolution
Catalase	Enzyme replacement in acatalasaemic mice
Cholinesterase	Assay of organophosphates
Chymotrypsin	Clotting of milk
Glucose oxidase	Glucose assay
Invertase	Glucose production
Kallikrein	Prekallikrein activation
Papain	Antibiotic degradation
	Beer chill proofing
Penicillin amidase	Penicillin hydrolysis
Peroxidase	Hydrogen peroxide analysis
Rennin	Clotting of milk
	Angiotensinogen activation
Streptokinase	Plasminogen activation
Thrombin	Fibrin degradation
Trypsin	Affinity chromatography of trypsin inhibitors
	Prothrombin activation
Urease	Urea assay
	Urea removal in artificial kidney

constant contact with their substrates which generally tend to stabilise the protein molecule. Improved thermal stability has been reported for lactate dehydrogenase attached to DEAE-cellulose, glucose oxidase on cellophane and ficin immobilised on carboxymethyl cellulose, but the opposite has been found for papain on *p*-aminobenzyl cellulose and glucose-6-phosphate dehydrogenase on collodion.

The rationalisation proposed for the observed cases of thermo-stabilisation is that attachment of the enzyme at several points on the matrix will restrict those conformational movements in the enzyme which would lead to its inactivation. Similar reasoning however can be put forward for the observed destabilisation, attachments restricting conformational flexibility could result in decreased probabilities of recovery after thermal perturbation.

Long term storage stability is promoted for some enzymes, for example, glucose oxidase is stabilised by attachment to collodion, but reduced inactivation can be effectively guaranteed only for proteases. Attachment of either soluble or insoluble macromolecules to trypsin, chymotrypsin, etc. sterically hinder the mutual approach of the enzyme molecules and so reduce their autoproteolysis.[11]

7.5.1 Immobilised Enzyme Engineering[15]

Several of the immobilised enzymes listed in Table 7.8 are currently undergoing active investigation in the pharmaceutical (e.g. penicillin amidase), sugar (e.g. amyloglucosidase), brewing (e.g. papain, amylase), dairy (e.g. β-galactosidase, rennin) and fine chemical (e.g. aminoacylase) industries. Many of these are still at the preliminary investigative or pilot plant stage, but four have proved successful on a commercial scale.

Penicillin acylases (EC 3.5.1.11), present intracellularly or secreted by many bacteria (e.g. *E. coli* and *Bacillus megatorium*) and fungi (e.g. *Penicillium chrysagenum* and *Fusarium semilectum*) catalyse the reversible interconversion of penicillin and 6-aminopenicillanic acid (6APA) eq. 7.33.

$$\text{R·CONH} \overset{\text{S}}{\underset{\text{O}}{\bigg|}} \text{COOH} \rightleftharpoons \text{NH}_2 \overset{\text{S}}{\underset{\text{O}}{\bigg|}} \text{COOH} + \text{R·COOH} \qquad (7.33)$$

Penicillin   6APA

$R = C_6H_5 \cdot CH(NH_2)-$ Ampicillin

$R = C_6H_5 \cdot OCH_2 -$ Phenoxymethyl penicillin

$R = C_6H_5 \cdot CH_2 -$ Benzyl penicillin

6APA is an important precursor in the formation of 7-aminocephalosporanic acid, and the semi-synthetic penicillins.

The various enzymes possess fairly broad specificities, hydrolysing cephalosporins, acylamino acids and some esters. In acid media the reverse direction predominates but at pH values above neutrality amidolysis is the favoured direction. The enzyme has been immobilised on triazinyl modified cellulose, cyanogen bromide activated sepharose and sephadex G 200 with little loss in activity or change in its kinetic parameters. The sephadex immobilised form is used in a batch and continuous process for 6APA production from benzyl penicillin.[16]

Sucrose, common sugar, is becoming increasingly expensive on the world markets and high fructose syrups (HFS), which contain almost equal concentrations of glucose and fructose are increasing in importance. Fructose is approximately twice as sweet as sucrose, and thus in the present financial climate the enzyme catalysed interconversion of D-glucose (from starch products) into D-fructose becomes economically feasible. Industrial processes for the production of HFS employ glucose isomerase adsorbed onto DEAE-cellulose and porous glass although one company has immobilised whole cells of streptomyces with high glucose isomerase activity to avoid costly purification of the enzyme.

Many individuals in the world's population lack the enzyme β-galactosidase in their intestinal mucosa and so suffer from the clinical condition, lactose intolerance, symptoms of which are severe gastrointestinal disorders. Milk is considered a complete food in the developing areas of Asia, Africa, Central America and the Middle East, precisely those areas where this syndrome is prevalent, but unfortunately one of its major components is lactose. Efforts have therefore been directed to breaking down milk lactose through enzyme action. Galactosidases, which catalyse the hydrolysis of lactose to D-glucose and D-galactose, have been immobilised with high retentions of activity on a wide variety of supports; cellulose acetate fibres, titania, controlled pore glass and metal oxides have given conjugates suitable for the conversion of milk into a more digestible food. One result of this action is the production of a sweeter, more energy rich food (due to the formation of glucose) from the rather bland tasting disaccharide.

This reaction has also been used to augment the limited supplies of glucose by the conversion of whey. Whey, the main component of which is lactose, was originally a discarded by-product of cheese production, but now it is a cheap source of the expensive sweeteners glucose and fructose.

Naturally occurring amino acids and those needed for biochemical investigations are of the L-configuration, whereas those synthesised chemically are racemic mixtures containing both stereoisomers. Applying the stereospecificity characteristics of an aminoacylase from *Aspergillus niger*, adsorbed onto DEAE-cellulose, Tosa *et al.*[17] have been able to effect the resolution of the two forms (eq. 7.34). The acylated racemate is made to flow down a column containing the immobilised enzyme.

$$\underset{\underset{\text{D,L}}{}}{R'CO\cdot NH\cdot CH(R)COOH} \xrightarrow{\text{Aminoacylase}} \underset{\underset{\text{D}}{}}{R'CO\cdot NH\cdot CH(R)COOH} + \underset{\underset{\text{L}}{}}{NH_2CH(R)COOH} + R'COOH \qquad (7.34)$$

After enzymic action the eluted products are separated by fractional crystallisation and the unchanged D-isomer further racemised. By this procedure, L-alanine, L-methionine, L-phenylalanine, L-tryptophan and L-valine have been produced. The immobilised enzyme packed into columns is reported to possess enhanced stability with very little leaching and contamination of the product.

Enzyme mediated production of other amino acids as a more economic alternative to present and traditional methods are under study. For example, the essential amino acid aspartic acid, widely used in medicine and as a food additive is usually produced by fermentation but has been reported[4] synthesised from ammonium fumarate by L-aspartase occluded in polyacrylamide gels.

7.5.2 Enzyme Electrodes

At various stages in their processes the food, pharmaceutical, clinical, environmental and biochemical industries, all require analytical measurements to be made. Ideally, analytical procedures should be selective, capable of determining one component in a complex mixture, automated so that routine measurements can be made simply and rapidly, and preferably non-destructive, using as small an amount of material as possible in order to facilitate continuous quality control monitoring. Some of these criteria describe the attributes of enzymes; their characteristics would make them suitable as components of systems for analysis of natural products, but their environmental sensitivity and difficulty in handling limits their routine usage.

A major advance in analytical enzymology was made by Hicks and Updike[18] who trapped glucose oxidase within a polyacrylamide gel and layered the film over the membrane of an oxygen electrode. This 'enzyme electrode' then combined the selectivity of the enzyme for its substrate, D-glucose, and the sensitivity of electrode potentiometric measurements. The resultant could alternatively be entitled a 'glucose electrode'. For a description of the oxygen electrode see Chapter 2. The potential at the cathode is found to be an exponential function of the oxygen concentration as it diffuses across the membrane. In the presence of glucose oxidase the oxygen flux is reduced as β-glucose, which diffuses into the gel, is oxidised to gluconic acid and hydrogen peroxide. When molecular oxygen is in rate-limiting excess and the glucose concentration below the apparent Michaelis constant of the immobilised enzyme, the oxygen reduction at the electrode is found to be proportional to the glucose concentration.

Since this first exposition of the 'enzyme electrode' concept, many others have been devised, based on various types of electrode, such as those specific for hydrogen ions, carbon dioxide, and monovalent cations. Some of the combined electrodes are listed in Table 7.9. All function in modes similar to that described above. Their selectivities are functions of those of the biological reaction and the electrode, but restrictions of substrate size and immobilisation procedure are also imposed. The sensitivity depends on that of the least sensitive partner but is usually in the range $10^{-1} - 10^{-5}$M, and the lifespan of the electrode is essentially that of the enzyme component. The glucose electrode of Updike and Hicks has been reported viable for over a year, but lifetimes of up to one month are more common for the others.

Enzyme electrodes as analytical tools have much future potential. As Clark (in reference 5) has pointed out, considering only an electrode sensitive to hydrogen peroxide, there are at least thirty enzyme/substrate pairs that produce this.

7.5.3 Insolubilised Enzymes in Biochemistry

Of particular relevance to the study of membranes are insoluble particles containing two or more enzymes catalysing consecutive reactions. Such multienzyme systems, in which metabolic intermediates will accumulate through restricted diffusion, are probably closer approximations to natural membraneous systems than single enzyme preparations. Mathematical analysis of two-enzyme membranes indicates that diffusional control can in fact lead to rate accelerations.[19] In homogeneous solution there is a lag period, during which time the intermediates must build up to their steady-

Table 7.9: Enzyme Electrodes

Substance assayed	Enzyme immobilised	Electrode base
Adenosine mono-phosphate	AMP deaminase	NH_4^+
Alcohols	Alcohol oxidase	Pt/O_2
D-Amino acids	D-Amino acid oxidase	NH_4
L-Amino acids	L-Amino acid oxidase	NH_4
Amygdalin	β-Glucosidase	CN^-
L-Asparagine	L-Asparaginase	NH_4^+
D-Glucose	Glucose oxidase	Pt/quinone
	Glucose oxidase	Pt/O_2
	Glucose oxidase/ catalase	Pt/H_2O_2
Glucose-6-phosphate	Alkaline phosphatase/ glucose oxidase	Pt/O_2
L-Glutamine	Glutaminase	NH_4
Lactate	Lactate dehydrogenase	Pt/ferri-cyanide
Penicillin	Penicillinase	pH
Phenylalanine	L-Amino acid oxidase/ peroxidase	Pt/H_2O_2
Phosphate	Alkaline phosphatase/ glucose oxidase	Pt/O_2
Tyrosine	Tyrosine decarboxylase	CO_2
Urea	Urease	CO_2
	Urease	NH_4^+
	Urease	pH
Uric acid	Uricase	Pt/O_2

state concentrations. In the binary membrane this lag period will be reduced because of a slower diffusion rate of the intermediates away from the immediate vicinities of the catalysts. These theoretical proposals have been amply verified by several workers. Mosbach and Mattiasson[20] for example, using co-immobilised hexokinase and glucose-6-phosphate dehydrogenase, attributed the absence of a lag phase and the increased initial velocities, compared to the unattached enzymes in solution, to the adjacency of the different enzyme molecules and to restricted external diffusion caused by the surrounding unstirred layers.

One particular function of a cellular or intracellular membrane is the transport of molecules against their concentration gradients. Thomas *et al.*[21] have mimicked this behaviour by reticulating layers of hexokinase and phosphatase on albumin and enclosing them within a membrane permeable to glucose but not to glucose-6-phosphate. This artificial membraneous system actively transported glucose against a glucose gradient, provided an energy supply was also coupled to the system, eq. 7.35.

$$\text{Glucose} \quad \begin{array}{|c|} \hline \text{Layer 1} \\ \text{Hexokinase} \\ \hline \xrightarrow{} \\ + \text{ATP} \\ \hline \end{array} \quad \text{Glucose-6-phosphate} \quad \begin{array}{|c|} \hline \text{Layer 2} \\ \text{Phosphatase} \\ \hline \xrightarrow{} \\ \hline \end{array}$$

$$\text{Glucose} + \text{P}_i \quad (7.35)$$

Glucose was taken up by the first layer and phosphorylated by ATP plus hexokinase to glucose-6-phosphate. This then diffused into the phosphatase layer where glucose was regenerated.

Insolubilised enzymes have been of particular value in the investigation of protein structure and activity; two examples illustrate their potential. One uncertainty encountered in the analysis of multi-subunit enzymes is the catalytic behaviour of the individual protomers and the extent to which their catalytic and protein properties are modified by agglomeration. Thermodynamically, dissociation is favoured by dilution but for many polymeric enzymes the extent of dilution necessary is often too large to allow meaningful extrapolation to higher concentration. One procedure whereby this has been overcome has been through coupling the enzyme to a carrier by only one of its subunits, removing the others by denaturation and then renaturing the attached monomer. Table 7.10 gives a list of the enzymes so treated and the catalytic activities of the isolated subunits.

Insolubilised enzymes can be used as affinity chromatography ligands in the same manner as those described in Chapter 1. One novel application which has possibilities for other enzymes was the isolation of an active site peptide from ribonuclease.[22] Ribonuclease was first affinity labelled with 5'-(4-diazophenylphosphoryl) uridine 2'(3')-phosphate; the uridine phosphate group directed the reagent to the active site and the diazo phenol formed the covalent attach-

Table 7.10: Immobilised Enzyme Subunits

Enzyme	Number of subunits	Immobilised subunit(s)
Aspartate transcarbamylase	$(AB_2)_3$	Active trimers (AB_2)
Aldolase	4	Active monomers
Glyceraldehyde-3-phosphate DH	4	Active dimers
Glycogen phosphorylase	4	Inactive monomers
Lactate DH	4	Inactive monomers bind NADH
Transaldolase	2	Active monomers
Triosephosphate isomerase	2	Active monomers

ment. The labelled enzyme was then digested with trypsin to a mixture of peptides, one of which contained an active site fragment. Instead of applying the usual electrophoretic and chromatographic techniques for its isolation and purification, Givol et al.[22] passed the digest down a sephadex affinity column containing bound ribonuclease. The uridine phosphate and hence the required active site fragment adsorbed to the enzyme on the column and thus could be recovered after washing through of the unwanted peptides.

7.6 REFERENCES

1. K. Aunstrup, 'Industrial Aspects of Biochemistry', in B. Spencer (ed.), *Fed. Europ. Biochem. Soc. Symp.* (Elsevier, North Holland, 1974), vol. 30, p. 23.
2. H. Wieland, *Enzymes in Food Processing* (Noyes Data Corp., Park Ridge, New Jersey, 1972).
3. J. R. Whitaker, *Food Related Enzymes* (Amer. Chem. Soc., 1973).
4. 'Immobilised Enzymes' in K. Mosbach (ed.), *Methods in Enzymology* (Academic Press, New York, 1976), vol. 44.
5. E. K. Pye and L. B. Wingard (eds.), *Enzyme Engineering* (Plenum Press, New York, 1974), vol. 2.
6. M. A. Mitz and L. J. Summaria, *Nature*, vol. 189 (1961), p. 576.
7. F. Wold, in C. H. W. Hirs (ed.), *Methods in Enzymology* (Academic Press, New York, 1967), vol. 11, p. 617.
8. K. Mosbach, B. Mattiasson and P. A. Srere, *Proc. Natl Acad. Sci.*, vol. 70 (1973), p. 2534.
9. T. M. S. Chang, Artificial Cells (Thomas Springfields, Illinois, 1972).
10. G. Goldstein, Y. Levin and E. Katchalski, *Biochem.*, vol. 3 (1964), p. 1913.

11. R. L. Foster, *Experientia*, vol. 31 (1975), p. 772.
12. E. Katchalski, I. Silman and R. Goldman, *Adv. Enzym.*, vol. 34 (1971), p. 445.
13. E. Katchalski, in P. Desnuelle, H. Neurath and M. Ottesen (eds.), *Structure-Function Relationships of Proteolytic Enzymes* (Munksgaard, Copenhagen, 1970), p. 198.
14. C. M. Wharton, E. M. Crook and K. Brocklehurst, *Europ. J. Biochem.*, vol. 6 (1968), p. 572.
15. A. Wiseman, *Handbook of Enzyme Biotechnology* (Ellis Horwood, Chichester, 1975).
16. S. Delin, B. Ekström, L. Nathorst-Westfelt, B. Sjöberg and H. Thalin, U.S. Patent 3736230 (1973).
17. T. Tosa, M. Mori and I. Chibata, *Enzymologia*, vol. 43 (1972), p. 213.
18. G. P. Hicks and S. J. Updike, *Nature*, vol. 214 (1967), p. 986.
19. R. Goldman and E. Katchalski, *J. Theor. Biol.*, vol. 32 (1971), p. 243.
20. K. Mosbach and B. Mattiasson, *Biochem. Biophys. Acta*, vol. 235 (1971), p. 253.
21. D. Thomas, D. Bourdillon, G. Broun and J. P. Kernevez, *Biochem.*, vol. 13 (1974), p. 2995.
22. D. Givol, Y. Weinstein, M. Gorecki and M. Wilchek, *Biochem. Biophys. Res. Commun.*, vol. 38 (1970), p. 825.

Appendix I: Key to Numbering and Classification of Enzymes

1. OXIDOREDUCTASES

1.1 Acting on the CH-OH group of donors

1.1.1	With NAD$^+$ or NADP$^+$ as acceptor
1.1.2	With a cytochrome as acceptor
1.1.3	With oxygen as acceptor
1.1.99	With other acceptors

1.2 Acting on the aldehyde or keto group of donors

1.2.1	With NAD$^+$ or NADP$^+$ as acceptor
1.2.2	With a cytochrome as acceptor
1.2.3	With oxygen as acceptor
1.2.4	With a disulphide compound as acceptor
1.2.7	With an iron-sulphur protein as acceptor
1.2.99	With other acceptors

1.3 Acting on CH-CH group of donors

1.3.1	With NAD$^+$ or NADP$^+$ as acceptor
1.3.2	With a cytochrome as acceptor
1.3.3	With oxygen as acceptor
1.3.7	With an iron-sulphur protein as acceptor
1.3.99	With other acceptors

1.4 Acting on the CH-NH$_2$ group of donors

1.4.1	With NAD$^+$ or NADP$^+$ as acceptor
1.4.3	With oxygen as acceptor
1.4.4	With a disulphide compound as acceptor
1.4.99	With other acceptors

1.5   *Acting on the CH-NH group of donors*

1.5.1	With NAD^+ or $NADP^+$ as acceptor
1.5.3	With oxygen as acceptor
1.5.99	With other acceptors

1.6   *Acting on NADH or NADPH*

1.6.1	With NAD^+ or $NADP^+$ as acceptor
1.6.2	With a cytochrome as acceptor
1.6.4	With a disulphide compound as acceptor
1.6.5	With a quinone or related compound as acceptor
1.6.6	With a nitrogenous group as acceptor
1.6.7	With an iron-sulphur protein as acceptor
1.6.99	With other acceptors

1.7   *Acting on other nitrogenous compounds as donors*

1.7.2	With a cytochrome as acceptor
1.7.3	With oxygen as acceptor
1.7.7	With an iron-sulphur protein as acceptor
1.7.99	With other acceptors

1.8   *Acting on a sulphur group of donors*

1.8.1	With NAD^+ or $NADP^+$ as acceptor
1.8.2	With a cytochrome as acceptor
1.8.3	With oxygen as acceptor
1.8.4	With a disulphide compound as acceptor
1.8.5	With a quinone or related compound as acceptor
1.8.6	With a nitrogenous group as acceptor
1.8.7	With an iron-sulphur protein as acceptor
1.8.99	With other acceptors

1.9   *Acting on a haem group of donors*

1.9.3	With oxygen as acceptor
1.9.6	With a nitrogenous group as acceptor
1.9.99	With other acceptors

1.10 Acting on diphenols and related substances as donors

    1.10.2   With a cytochrome as acceptor
    1.10.3   With oxygen as acceptor

1.11 Acting on hydrogen peroxide as acceptor

1.12 Acting on hydrogen as donor

    1.12.1   With NAD^+ or $NADP^+$ as acceptor
    1.12.2   With a cytochrome as acceptor
    1.12.7   With an iron-sulphur protein as acceptor

1.13 Acting on single donors with incorporation of molecular oxygen (oxygenases)

    1.13.11   With incorporation of two atoms of oxygen
    1.13.12   With incorporation of one atom of oxygen (internal monooxygenases or internal mixed function oxidases)
    1.13.99   Miscellaneous (requires further characterisation)

1.14 Acting on paired donors with incorporation of molecular oxygen

    1.14.11   With 2-oxoglutarate as one donor, and incorporation of one atom each of oxygen into both donors
    1.14.12   With NADH or NADPH as one donor, and incorporation of two atoms of oxygen into one donor
    1.14.13   With NADH or NADPH as one donor, and incorporation of one atom of oxygen
    1.14.14   With reduced flavin or flavoprotein as one donor, and incorporation of one atom of oxygen
    1.14.15   With a reduced iron-sulphur protein as one donor, and incorporation of one atom of oxygen
    1.14.16   With reduced pteridine as one donor, and incorporation of one atom of oxygen
    1.14.17   With ascorbate as one donor, and incorporation of one atom of oxygen
    1.14.18   With another compound as one donor, and incorporation of one atom of oxygen
    1.14.99   Miscellaneous (requires further characterisation)

1.15   Acting on superoxide radicals as acceptor

1.16 Oxidizing metal ions

   1.16.3   With oxygen as acceptor

1.17 Acting on CH_2-groups

   1.17.1   With NAD^+ or $NADP^+$ as acceptor
   1.17.4   With a disulphide compound as acceptor

2.   **TRANSFERASES**

2.1 Transferring one-carbon groups

   2.1.1   Methyltransferases
   2.1.2   Hydroxymethyl-, formyl- and related transferases
   2.1.3   Carboxyl- and carbamoyltransferases
   2.1.4   Amidinotransferases

2.2 Transferring aldehyde or ketonic residues

2.3 Acyltransferases

   2.3.1   Acyltransferases
   2.3.2   Aminoacyltransferases

2.4 Glycosyltransferases

   2.4.1   Hexosyltransferases
   2.4.2   Pentosyltransferases
   2.4.99   Transferring other glycosyl groups

2.5 Transferring alkyl or aryl groups, other than methyl groups

2.6 Transferring nitrogenous groups

   2.6.1   Aminotransferases
   2.6.3   Oximinotransferases

2.7   *Transferring phosphorus-containing groups*

    2.7.1   Phosphotransferases with an alcohol group as acceptor
    2.7.2   Phosphotransferases with a carboxyl group as acceptor
    2.7.3   Phosphotransferases with a nitrogenous group as acceptor
    2.7.4   Phosphotransferases with a phospho-group as acceptor
    2.7.5   Phosphotransferases with regeneration of donors (apparently catalysing intramolecular transfers)
    2.7.6   Diphosphotransferases
    2.7.7   Nucleotidyltransferases
    2.7.8   Transferases for other substituted phospho-groups
    2.7.9   Phosphotransferases with paired acceptors

2.8   *Transferring sulphur-containing groups*

    2.8.1   Sulphurtransferases
    2.8.2   Sulphotransferases
    2.8.3   CoA-transferases

3.   HYDROLASES

3.1   *Acting on ester bonds*

    3.1.1   Carboxylic ester hydrolases
    3.1.2   Thiolester hydrolases
    3.1.3   Phosphoric monoester hydrolases
    3.1.4   Phosphoric diester hydrolases
    3.1.5   Triphosphoric monoester hydrolases
    3.1.6   Sulphuric ester hydrolases
    3.1.7   Diphosphoric monoester hydrolases

3.2   *Acting on glycosyl compounds*

    3.2.1   Hydrolysing O-glycosyl compounds
    3.2.2   Hydrolysing N-glycosyl compounds
    3.2.3   Hydrolysing S-glycosyl compounds

3.3   *Acting on ether bonds*

3.3.1   Thioether hydrolases
3.3.2   Ether hydrolases

3.4   *Acting on peptide bonds (peptide hydrolases)*

3.4.11   α-Aminoacylpeptide hydrolases
3.4.12   Peptidylamino-acid or acylamino-acid hydrolases
3.4.13   Dipeptide hydrolases
3.4.14   Dipeptidylpeptide hydrolases
3.4.15   Peptidyldipeptide hydrolases
3.4.21   Serine proteinases
3.4.22   SH-proteinases
3.4.23   Acid proteinases
3.4.24   Metalloproteinases
3.4.99   Proteinases of unknown catalytic mechanism

3.5   *Acting on carbon-nitrogen bonds, other than peptide bonds*

3.5.1   In linear amides
3.5.2   In cyclic amides
3.5.3   In linear amidines
3.5.4   In cyclic amidines
3.5.5   In nitriles
3.5.99   In other compounds

3.6   *Acting on acid anhydrides*

3.6.1   In phosphoryl-containing anhydrides
3.6.2   In sulphonyl-containing anhydrides

3.7   *Acting on carbon-carbon bonds*

3.7.1   In ketonic substances

3.8   *Acting on halide bonds*

3.8.1   In C-halide compounds
3.8.2   In P-halide compounds

3.9 Acting on phosphorus-nitrogen bonds

3.10 Acting on sulphur-nitrogen bonds

3.11 Acting on carbon-phosphorus bonds

4. LYASES

4.1 Carbon-carbon lyases

    4.1.1   Carboxy-lyases
    4.1.2   Aldehyde-lyases
    4.1.3   Oxo-acid-lyases
    4.1.99   Other carbon-carbon lyases

4.2 Carbon-oxygen lyases

    4.2.1   Hydro-lyases
    4.2.2   Acting on polysaccharides
    4.2.99   Other carbon-oxygen lyases

4.3 Carbon-nitrogen lyases

    4.3.1   Ammonia-lyases
    4.3.2   Amidine-lyases

4.4 Carbon-sulphur lyases

4.5 Carbon-halide lyases

4.6 Phosphorus-oxygen lyases

4.99 Other lyases

5.   ISOMERASES

5.1   Racemases and epimerases

    5.1.1   Acting on amino acids and derivatives
    5.1.2   Acting on hydroxy acids and derivatives
    5.1.3   Acting on carbohydrates and derivatives
    5.1.99   Acting on other compounds

5.2   Cis-trans isomerases

5.3   Intramolecular oxidoreductases

    5.3.1   Interconverting aldoses and ketoses
    5.3.2   Interconverting keto- and enol-groups
    5.3.3   Transposing $C=C$ bonds
    5.3.4   Transposing S-S bonds
    5.3.99   Other intramolecular oxidoreductases

5.4   Intramolecular transferases

    5.4.1   Transferring acyl groups
    5.4.2   Transferring phosphoryl groups
    5.4.3   Transferring amino groups
    5.4.99   Transferring other groups

5.5   Intramolecular lyases

5.99   Other isomerases

6.   LIGASES (SYNTHETASES)

6.1   Forming carbon-oxygen bonds

    6.1.1   Ligases forming aminoacyl-tRNA and related compounds

6.2   *Forming carbon-sulphur bonds*

    6.2.1   Acid-thiol ligases

6.3   *Forming carbon-nitrogen bonds*

    6.3.1   Acid-ammonia ligases (amide synthetases)
    6.3.2   Acid-amino-acid ligases (peptide synthetases) .
    6.3.3 .   Cyclo-ligases
    6.3.4   Other carbon-nitrogen ligases
    6.3.5   Carbon-nitrogen ligases with glutamine as amido-*N*-donor

6.4   *Forming carbon-carbon bonds*

6.5   *Forming phosphate ester bonds*

Appendix II: Tutorial Questions in Enzyme Kinetics

The following questions are taken from the author's tutorial programme on enzyme kinetics.

1. In the treatment of a metabolic disorder, it is required to inhibit non-competitively an enzyme catalysed reaction. The following data were obtained with the inhibitor:

Inhibitor conc. (mM)	0.01	0.02	0.03	0.04	0.05
Reciprocal initial rate (s/μM)	0.2	0.3	0.4	0.5	0.6

Calculate the concentration of inhibitor required at the site of action to reduce the maximum velocity of the reaction by a factor of ten.

Answer: I = 0.09 mM

2. The following results were obtained for an enzyme-catalysed reaction in the presence of a competitive inhibitor:

At substrate concn. of 0.5 mM

Concentration of inhibitor (mM)	0	0.3	0.6	1.0
Reciprocal initial reaction velocity (s/μM)	0.833	1.045	1.256	1.536

At substrate concn. of 1.05 mM

Concentration of inhibitor (mM)	0	0.3	0.6	1.0
Reciprocal initial reaction velocity (s/μM)	0.500	0.600	0.700	0.833

Calculate the enzyme-inhibitor dissociation constant (K_i), the maximum initial reaction velocity (V) and the enzyme-substrate dissociation constant (K_m).

$$\text{Answer: } K_i = 0.9 \text{ mM}$$
$$V = 5.0 \ \mu\text{M/s}$$
$$K_m = 1.6 \text{ mM}$$

3. From each of two organs within the same species, enzymes were isolated which specifically catalysed the same reaction, exhibiting the substrate concentration dependent velocities below:

Reciprocal substrate concentration (mmol^{-1})		2.5	4.5	6.0	9.0
Reciprocal velocity (s/m mol)	Enzyme 1	6.0	7.6	8.8	11.3
	Enzyme 2	10.4	13.3	15.3	19.5

At the substrate concentrations tabulated, addition of 0.2 mM reaction product caused a 25 percent decrease in each of the rates of catalysis by Enzyme 1 but *no* effect was observed on Enzyme 2.
   (i) Obtain numerical values for the parameters which characterise the uninhibited enzyme catalysed reactions.
   (ii) What type of inhibition is shown by the product and what metabolic significance could this have for the enzyme?
   (iii) What conclusions can be drawn concerning the relationship to each other of the two enzymes?

Answer:
   (i) $K_m = 0.2$ mM; $V_{max} = 0.25, 0.143$ mM s^{-1}; $K_i = 0.57$ mM
   (ii) Noncompetitive
   (iii) Isoenzymes

4. An enzyme inhibitor, with $K_i = 1.9$ mM is available as a crude product of unknown concentration. When added to an enzyme-substrate system at crude product concentrations of 2 g/l and 6g/l the following results were obtained:

Reciprocal substrate conc. (l/g)	0.2	0.4	0.6	0.8	1.0
Reciprocal initial rate (s/mg) at inhibitor conc. I	0.34	0.4	0.45	0.5	0.55
Reciprocal initial rate (s/mg) at inhibitor conc. 3 I	0.4	0.5	0.6	0.7	0.8

Calculate the concentration of active inhibitor in the crude product, and determine the type of inhibition and the Michaelis constant, K_m, of the substrate-enzyme complex.

Answer: I = 0.8 mM
Competitive
$K_m = 0.50$ mM

Author Index

Subject Index

Enzyme Index

Enzymes referred to in the text are indexed in order of their 'recommended' names, page 40 and Appendix I.